Rhizosphere Biology

Series Editor

Anil Kumar Sharma, Biological Sciences, CBSH, G.B. Pant University of Agriculture & Technology, Pantnagar, Uttarakhand, India

The Series **Rhizosphere Biology**, emphasizes on the different aspects of Rhizosphere. Major increase in agricultural productivity, to meet growing food demands of human population is imperative, to survive in the future. Along with methods of crop improvement, an understanding of the rhizosphere biology, and the ways to manipulate it, could be an innovative strategy to deal with this demand of increasing productivity. This Series would provide comprehensive information for researchers, and encompass all aspects in field of rhizosphere biology. It would comprise of topics ranging from the classical studies to the most advanced application being done in the field. Rhizoshpere is a dynamic environment, and a series of processes take place to create a congenial environment for plant to grow and survive. There are factors which might hamper the growth of plants, resulting in productivity loss, but, the mechanisms are not very clear. Understanding the rhizosphere is needed, in order to create opportunities for researchers to come up with robust strategies to exploit the rhizosphere for sustainable agriculture.

There are titles already available in the market in the broad area of rhizosphere biology, but there is a major lack of information as to the functions and future applications of this field. These titles have not given all the up-to-date information required by the today's researchers and therefore, this Series aims to fill out those gaps.

Puneet Singh Chauhan
Shri Krishna Tewari • Sankalp Misra
Editors

Plant-Microbe Interaction and Stress Management

Springer

Editors
Puneet Singh Chauhan
Microbial Technology Division
CSIR- National Botanical Research Institute
Lucknow, Uttar Pradesh, India

Shri Krishna Tewari
Plant Conservation & Agrotech Division
CSIR- National Botanical Research Institute
Lucknow, Uttar Pradesh, India

Sankalp Misra
Faculty of Biosciences
Institute of Biosciences and Technology,
Shri Ramswaroop Memorial University
Barabanki, Uttar Pradesh, India

ISSN 2523-8442 ISSN 2523-8450 (electronic)
Rhizosphere Biology
ISBN 978-981-97-4238-7 ISBN 978-981-97-4239-4 (eBook)
https://doi.org/10.1007/978-981-97-4239-4

This Springer imprint is published by the registered company Springer Nature Singapore Pte Ltd.
The registered company address is: 152 Beach Road, #21-01/04 Gateway East, Singapore 189721, Singapore

Preface

The search for robust and sustainable crop production techniques is a top priority in the quickly developing sector of agriculture, where microbial interaction plays a critical role as an innovation catalyst. The interactions between plants and microorganisms are diverse and complex, ranging from mutualistic to parasitic. This book *Plant-Microbe Interaction and Stress Management* unravels the intricate mechanisms underlying the interactions between microbes and plants under stress conditions. This extensive volume provides a comprehensive examination of the applications of microbial interactions in boosting resilient plant development under a range of environmental conditions by bringing together the most recent findings, ideas, and innovations in the field.

Chapter 1 summarizes the cross-dynamics of plant–microbe interaction that can be applied in plant stress management for sustainable agricultural programs. Research on extending the shelf life of the microbial inoculum, scaling up fermentation, releasing microbes or their metabolites in the field, and bolstering legislative policies that can assist plant growth promoting microbes (PGPMs) in completing their transition from laboratory to field offer a cost-effective and environmentally beneficial method of sustainable agriculture.

Chapter 2 examines the current understanding of plant stress management when PGPM is present in unfavorable environmental settings. It also looks at the mechanism underlying the plant–PGPM interaction and how it might be applied to sustainable agriculture.

Chapter 3 highlights the ecological significance and potential of the phyllosphere microbiome in agricultural and environmental management. Understanding how microbes and plants interact in the phyllosphere makes it easier to use microbial resources to boost plant resilience and encourage ecosystem sustainability.

Chapter 4 embarks on an enlightening exploration of the impacts of waterlogging and drought on plant physiology, plant responses to climate change, and molecular pathways. Moreover, the mechanisms underlying these modifications are explained. It also explores how microorganisms that promote plant growth can enhance a plant's ability to withstand waterlogging and drought conditions.

Chapter 5 investigates the basic mechanism of plants' response to high salt concentrations and the development of effective and modern techniques for reducing salinity. It also summarizes the principal strategies for reducing the negative effects of salt on plants and provides a critical evaluation of the major advancements in this field.

Chapter 6 extends our understanding of the microbes associated with the plant's rhizosphere which play a crucial role in metal mobilization, uptake, and detoxification. Understanding plant–microbe interactions under metal stress would help in developing better agronomic and environmental practices for detoxifying metal toxicity to promote better plant growth for more productivity and environmental sustainability.

Chapter 7 lays the groundwork for creating focused and long-lasting approaches for managing bacterial diseases and supporting the vital function of both above-and below-ground beneficial biofilms. The comprehension of synergistic linkages in the plant microbiome environment and their significance for sustainable agriculture is bolstered by the insights obtained from beneficial biofilms.

Chapter 8 accentuates the important characteristics of the rhizospheric microbiome, important methods used to characterize microbial diversity, and their use in disease suppression with their mode of action. An updated knowledge framework in the field of linked rhizospheric microbiomes might be provided by analyzing the facets of the microbiome and discussing their application.

Chapter 9 provides insight into the top 10 plant viruses ranked based on their prevalence and the economic damages they inflict on their host plants. This chapter also deals with their size, shape, mode of transmission, genomic organization, replication, and movement in host cells to understand the nature of their complexity.

Chapter 10 serves as an invaluable resource for researchers and practitioners keen on leveraging insects as biotic stress in agriculture. Plants have evolved to counter the adverse effects of insects as well as to gain from them; such interaction is important for a dynamic ecosystem.

Chapter 11 discusses major antagonistic approaches toward phytopathogens emphasizing microbial biological control agents and their mode of action. In addition, this chapter also addresses the potential of microbial engineering and unravels the perspectives and possibilities of microbes as biocontrol agents, thus offering new avenues for innovation in agriculture.

Chapter 12 focuses on the cutting-edge methods and technologies that have created new opportunities for gaining a precise understanding of the interactions that occur between PGPRs and the host plant. Moreover, specific perspectives on recent technology to investigate the hitherto unknown aspect of soil microbiome have also been discussed.

Chapter 13 explores the comprehensive effects of sodic soil on soil characteristics, plant development, and the mechanisms by which bacteria promote tolerance in plants under sodic environments. Achieving the best crop production and yield under sodic stress conditions requires an understanding of the sodic environment and the mechanisms of microorganisms that promote plant growth in a sodic-tolerant manner.

Chapter 14 outlines the significance of endophytic microorganisms in comprehending plant–microbe interactions by succinctly summarizing advanced technologies and instruments for interpreting the genetic composition and ecological roles of these bacteria.

Chapter 15 concentrates on various plant biotic stressors, with a focus on fungal infections in particular. This chapter also explores the potential of exogenous administration of specific microbial agents to alleviate the detrimental effects of biotic stress on plant growth.

Throughout this book, we have included contributions from leading experts in the field, who have provided in-depth reviews of the current state of knowledge and future directions for research. We hope this book will be a valuable resource for researchers, teachers, and practitioners, fostering a deeper understanding of the intricate and mutually advantageous relationships between microorganisms and plants for sustainable agriculture.

Lastly, we would like to acknowledge the support and encouragement from the publishers, who recognized the importance of this topic and provided us with the opportunity to compile and share this valuable information with a wider audience.

Lucknow, Uttar Pradesh, India — Puneet Singh Chauhan
Lucknow, Uttar Pradesh, India — Shri Krishna Tewari
Barabanki, Uttar Pradesh, India — Sankalp Misra

Contents

Editors and Contributors

About the Editors

Puneet Singh Chauhan is a Senior Principal Scientist in Microbial Technology Division at CSIR-National Botanical Research Institute, Lucknow, India. He completed his Ph.D. from CSIR-NBRI, Lucknow, and was awarded with Postdoctoral Research Fellowship from Brain Korea 21 project, Chungbuk National University, Cheongju, South Korea. His main research interests include plant–microbe interactions, microbial ecology, and soil metagenomics. He is a member of the editorial board of several leading scientific journals, has published more than 100 SCI research and review articles, and has authored 20 book chapters. He has been awarded with Biotech Research Excellence Award from Biotech Research Society and The Society of Life Sciences and Innovative Young Scientist Award from Asian PGPR Society for Sustainable Agriculture, Hyderabad, India.

Shri Krishna Tewari is Chief Scientist and Professor, Academy of Scientific and Innovative Research (AcSIR) at CSIR-National Botanical Research Institute, Lucknow, India. He is Head of Botanic Garden and Plant Conservation and Agrotechnology Divisions of the institute. Having completed his Ph.D. from G.B. Pant University of Agriculture and Technology, Pantnagar, he joined CSIR-NBRI and worked in the research areas of conservation, propagation, and agrotechnology of economically important plants. His research interests include amelioration and management of sodic soil, nontraditional economic plants, bio-composting, and extension of green technologies for rural development. He has received several national awards, including the "CAIRD-2011" award from CSIR. He has published over 100 research papers in reputed journals, 10 books, 11 review articles, 25 chapters in books, 22 extension bulletins, and more than 100 popular articles.

Sankalp Misra is an Assistant Professor at Shri Ramswaroop Memorial University (SRMU), Uttar Pradesh, India. He completed his Ph.D. in Biological

Sciences from CSIR-National Botanical Research Institute (NBRI), Lucknow, and was awarded by the Academy of Scientific and Innovative Research (AcSIR). He has qualified for CSIR-UGC NET and has been awarded a Senior Research Fellowship (SRF) and Research Associateship (RA) from Indian Council of Medical Research (ICMR), New Delhi, India. He has been felicitated with the Young Researcher Award from the Institute of Scholars and SRMU. He has also received Best Poster Awards at international and national conferences. He holds a Life Membership of "The Association of Microbiologists of India (AMI)." He has published more than 15 SCI research articles and authored more than 10 book chapters. He has received research grants from different funding agencies including ICMR, UPCST, and SRMU. His research interests include sustainable agriculture, abiotic stress resilient agriculture, and biofertilizer-based technologies for improving soil fertility and crop productivity.

Contributors

Rangasamy Anandham Department of Agricultural Microbiology, Tamil Nadu Agricultural University, Coimbatore, Tamil Nadu, India

The Institute of Ecology and Earth Sciences, University of Tartu, Tartu, Estonia

Anjali Chaudhary CSIR-Institute of Himalayan Bioresource Technology, Palampur, India

Academy of Scientific and Innovative Research (AcSIR), Ghaziabad, India

Vasvi Chaudhry Microbial Interactions in Plant Ecosystems, Center for Plant Molecular Biology, University of Tübingen, Tübingen, Germany

Updesh Chauhan Faculty of Bio-Sciences, Institute of Bio-Sciences and Technology, Shri Ramswaroop Memorial University, Barabanki, Uttar Pradesh, India

Anand Dave Department of Microbiology, Seventh Day Adventist Arts and Science College, Ahmedabad, Gujrat, India

Vinay Kumar Dhiman Department of Basic Sciences, College of Forestry, Dr. Yashwant Singh Parmar University of Horticulture and Forestry, Nauni, Solan, India

Vivek Kumar Dhiman Department of Biotechnology, College of Horticulture, Dr. Yashwant Singh Parmar University of Horticulture and Forestry, Nauni, Solan, India

Namo Dubey CSIR-Institute of Himalayan Bioresource Technology, Palampur, India

Academy of Scientific and Innovative Research (AcSIR), Ghaziabad, India

Smriti Gaur Department of Biotechnology, Bennett University, Greater Noida, Uttar Pradesh, India

Daniel Gómez-Pérez Faculty of Biology, Center for Organismic Adaptation, LMU Munich, Planegg-Martinsried, Germany

Anmol Gupta Integral University, Lucknow, Uttar Pradesh, India

Garima Gupta Faculty of Bio-Sciences, Institute of Bio-Sciences and Technology, Shri Ramswaroop Memorial University, Barabanki, Uttar Pradesh, India

Microbial Technology Division, CSIR-National Botanical Research Institute, Lucknow, Uttar Pradesh, India

Govind Gupta Department of Chemistry, Environmental Science Discipline, Manipal University, Jaipur, Rajasthan, India

Indu CSIR-Institute of Himalayan Bioresource Technology, Palampur, India

K. Jayachandran School of Biosciences, Mahatma Gandhi University, Kottayam, Kerala, India

Siya Kamat Department of Biochemistry, Indian Institute of Science, Bengaluru, India

Rupali Katoch CSIR-Institute of Himalayan Bioresource Technology, Palampur, India

Ramasamy Krishnamoorthy Department of Crop Management, Vanavarayar Institute of Agriculture, Pollachi, Tamil Nadu, India

Manoj Kumar Institute of Plant Sciences, Agricultural Research Organization, Volcani Center, Rishon LeZion, Israel

Susheel Kumar Plant Molecular Virology Laboratory, Molecular Biology and Biotechnology Division, CSIR-National Botanical Research Institute, Lucknow, Uttar Pradesh, India

Madhuree Kumari Department of Biochemistry, Indian Institute of Science, Bengaluru, India

Sivakumar Madhan Department of Agricultural Microbiology, Tamil Nadu Agricultural University, Coimbatore, Tamil Nadu, India

Kamna Madheshiya Faculty of Bio-Sciences, Institute of Bio-Sciences and Technology, Shri Ramswaroop Memorial University, Barabanki, Uttar Pradesh, India

Ariyan Manikandan Department of Agricultural Microbiology, Tamil Nadu Agricultural University, Coimbatore, Tamil Nadu, India

The Institute of Ecology and Earth Sciences, University of Tartu, Tartu, Estonia

Dilip Kumar Maurya Plant Molecular Virology Laboratory, Molecular Biology and Biotechnology Division, CSIR-National Botanical Research Institute, Lucknow, Uttar Pradesh, India

Prabhat Kumar Maurya Plant Molecular Virology Laboratory, Molecular Biology and Biotechnology Division, CSIR-National Botanical Research Institute, Lucknow, Uttar Pradesh, India

Suraj Kumar Modi School of Life Sciences Pharmacy and Chemistry, Biomolecular Sciences Department, Kingston University, London, UK

Himanshu Pandey PG Department of Agriculture, Khalsa College, Amritsar, Punjab, India

Minakshi Pandey Faculty of Biosciences, Institute of Biosciences and Technology, Shri Ramswaroop Memorial University, Barabanki, Uttar Pradesh, India

E. K. Radhakrishnan School of Biosciences, Mahatma Gandhi University, Kottayam, Kerala, India

Rajasekaran Raghu Department of Plant Biotechnology, Tamil Nadu Agricultural University, Coimbatore, Tamil Nadu, India

S. Vijay Anand Raj Plant Molecular Virology Laboratory, Molecular Biology and Biotechnology Division, CSIR-National Botanical Research Institute, Lucknow, Uttar Pradesh, India

S. K. Raj Sector-G, Janki Puram, Lucknow, Uttar Pradesh, India

Archana Rathore Plant Molecular Virology Laboratory, Molecular Biology and Biotechnology Division, CSIR-National Botanical Research Institute, Lucknow, Uttar Pradesh, India

Tina Roy CSIR-Institute of Himalayan Bioresource Technology, Palampur, India

Monja Schmid Microbial Interactions in Plant Ecosystems, Center for Plant Molecular Biology, University of Tübingen, Tübingen, Germany

Duraisamy Selvi Department of Soil Science and Agricultural Chemistry, Tamil Nadu Agricultural University, Coimbatore, Tamil Nadu, India

Murugaiyan Senthilkumar Department of Agricultural Microbiology, Tamil Nadu Agricultural University, Coimbatore, Tamil Nadu, India

Avinash Sharma Faculty of Agricultural Sciences, Arunachal University of Studies, Namsai, Arunachal Pradesh, India

Riddhi Shrivastava Department of Chemistry, Environmental Science Discipline, Manipal University, Jaipur, Rajasthan, India

Poornima College of Engineering, Jaipur, Rajasthan, India

Devendra Singh Faculty of Biotechnology, Institute of Biosciences and Technology, Shri Ramswaroop Memorial University, Barabanki, Uttar Pradesh, India

Durgesh Singh Faculty of Bio-Sciences, Institute of Bio-Sciences and Technology, Shri Ramswaroop Memorial University, Barabanki, Uttar Pradesh, India

Kunal Singh CSIR-Institute of Himalayan Bioresource Technology, Palampur, India

Academy of Scientific and Innovative Research (AcSIR), Ghaziabad, India

Naveen Kumar Singh Department of Chemistry, Environmental Science Discipline, Manipal University, Jaipur, Rajasthan, India

Poonam C. Singh Microbial Technology Division, CSIR-National Botanical Research Institute, Rana Pratap Marg, Lucknow, India

Academy of Scientific and Innovative Research (AcSIR), Ghaziabad, India

Puneet Singh Chauhan Microbial Technology Division, CSIR-National Botanical Research Institute, Lucknow, Uttar Pradesh, India

Padinjarakavil Soumya School of Biosciences, Mahatma Gandhi University, Kottayam, Kerala, India

Subramanium Thiyageshwari Department of Soil Science and Agricultural Chemistry, Tamil Nadu Agricultural University, Coimbatore, Tamil Nadu, India

Shalini Tiwari Department of Biochemistry and Molecular Biology, Oklahoma State University, Stillwater, OK, USA

Priyam Vandana Division of Crop Protection, ICAR-Indian Institute of Sugarcane Research, Lucknow, Uttar Pradesh, India

Sunil Kumar Verma Faculty of Biotechnology, Institute of Biosciences and Technology, Shri Ramswaroop Memorial University, Barabanki, Uttar Pradesh, India

Tanu Vishwakarma Plant Molecular Virology Laboratory, Molecular Biology and Biotechnology Division, CSIR-National Botanical Research Institute, Lucknow, Uttar Pradesh, India

Kanchan Yadav CSIR-Institute of Himalayan Bioresource Technology, Palampur, India

Academy of Scientific and Innovative Research (AcSIR), Ghaziabad, India

Pooja Yadav CSIR-Institute of Himalayan Bioresource Technology, Palampur, India

Academy of Scientific and Innovative Research (AcSIR), Ghaziabad, India

Udit Yadav Dr. B. Lal Institute of Biotechnology, Jaipur, Rajasthan, India

Gunasekaran Yazhini Department of Soil Science and Agricultural Chemistry, Tamil Nadu Agricultural University, Coimbatore, Tamil Nadu, India

Leonie M. Zott Microbial Interactions in Plant Ecosystems, Center for Plant Molecular Biology, University of Tübingen, Tübingen, Germany

Chapter 1
Plant–Microbe Interaction: Stress Management for Sustainable Agriculture

Siya Kamat, Suraj Kumar Modi, Smriti Gaur, and Madhuree Kumari

Abstract The current agricultural practices demonstrate a high dependence on agrochemicals to cope up biotic and abiotic stress, which in turn results in a negative impact on environment and increases the economic burden on farmers. Plant growth-promoting microbes (PGPMs) have emerged as an eco-friendly and economic approach to combat the multiple stress faced by plants. Aboveground and belowground interactions and colonization of PGPMs can contribute to increased plant growth as well as a heathy ecosystem. Plant growth-promoting bacteria, viz. *Pseudomonas* sp., *Bacillus* sp., *Aneurinibacillu* sp., plant growth-promoting fungi, viz. *Trichoderma* sp., and plant growth-promoting actinomycetes are the most studied PGPMs in the literature used for sustainable agriculture. The main mechanisms employed by PGPMs that can be exploited for mitigating stress in plants are plant growth promotion, availability of nutrients, production of secondary metabolites, enzymes and hormones, and modulation of plant oxidative stress and plant defense. Uses of a single PGPM or the consortia of microbes have widely been reported in the literature for ameliorating plants' biotic and abiotic stress in greenhouse conditions. Despite the potential of PGPMs in sustainable agriculture, there are many hurdles that still need to be overcome. Research in sustainable release of microbes or their metabolites in the field condition, increasing shelf-life of the microbial inoculum, scaling up of fermentation, and supporting governmental policies can help the PGPMs complete their journey from lab to land and provide an eco-friendly and economic approach for sustainable agriculture.

Keywords PGPMs · Stress · Sustainable agriculture · Mechanisms · Challenges

---

S. Kamat · M. Kumari (✉)
Department of Biochemistry, Indian Institute of Science, Bengaluru, India

S. K. Modi
School of Life Sciences Pharmacy and Chemistry, Biomolecular Sciences Department, Kingston University, London, UK

S. Gaur
Department of Biotechnology, Bennett University, Greater Noida, Uttar Pradesh, India

P. Singh Chauhan et al. (eds.), *Plant-Microbe Interaction and Stress Management*, Rhizosphere Biology,
https://doi.org/10.1007/978-981-97-4239-4_1

1.1 Introduction

Modern agriculture promises to support the food security of the rapidly growing world population. However, the agricultural practices under this modern scheme are characterized by monocultures, indiscriminate application of pesticides, chemical fertilizers—all of which contribute to nitrous oxide, greenhouse gas emission, and major biodiversity loss (Cappelli et al. 2022; Malgioglio et al. 2022). Regarding agriculture, being the backbone of several developing nations, the major challenge is to meet the food demands of the growing population within the confines of limited resources (freshwater, arable land, human labor) judiciously. Sustainable agriculture aims to meet these challenges with innovative techniques and optimize the balance between agriculture and modernity (Pandey 2018). Although sustainable agriculture is far from mainstream today, almost all countries aspire to progress toward this goal and therefore have agriculture policies in this direction. The most popular sustainable agriculture practices include permaculture (vermicompost), organic and natural farming, crop rotation, hydroponics, agroforestry, vertical farming, and integrated pest management (Gupta et al. 2021). Plant health depends on interactions with the several macro-and micro communities. While most of these practices focus aboveground, it is now established that the critical factor is plant–microbe interaction—both above and belowground (Cappelli et al. 2022; MacLaren et al. 2022). To employ microbes in sustainable agriculture, it is necessary to decipher the microbial and plant diversity, and functionality within the soil and plant microbiome. In this chapter, we aim to summarize the cross-dynamics of plant–microbe interaction that can be applied in plant stress management for sustainable agricultural programs.

1.1.1 Plant–Microbe Interaction

Plants and microbes interact throughout their life cycle. Spatially, the interaction occurs belowground, aboveground, and within the plant—all contributing to a healthy ecosystem. While most interactions are mediated by the plant, some occur by chance. For example, benzoxazinoids are secondary metabolites released by grasses that regulate plant health and plant–microbe integrations below-and aboveground (Cotton et al. 2019). The integration of several multiomic techniques has made it possible to investigate the beneficial plant–microbe interactions (Gamalero et al. 2022).

The spatially separated plant–microbe interactions are:

Aboveground Interaction: Microbes that colonize leaves, flowers, seeds, fruit, and stem contribute to this category. The epiphytes or phyllo-spheric microbes are influenced by the environment and therefore unstable. These microbes are often pathogenic and derive nutrition from plants. Endophytes, on the other hand, live through internal tissues through symbiosis (Nadarajah and Abdul Rahman 2021).

Chaudhry et al. (2021) summarized the properties of leaf surface and its influence on leaf-colonizing microbiota. The leaf microbiome of tomato plant could promote plant health and growth (Romero et al. 2014). Biotic and abiotic stimuli contribute to immunity priming in plant via shaping the microbial communities without much effect on plant growth. These organisms eventually trigger the host plant's basal defense and enhance resistance to phytopathogens. Microbes produce signaling molecules such as (R)-β-homoserine, β-aminobutyric acid that primes production of jasmonic acid, salicylic acid, ethylene, and brassinosteroids, thus protecting the plant from biotrophic and necrotrophic pathogens (Buswell et al. 2018). Since the aboveground microbes colonizing plants are highly adaptable to environmental fluctuations and stresses such as insects, phytopathogens, herbivores, and climate change, they represent an enormous pool of useful genetic traits.

Belowground Interaction: In plant roots, its exudates to soil, and decaying material determines the rhizospheric community or the rhizobiome. The rhizobiome in turn influences plant health, nutrient availability, and the entry or protection from pathogens. Rhizobiome is a spatial interaction that contains several bacteria and fungi drawn toward plant roots. Gu et al. (2020) investigated 80 rhizobiomes in which 2150 bacterial species [Proteobacteria (50%), Firmicutes (24%), Bacteroidetes (18%), and Actinobacteria (8%)] collectively outcompeted plant pathogen *Ralstonia solanacearum*—soilborne bacterium that causes bacterial wilt. The rhizobiome secreted siderophores—creating a competition for iron in the soil, thus suppressing competitors and protecting tomato plants. Other ways in which the rhizobiome contributes to plant health include antibiotic production, nitrogen fixation, polysaccharide degradation, and quorum sensing (Pollak and Cordero 2020).

1.1.2 The Protective Effects of Plant–Microbe Interaction

Crops are constantly exposed to a multitude of environmental stresses-biotic and abiotic. Intensive modern agricultural practices, increasing temperatures, and climate change have further aggravated loss of crops and agricultural wealth worldwide. The use of genetic engineering technologies to make resistant plant varieties is a long-term goal, and several nations do not accept these solutions. Plants in nature live with several microbial populations, and therefore, one can hypothesize that microbes can protect or help plants in biotic or abiotic stresses (Zhang et al. 2022).

In such a scenario, the use of plant microbes can be an economic strategy toward sustainable agriculture goals. Arbuscular mycorrhizal fungi and plant growth-promoting rhizobacteria have been investigated extensively for elevating plant stress tolerance and productivity (Inbaraj 2021). Figure 1.1 provides a description of how the plant–microbial communities are linked to sustainable agriculture. Plants routinely encounter abiotic and biotic stress, which can influence the microbial communities living within and around plant—above-and belowground. These stress-conditioned microbial communities have evolved to improve the plant growth

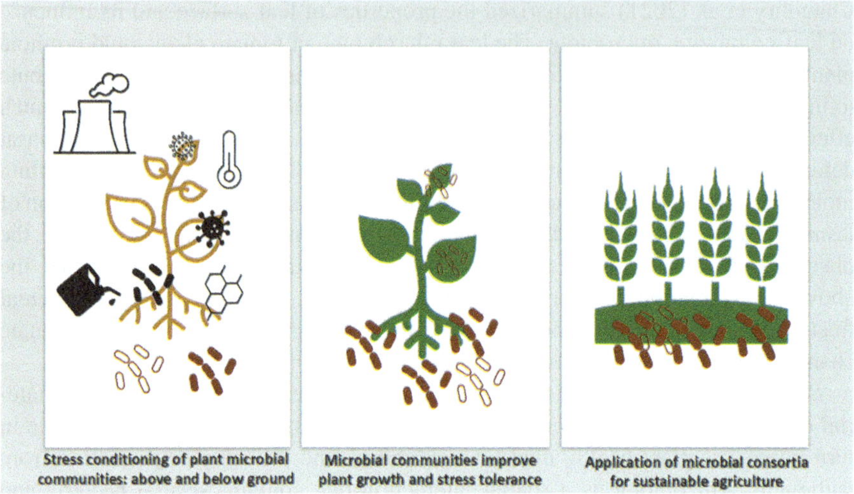

Fig. 1.1 Role of plant–microbial communities in sustainable agriculture

and stress tolerance. The application of these microbial cultures in fields can improve the productivity and provide sustainable solutions for agricultural economy.

1.1.3 Biotic Stress

Crop plants often contract pests, pathogens, and parasites, resulting in biotic stress. The infection results in vascular wilts, chlorosis, stunted growth, cankers, etc. (Hashem et al. 2019). Some examples of pathogens that cause serious diseases in crop plants are described here.

Artichoke species of the central and Mediterranean area is susceptible to several fungal infections. However, *Verticillium dahliae* Kleb. is the most infective soil-borne pathogen that causes chlorosis, stunted growth, wilting, vascular discoloration, and plant death. The microsclerotia of *V. dahlia* live in the soil for 13 years and lead to the propagation of infection through infested soil (Armengol et al. 2005). *Mycosphaerella graminicola* is a deadly fungal pathogen that causes septoria leaf blotch disease in bread and durum wheat crops (Palmer and Skinner 2002). *Pseudomonas* strains are notable for their diverse interactions with plants. One such strain is *Pseudomonas syringae* pv. *tomato* DC300 that causes bacterial infection in several plants, including tomato, and *Arabidopsis thaliana* (Worley et al. 2013). Saffron or *Crocus sativus* is an economically important crop often used as a spice, dye, and for medicinal purpose. It is often infected with fungal pathogens *Fusarium oxysporum* and *Rhizopus oryzae* causing corm rot disease (Mirghasempour et al. 2022). β-Proteobacterium *Ralstonia solanacearum* is a soilborne bacterial pathogen that causes bacterial wilt in *Capsicum annuum* L. Global warming has resulted in a continuous spread of this infection (Lee et al. 2022). Recent studies have detected cross-pathogenic bacterial pathogens—such as *Rhizobium*, *Pantoea*, *Erwinia*, *Burkholderia*, and *Pseudomonas*—that can cause infections in plants and animals/humans (Kizheva et al. 2022). Necrotrophic fungus *Botrytis cinerea* is a deadly pathogen that infects postharvest pepper fruit and causes gray mold, leading to huge economic losses (Wang et al. 2022). Apple plants are often infected with *Colletotrichum* species (*C. alienum*, *C. fructicola*, *C. gloeosporioides* sensu stricto, *C. nymphaeae*, *C. siamense,* and one new species, *C. orientalis*) (Chen et al. 2022). In a recent study, new pathogenic fungal strains *Neofusicoccum australe, N. crypto australe/stellenboschiana*, *N. luteum*, *N. parvum*, and *Lasiodiplodia brasiliensis* were detected that cause dieback and stem-end rot diseases in avocado plants in the Canary Islands of Spain (Hernández et al. 2023).

1.1.4 Abiotic Stress

While a lot is discussed about the biotic stress and abiotic stress individually, very little is known about how abiotic stress affects the plant's ability to combat infections or how stress changes the plant's microbiome. Bhandari et al. (2023) studied the effect of elevated tropospheric ozone (O_3)—alone and in combination with a bacterial infection (*Xanthomonas perforans*)—on susceptible and resistant varieties of pepper cultivars, and their associated microbiome. It was observed that O_3 stress elevated the severity of *X. perforans* infection only in the resistant pepper cultivar. This was due to the dramatic shift in the microbiome structure, which indicates that the microbial communities respond distinctly to biotic and abiotic stress. The endophytic fungi of wheat plants sampled from North China could protect the host wheat plant from several abiotic stresses such as drought, temperature tolerant, salinity, and heavy metal and could promote plant growth. Tomato plants are severely affected due to saline soils and quality of irrigation methods, both of which result in considerable produce losses. Muthuraja and Muthukumar (2022) sampled non-mycorrhizal bacteria—*Aspergillus violaceofuscus* and *Bacillus licheniformis* from saxicolous habitat—and proved the usage of this coculture to withstand salinity and promote growth in tomato plants. In another study, microbial consortium comprised of *Enterobacter* sp., *Bacillus* sp., *Achromobacter* sp., and *Delftia* sp. could alleviate salinity stress and promote growth of tomato plants (Kapadia et al. 2021). Jiao et al. (2016) discovered melatonin-producing endophytic bacterium, *B. amyloliquefaciens* SB-9, from grapevine roots. The endophyte colonization and melatonin production could counteract the abiotic stress—salt and drought—protecting and promoting the growth of grapevine. Plant growth-promoting fungi are also investigated as inducers of heavy metal detoxification by seed priming and biofortification (Geetha et al. 2023).

1.2 Plant–Microbe Interaction Can Mitigate Biotic and Abiotic Stress

Plant interaction with the microbial life is a diverse ecosystem wherein different interactions occur: symbiosis, mutualism, parasitic, etc. Plant microbiome—which includes endophytes, plant growth-promoting rhizobia, and mycorrhizal fungi—is capable of alleviating the biotic and abiotic stress and improve plant growth. These microorganisms are critical players of plants' defense and well-being. Therefore, they offer sustainable solutions to increase agricultural productivity (Munir et al. 2022). Table 1.1 describes recent examples of the potential use of microbes in mitigating biotic and abiotic stress in plants.

1.3 Mechanisms Employed by Plants for Overcoming Stress

Plants have evolved and acquired immense mechanisms to tackle, avoid, and encounter several biotic and abiotic stresses. Failing to have proper strategies can threaten the survival of plants and essentially affect the productivity of plants and crops. Plants are known to establish symbiotic relationships with multiple microbes, which protect and profit from each other in multiple ways. The key strategies and mechanisms evolved by the plant species for stress management are the following.

1.3.1 Plant Growth Promotion

Plants have established symbiotic relationships with various microbes to tackle biotic and abiotic stress followed by promoting sustainable strategies. For example, plant growth-promoting rhizobacteria (PGPR) is one of the most common and important plant-associated bacteria that have been extensively utilized as it promotes plant growth and promises sustainable agriculture. Additionally, it enhances plant–microbe interaction with a nominal adverse effect on the soil microbiota. Studies have shown that phytohormone melatonin produced by the native PGPR enhances the tolerance potential of plants from various biotic and abiotic stress. Jofre et al. (2023) evaluated the production of melatonin by two native PGPR bacteria *Enterobacter* 64S1 and *Pseudomonas* 42P4 in the *Arabidopsis thaliana* plant. Moreover, these two bacteria could enhance endogenous melatonin content even when *Arabidopsis thaliana* was exposed to grow under drought conditions. Mechanistic analysis suggested that these PGPR native strains could enhance the tolerance level of plants in drought by preventing membrane damage, and oxidative stress, and ultimately promoting plant growth (Jofre et al. 2023). Besides rhizobacteria, several other endophytic bacterial species play a critical role in providing

Table 1.1 Recent examples of the potential use of microbes in mitigating biotic and abiotic stress in plants

Biotic stress			
Pathogen	Host plant	Plant growth-promoting microbe	Reference
Fusarium spp.	Saffron	*P. Aeruginosa* YY322	Mirghasempour et al. (2022)
Botryosphaeria dothidea	Apple plant	*Bacillus amyloliquefaciens* FX2	Li et al. (2022a, b)
Blumeria graminis	*Achnatherum inebrians*	*Epichloë* endophyte	Zhu et al. (2022)
Fusarium oxysporum	Tomato	Endophyte: *Beauveria bassiana*	Nchu et al. (2022)
Fungal pathogens: *Fusarium oxysporum* f.sp. *cicero, Rhizoctonia solani, Sclerotium rolfsii*	Chickpea	*Bacillus subtilis*	Mageshwaran et al. (2022)
Golovinomyces cichoracearum: Powdery mildew disease	Tobacco	*Seimatosporium* sp. M7SB41	Zhao et al. (2023)
Fusarium oxysporum, Rhizoctonia solani, Pythium aphanidermatum	*Pisum sativum*	Endophytes isolated from *Ocimum sanctum* Linn: *Pseudomonas aeruginosa* OS_12, *Aneurinibacillus aneurinilyticus* OS_25	Gupta et al. (2022)

Abiotic stress			
Stress	Host plant	Plant growth-promoting microbe	Reference
Salinity	*Hibiscus hamabo*	Arbuscular mycorrhizal fungi	Yuan et al. (2019)
Heavy metals, salinity, drought, and temperature	Wheat	Wheat endophytic fungi: *Trichoderma* strains	Ripa et al. (2019)
Salinity	Tomato	*Bacillus spizizenii* FMH_{45}	Masmoudi et al. (2021)
Salinity	Tomato	Co-inoculation: *Bacillus licheniformis* and *Aspergillus violaceofuscus*	Muthuraja and Muthukumar (2022)
Salinity and drought	Grapevine	*Bacillus amyloliquefaciens* SB-9	Jiao et al. (2016)
Metal stress	Soyabean	*Plant growth-promoting rhizobacteria—Acinetobacter beijerinckii* and *Raoultella planticola*	Hussain et al. (2023)

resources and nutrients needed for plant growth, thus promoting plant growth and modulating the development of plants (Ma et al. 2016; Santoyo et al. 2016).

1.3.2 Root Colonization and Availability of Nutrients

Plant roots are one of the best parts for colonization as they remain exposed to soil and offer a wide range of interaction and inhabiting soil microorganisms including the rhizosphere (Saeed et al. 2021). Synek et al. (2021) showed the colonization and inhabiting of endophyte *Enterobacter* sp. SA187 in various plants, including, wheat, barley, and *Arabidopsis*. Essentially, this helps in plant growth and strengthen plant-tolerating potential under an unfavorable environment (Synek et al. 2021). Many bacterial colonization offer the availability of essential nutrients and supplements required for plant growth and utility. PGPR colonizes the rhizosphere of an agronomically important crop *Azospirillum brasilense,* followed by modulating the root architecture and enhancing the bioavailability of nutrients (Pii et al. 2015). Another recent study demonstrated that *Pseudomonas putida* KT2440 is an excellent root colonizer when colonized soybean and corn plants resulted in enhanced seed germination and stem length under elevated saline conditions compared with uninoculated plant (Costa-Gutierrez et al. 2020).

1.3.3 Secretion of Phytohormones and Enzymes

Plants survival under biotic and abiotic stress becomes feasible as the symbiotic association of plants with microbes helps not only in their growth but also provides protection from different phytopathogens, including infectious bacteria, fungi, and viruses. Mechanistically symbiotic bacteria and PGPR induce secretion of phytohormones and upregulated enzymes required by plants for their overall development. For example, *Bacillus subtilis* is well known for playing a significant role in improving tolerance potential to biotic stresses. Moreover, it degrades the precursor of ethylene ACC and helps plants grow and maintain their normal growth under stressful conditions (Glick et al. 2007; Hashem et al. 2019). Additionally, *Bacillus* spp. is also known to secrete siderophore exopolysaccharides that downregulate the movement of toxic ions and hence maintain the ionic balance. Interestingly, it enhances the water transportation across plant tissues and essentially protect from pathogenic microbes (Radhakrishnan et al. 2017). *Bacillus subtilis* possess chemoreceptors that enable them to recognize the right environment to interact.

1.3.4 Secretion of Secondary Metabolites and Volatile Organic Compounds

Plants produce a wide range of secondary metabolites and volatile organic compounds in response to the stress caused due to biotic and abiotic factors. Environmental factors, including humidity, temperature, and CO_2 content, can influence the production of secondary metabolite production. Certain environmental conditions, including high salinity, freezing temperature, drought, and mineral content in soil, essentially can cause adverse effects on plants. However, various volatile organic compounds and secondary metabolites encounter the adversity and allow them not only to stand but also maintain homeostasis (Ramakrishna and Ravishankar 2011).

Phytoalexins are one of the secondary metabolites usually present in trace amounts in healthy orchids, and when they are exposed to pathogenic fungi they significantly upregulate the genes encoding phytoalexin enzymes (Reinecke and Kindl 1994). Similarly, terpenes have multiple functions in plants, including protection against various phytopathogens and herbivores. As per previous studies, hemiterpenes enhance thermotolerance and photosynthesis rate in some of the plant species. Moreover, it has been demonstrated that when oak (*Quercus ilex)* leaves are fumigated with monoterpenes, they enhance the thermotolerance of the plant (Bartwal et al. 2013). Snowdrop lectin produced in *Triticum aestivum*, *Oryza sativa*, and *Arabidopsis* spp. potentially works as insecticidal for insects, including *Nilaparvata lugens*, Aphids, *Pieris rapae*, and *Spodoptera littoralis* (Saha et al. 2006). There are many more secondary metabolites known to protect plants and promote their growth and play a crucial role in maintaining homeostasis (Al-Khayri et al. 2023).

1.3.5 Reduces Oxidative Stress

A plant's ability to respond to the extreme stress condition is considered a crucial function. Understanding plant's strategies for tackling oxidative stress is important as it significantly regulates the plant's health and development. Several environmental factors, including UV stress, herbicide action, pathogens invasion (hypersensitive reactions), and unavailability of enough oxygen, can induce oxidative stress in plants. Besides environmental factors, plant's developmental transitions such as seed maturation induce reactive oxygen species (ROS). However, plants have evolved themselves and defensively act against ROS by activating genes associated with antioxidant molecule synthesis. Several studies have reported that chloroplast and peroxisomes are more prone and attacked by ROS (Grene 2002; Blokhina et al. 2003; Saha et al. 2006; Al-Khayri et al. 2023). To neutralize the effect of ROS and their effect, plants respond with upregulated synthesis of certain enzymes, including superoxidase, monodehydroascorbate reductase, and ascorbate peroxidase (Mullineaux et al. 2000).

1.3.6 Modulation of Plant Defense System

Plants are subjected to a plethora of foreign invaders throughout life that potentially impact their health and productivity. However, long exposure to such invaders allows plants to modulate their immune system and resist and respond against pathogens. Plant's first line of defense are equipped with pattern recognition receptor that potentially identify pathogen-associated marker patterns (PAMPs), known as PAMPs-triggered immunity (PTI) (Monaghan and Zipfel 2012). Moreover, plants are also acquired for the rapid and transient perturbations of calcium ion (Ca^{2+}) concentration, which activate a set of gene activation to generate requisite response to invaders (Reddy et al. 2011). Additionally, in response to the biotic stresses or invasion, PTI stimulate the expression of Ca^{2+} ion channels, resulting in an increase in cytoplasmic Ca^{2+} concentration and ultimately make an unfavorable environment to survive the pathogens and invaders.

1.4 Process of Microbial Inoculation: From Lab to Land

Environmental stresses, including biotic and abiotic factors, directly impact the productivity of plants and crops. Therefore, it has been imperative to develop strategies that can globally increase the plants' utility and enhance the agricultural food productivity in more economical and sustainable ways (Lopes et al. 2021). Besides limiting the exposure to chemical fertilizers, pesticides, and insecticides, it is also important to plan and uplift the idea of using beneficial microorganism and colonizing and integrating them with plant system to not only reduce the exposure to toxic and harmful chemical to the environment but also be ecofriendly, a sustainable approach to enhance the crops' productivity (Khan et al. 2020). Extreme environment can be a good source for selecting, isolating, and utilizing the beneficial microorganisms. For example, Inostroza et al. (2017) showed that microbial consortia taken from the rhizosphere of *Atriplex* sp., when grown in soil with undisturbed arid Chilean ecosystem, followed by wheat seedling result in the enhanced growth even in dry soil with limited phosphorus content (Inostroza et al. 2017). Environments like halophytic area can be a potent source for identifying beneficial microorganisms (Salwan et al. 2019). An exhaustive study by Mayak et al. (2004) led to the finding of 115 endophytes isolates from different crop wild relative (CWR) halophytes, which includes *Matthiola tricuspidata, Cakile maritima*, and *Crithmum maritimum*. Interestingly, many of the strains prevent the invasion of phytopathogens and a few were also acting as an antagonist for human pathogens like *Aspergillus fumigatus* (Mayak et al. 2004). Furthermore, some of the strains complemented with enhanced plant growth and uplifted salinity tolerance. Studies have also reported several other microorganisms associated with PGPM such as *Azospirillum, Bacillus, Enterobacter, Burkholderia, Flavobacterium, Clostridium, Pseudomonas, Frankia, Trichoderma,* and *Rhizobium* that have efficiently played imperative roles toward establishing a sustainable agriculture and improved plant productivity (Abhilash et al. 2016; Van Oosten et al. 2017; Gouda et al. 2018). These PGPM establish a symbiotic relationship with plants and help them by modulating phytohormone production, enhancing nutrient availability of soil and acting as barriers to pathogens (Asghari et al. 2020). Moreover, it can be colonized through different inoculation methods, including roots, soil, and foliar inoculation, which will have adequate interaction between plants and microorganisms. Among these methods, seed inoculations are the most common while foliar is the least. Interestingly, seed inoculation method is potentially used as an alternative to chemical treatment. Seed inoculation implies the immersing of the seed in microbial solution of known concentration (Lopes et al. 2018). The seed germination process provides abundant amino acids and carbohydrates in the form of seed exudates (Ahemad and Kibret 2014). Associated microorganisms utilize them as a source of nutrients and colonize and emerge into them when inoculated seed get transferred into the soil (Ammor et al. 2008). A study showed that seed inoculation with *Rhizobial* and *Bacillus* sp. enhanced the biomass production of *Cicer arietinum* (Khan et al. 2019) and *Oryza sativa* (Ullah et al. 2017), respectively. Another similar study reported that seed inoculation with *Burkholderia phytofirmans* led to enhanced phytoremediation of organic pollutants, including hydrocarbons (Afzal et al. 2013). Similarly, *Pseudomonas fluorescens* essentially increase the vigor biomass and enhance the resistance potential to water stress of *Virgin radiate* (Saravanakumar 2012).

Like seed inoculation, root inoculation method implies immersing roots in a microorganism solution. Post inoculation, seedling is planted to proper substrate and allowing size standardization. Additional advantages of this method are that inoculum is directly exposed to the host root and improves root colonization. Studies have shown that in plant species with asexual propagation, PGPM inoculated through the root method could show the enhanced ability of phytohormone synthesis such as auxin. Moreover, it also showed preventive activities against phytopathogens (Ahemad and Kibret 2014; Gouda et al. 2018). A study demonstrated that root inoculation with *Burkholderia* sp. enhanced *Vitis vinifera plant* growth and yield by elevating the tolerance level tolerance to low-temperature, modified carbohydrate metabolism (Fernandez et al. 2012). Another method of inoculation is soil inoculation, where PGPM are directly introduced into the soil by drenching, soil incorporation (mixed in the substrate), or microcapsule (Lopes et al. 2018). A study by Li et al. (2022a, b) demonstrated that when soil was inoculated with *Bacillus* it resulted in better nodulation and growth of *Cicer arietinum* (Li et al. 2022a, b). Another study by Hernandez-Montiel (2017) showed that the growth and productivity of *Lycopersicon esculentum* were significantly enhanced when soil was inoculated using *Pseudomonas putida* incorporated into the microcapsule (Hernández-Montiel et al. 2017). Additional advantages of delivering *Pseudomonas putida* in microcapsule are gradual release, greater viability, protection, and stability of roots. These results suggest that the inoculation method essentially influences the growth of plants and productivity of crops when beneficial microorganisms are utilized in a certain way.

The basic roadmap to the commercialization of PGPRs is not stringently specified but includes the basic orderly guidelines (Backer et al. 2018):

(a) Extraction of the bacterial species from the plant.
(b) Growth and screening of the microbes (single or multiple).
(c) Selection of the appropriate field for the growth of the crop within a certain environmental frame.
(d) Trail and testing of the best possible combinations of the bacterial strains.
(e) Planning of the prime field management applications.
(f) Purification of the final product.
(g) Checking the toxicity effects.
(h) Analysis of the finest form of delivery of the product.
(i) Ethical and regulatory approvals.
(j) Marketing of the product.

To formulate PGPR-based products, the composition should be such that it allows even dispersal in the field such as inoculums in the legume industry are mainly subjected to solid-based carriers, for example, sterilized peat (Bashan et al. 2014). This carrier system includes the inoculated cells, which are stuck to the surface of the seed during sowing in the field. Alternatively, alginate-based solid carriers have also been used due to sustainability issues of the peat (Bashan 2016). Other sources as potential candidates include biochar due to the flexibility to alter the nutrient and porous nature of the material (Głodowska et al. 2016).

1.5 Challenges, Limitations, and Future Perspective

Agriculture sector has been thriving over the years with the enormous use of chemical fertilizers and pesticides, which has led to a steep increase in the destruction of good soil microflora, causing detrimental effects to the environment as well as to the farmers. In search for the production of the beneficial crop, recent methods focus on the plant–microbe interaction to improve the yield and variety of the crop plants. The good microbes produce certain phytohormones, including auxins, gibberellins, abscisic acid, and cytokinins (Fahad et al. 2015). PGPMs can prove to be good alternatives that can strengthen the rhizosphere of the plants by naturally secreting the organic compounds and certain growth hormones required for viability of plant tissues. The plant–microbe symbiosis is triggered by the secretion and stimulation of favorable compounds required for plant growth. The major microbes benefiting the soil rhizosphere include *Pseudomonas* sp., *Bacillus* sp., *Rhizobium* sp., *Klebsiella,* and *Clostridium* sp., amongst others (Abhilash et al. 2016; Gouda et al. 2018; Van Oosten et al. 2017).

However, one of the major challenges faced by the overall composition of the rhizomicrobiome is the continued growth and maintenance of the live bacterial cells, that is, their shelf life. Studies report that the spore-forming Gram-positive bacteria maintain longer shelf life than the Gram-negative bacterial species (Schoebitz et al. 2013). Also, a big challenge is to eradicate the crosstalk between different bacterial species, which may lead to the development of biofilms (Ajijah et al. 2023; Karimi et al. 2022). Some of the interspecies interactions may prove to be fatal due to the matrix forming proteins and enzymes. Moreover, studies report the pathogenicity of a few microbial species like *Pseudomonas* sp. and *Burkholderia* sp. to humans (Kumar et al. 2013). The plant–microbe interaction serves as a mutual relationship, with the plant serving as a nutrient source and the microbes enhancing the growth of the plant. However, environmental changes play an important role in maintaining this plant–microbe interaction. The humid conditions within the soil and the extreme temperature fluctuations due to heat or cold and salinity make it difficult for the microbial inoculum to grow and proliferate (Bailey-Serres et al. 2019). Also, the other limitation is that the plants are exposed to pest attacks, which when healed are most likely to re-emerge due to climate changes. The constant climate changes have limited the yield from plants, leading to the adaptation of more synthetic ways for a highly viable crop production (Moore et al. 2017). The strategies to enhance the resilience and sustainable growth of plants will be beneficial with their localized implementation in agriculture fields.

Studies have proven to be impactful in using PGPMs as a method to safeguard the ecosystem by reducing the usage of chemicals and improving the productivity of the plants. Certain PGPRs, including that of *Pseudomonas, Rhizobium,* and *Azotobacter* species, amongst others, are being utilized commercially for large-scale production (Parray et al. 2016). Other approaches to enhance the utility and mass production of PGPRs are by modulating their usage through new technologies. Nanoparticle bead encapsulation is being done for the refinement of PGPRs (de Moraes et al. 2021; Nayana et al. 2020). The microbiome of the plant seeds can be modulated by using the microbial inoculation within the seeds for enhancement of the plant growth (Mitter et al. 2017). Future studies can be directed toward the use of microbial consortium with many advantages. Certain plant microbes can help provide essential nutrients to the plant growth and improve the fertility of the soil. The microbes used as PGPRs can also act as decomposers of toxic waste and help in getting rid of the accumulation of harmful waste products. These PGPRs can also be beneficial by supplementing the plant system with vital products that may help boost the immunity of the plants. Another important aspect could be the shielding effect from the environmental stresses. The future holds promise in the field of crop improvement with the help of plant growth-promoting microbes. However, economical techniques using microbes for more effective utilization by the farmers are needed. The manufacturers can include government-aided norms for regular and timely delivery of a fresh bacterial slot with higher stability and shelf life (Bailey-Serres et al. 2019). Also, proper guidelines and training can be given to the farmers by the communal administration to use the biofertilizers, including microbes, effectively. The government initiatives to help localize the use of microbes by small-scale setups by the farmers can help them maintain a systematic and consistent production of the inoculum (Vishwakarma et al. 2020). These measures can certainly assist in improving cell viability and stability of the bacterial inoculum to be used in the future.

1.6 Conclusions

With a continuously increasing population and climate change, there is an urgent need to search for sustainable substitutes for agrochemicals. Biofertilizers and biopesticides are eco-friendly and economic alternatives to toxic chemicals for stress management and plant growth promotion. Despite some limitations, research should be focused on the effective and timely transition of PGPMs from lab to land for stress management in sustainable agriculture.

References

Abhilash PC, Dubey RK, Tripathi V, Gupta VK, Singh HB (2016) Plant growth-promoting microorganisms for environmental sustainability. Trends Biotechnol 34:847–850. https://doi.org/10.1016/J.TIBTECH.2016.05.005

Afzal M, Khan S, Iqbal S, Mirza MS, Khan QM (2013) Inoculation method affects colonization and activity of *Burkholderia phytofirmans* PsJN during phytoremediation of diesel-contaminated soil. Int Biodeterior Biodegradation 85:331–336. https://doi.org/10.1016/J.IBIOD.2013.08.022

Ahemad M, Kibret M (2014) Mechanisms and applications of plant growth promoting rhizobacteria: current perspective. J King Saud Univ Sci 26:1–20. https://doi.org/10.1016/J.JKSUS.2013.05.001

Ajijah N, Fiodor A, Pandey AK, Rana A, Pranaw K (2023) Plant growth-promoting bacteria (PGPB) with biofilm-forming ability: a multifaceted agent for sustainable agriculture. Diversity 15:112. https://doi.org/10.3390/D15010112

Al-Khayri JM, Rashmi R, Toppo V, Chole PB, Banadka A, Sudheer WN, Nagella P, Shehata WF, Al-Mssallem MQ, Alessa FM, Almaghasla MI (2023) Plant secondary metabolites: the weapons for biotic stress management. Meta 13:716. https://doi.org/10.3390/METABO13060716

Ammor MS, Michaelidis C, Nychas GJE (2008) Insights into the role of quorum sensing in food spoilage. J Food Prot 71:1510–1525. https://doi.org/10.4315/0362-028X-71.7.1510

Armengol J, Berbegal M, Giménez-Jaime A, Romero S, Beltrán R, Vicent A, Ortega A, García-Jiménez J (2005) Incidence of Verticillium wilt of artichoke in Eastern Spain and role of inoculum sources on crop infection. Phytoparasitica 33:397–405. https://doi.org/10.1007/BF02981308

Asghari B, Khademian R, Sedaghati B (2020) Plant growth promoting rhizobacteria (PGPR) confer drought resistance and stimulate biosynthesis of secondary metabolites in pennyroyal (*Mentha pulegium* L.) under water shortage condition. Sci Hortic 263:109132. https://doi.org/10.1016/J.SCIENTA.2019.109132

Backer R, Rokem JS, Ilangumaran G, Lamont J, Praslickova D, Ricci E, Subramanian S, Smith DL (2018) Plant growth-promoting rhizobacteria: context, mechanisms of action, and roadmap to commercialization of biostimulants for sustainable agriculture. Front Plant Sci 871:402666. https://doi.org/10.3389/FPLS.2018.01473/BIBTEX

Bailey-Serres J, Parker JE, Ainsworth EA, Oldroyd GED, Schroeder JI (2019) Genetic strategies for improving crop yields. Nature 575:109. https://doi.org/10.1038/S41586-019-1679-0

Bartwal A, Mall R, Lohani P, Guru SK, Arora S (2013) Role of secondary metabolites and Brassinosteroids in plant defense against environmental stresses. J Plant Growth Regul 32:216–232. https://doi.org/10.1007/S00344-012-9272-X

Bashan N (2016) Inoculant formulations are essential for successful inoculation with plant growth-promoting bacteria and business opportunities. Indian Phytopathol 69:739–743

Bashan Y, De -Bashan LE, Prabhu SR, Hernandez JP (2014) Advances in plant growth-promoting bacterial inoculant technology: formulations and practical perspectives (1998-2013). Plant Soil 378:1–33. https://doi.org/10.1007/S11104-013-1956-X

Bhandari R, Sanz-Saez A, Leisner CP, Potnis N (2023) Xanthomonas infection and ozone stress distinctly influence the microbial community structure and interactions in the pepper phyllosphere. ISME Commun 3:24. https://doi.org/10.1038/s43705-023-00232-w

Blokhina O, Virolainen E, Fagerstedt KV (2003) Antioxidants, oxidative damage and oxygen deprivation stress: a review. Ann Bot 91:179–194. https://doi.org/10.1093/AOB/MCF118

Buswell W, Schwarzenbacher RE, Luna E, Sellwood M, Chen B, Flors V, Pétriacq P, Ton J (2018) Chemical priming of immunity without costs to plant growth. New Phytol 218:1205–1216. https://doi.org/10.1111/nph.15062

Cappelli SL, Domeignoz-Horta LA, Loaiza V, Laine AL (2022) Plant biodiversity promotes sustainable agriculture directly and via belowground effects. Trends Plant Sci 27:674–687. https://doi.org/10.1016/j.tplants.2022.02.003

Chaudhry V, Runge P, Sengupta P, Doehlemann G, Parker JE, Kemen E (2021) Shaping the leaf microbiota: plant–microbe–microbe interactions. J Exp Bot 72:36–56. https://doi.org/10.1093/jxb/eraa417

Chen Y, Fu D, Wang W, Gleason ML, Zhang R, Liang X, Sun G (2022) Diversity of Colletotrichum species causing apple bitter rot and Glomerella leaf spot in China. J Fungi 8:740. https://doi.org/10.3390/jof8070740

Costa-Gutierrez SB, Lami MJ, Santo MC, Zenoff AM, Vincent PA, Molina-Henares MA, Espinosa-Urgel M, de Cristóbal RE (2020) Plant growth promotion by *Pseudomonas putida* KT2440 under saline stress: role of ept A. Appl Microbiol Biotechnol 104:4577–4592. https://doi.org/10.1007/S00253-020-10516-Z

Cotton TA, Pétriacq P, Cameron DD, Meselmani MA, Schwarzenbacher R, Rolfe SA, Ton J (2019) Metabolic regulation of the maize Rhizobiome by benzoxazinoids. ISME J 13:1647–1658. https://doi.org/10.1038/s41396-019-0375-2

de Moraes ACP, Ribeiro LDS, de Camargo ER, Lacava PT (2021) The potential of nanomaterials associated with plant growth-promoting bacteria in agriculture. 3 Biotech 11:318. https://doi.org/10.1007/S13205-021-02870-0

Fahad S, Hussain S, Bano A, Saud S, Hassan S, Shan D, Khan FA, Khan F, Chen Y, Wu C, Tabassum MA, Chun MX, Afzal M, Jan A, Jan MT, Huang J (2015) Potential role of phytohormones and plant growth-promoting rhizobacteria in abiotic stresses: consequences for changing environment. Environ Sci Pollut Res 22:4907–4921. https://doi.org/10.1007/S11356-014-3754-2

Fernandez O, Theocharis A, Bordiec S, Feil R, Jacquens L, Clément C, Fontaine F, Barka EA (2012) *Burkholderia phytofirmans* PsJN acclimates grapevine to cold by modulating carbohydrate metabolism. Mol Plant-Microbe Interact 25:496–504. https://doi.org/10.1094/MPMI-09-11-0245

Gamalero E, Bona E, Glick BR (2022) Current techniques to study beneficial plant-microbe interactions. Microorganisms 10:1380. https://doi.org/10.3390/microorganisms10071380

Geetha N, Sunilkumar CR, Bhavya G, Nandini B, Abhijith P, Satapute P, Shetty HS, Govarthanan M, Jogaiah S (2023) Warhorses in soil bioremediation: seed biopriming with PGPF secretome to phytostimulate crop health under heavy metal stress. Environ Res 216:114498. https://doi.org/10.1016/j.envres.2022.114498

Glick WH, Miller CC, Cardinal LB (2007) Making a life in the field of organization science. J Organ Behav 28:817–835. https://doi.org/10.1002/JOB.455

Głodowska M, Husk B, Schwinghamer T, Smith D (2016) Biochar is a growth-promoting alternative to peat moss for the inoculation of corn with a pseudomonad. Agron Sustain Dev 36:1–10. https://doi.org/10.1007/S13593-016-0356-Z

Gouda S, Kerry RG, Das G, Paramithiotis S, Shin HS, Patra JK (2018) Revitalization of plant growth promoting Rhizobacteria for sustainable development in agriculture. Microbiol Res 206:131–140. https://doi.org/10.1016/J.MICRES.2017.08.016

Grene R (2002) Oxidative stress and acclimation mechanisms in plants. Arabidopsis Book 1:e0036. https://doi.org/10.1199/TAB.0036.1

Gu S, Wei Z, Shao Z, Friman VP, Cao K, Yang T, Kramer J, Wang X, Li M, Mei X, Xu Y (2020) Competition for iron drives phytopathogen control by natural rhizosphere microbiomes. Nat Microbiol 5:1002–1010. https://doi.org/10.1038/s41564-020-0719-8

Gupta N, Pradhan S, Jain A, Patel N (2021) Sustainable agriculture in India 2021. CEEW report. 122. https://www.ceew.in/sites/default/files/CEEWFOLU-Sustainable-Agriculture-in-India-2021-20Apr21.Pdf

Gupta S, Pandey S, Sharma S (2022) Decoding the plant growth promotion and antagonistic potential of bacterial endophytes from *Ocimum sanctum* Linn. Against root rot pathogen Fusarium oxysporum in *Pisum sativum*. Front. Plant Sci 13:813686. https://doi.org/10.3389/fpls.2022.813686

Hashem A, Tabassum B, Abd Allah EF (2019) *Bacillus subtilis*: a plant-growth promoting rhizobacterium that also impacts biotic stress. Saudi J Biol Sci 26:1291–1297. https://doi.org/10.1016/j.sjbs.2019.05.004

Hernández D, García-Pérez O, Perera S, González-Carracedo MA, Rodríguez-Pérez A, Siverio F (2023) Fungal pathogens associated with aerial symptoms of Avocado (*Persea americana* Mill.) in Tenerife (Canary Islands, Spain) focused on species of the family Botryosphaeriaceae. Microorganisms 11:585. https://doi.org/10.3390/microorganisms11030585

Hernández-Montiel LG, Chiquito Contreras CJ, Murillo Amador B, Vidal Hernández L, Quiñones Aguilar EE, Chiquito Contreras RG (2017) Efficiency of two inoculation methods of pseudomonas putida on growth and yield of tomato plants. J Soil Sci Plant Nutr 17:1003–1012. https://doi.org/10.4067/S0718-95162017000400012

Hussain A, Shah M, Hamayun M, Iqbal A, Qadir M, Alataway A, Dewidar AZ, Elansary HO, Lee IJ (2023) Phytohormones producing rhizobacteria alleviate heavy metals stress in soybean through multilayered response. Microbiol Res 266:127237. https://doi.org/10.1016/j.micres.2022.127237

Inbaraj MP (2021) Plant-microbe interactions in alleviating abiotic stress—a mini review. Front Agron 3:667903. https://doi.org/10.3389/fagro.2021.667903

Inostroza NG, Barra PJ, Wick LY, Mora ML, Jorquera MA (2017) Effect of rhizobacterial consortia from undisturbed arid-and agro-ecosystems on wheat growth under different conditions. Lett Appl Microbiol 64:158–163. https://doi.org/10.1111/LAM.12697

Jiao J, Ma Y, Chen S, Liu C, Song Y, Qin Y, Yuan C, Liu Y (2016) Melatonin-producing endophytic bacteria from grapevine roots promote the abiotic stress-induced production of endogenous melatonin in their hosts. Front Plant Sci 7:1387. https://doi.org/10.3389/fpls.2016.01387

Jofre MF, Mammana SB, Appiolaza ML, Silva MF, Gomez FJ, Cohen AC (2023) Melatonin production by rhizobacteria native strains: towards sustainable plant growth promotion strategies. Physiol Plant 175:e13852. https://doi.org/10.1111/PPL.13852

Kapadia C, Sayyed RZ, El Enshasy HA, Vaidya H, Sharma D, Patel N, Malek RA, Syed A, Elgorban AM, Ahmad K, Zuan ATK (2021) Halotolerant microbial consortia for sustainable mitigation of salinity stress, growth promotion, and mineral uptake in tomato plants and soil nutrient enrichment. Sustain For 13:8369. https://doi.org/10.3390/su13158369

Karimi E, Aliasgharzad N, Esfandiari E, Hassanpouraghdam MB, Neu TR, Buscot F, Reitz T, Breitkreuz C, Tarkka MT (2022) Biofilm forming rhizobacteria affect the physiological and biochemical responses of wheat to drought. AMB Express 12:93. https://doi.org/10.1186/S13568-022-01432-8

Khan N, Bano A, Rahman MA, Guo J, Kang Z, Babar MA (2019) Comparative physiological and metabolic analysis reveals a complex mechanism involved in drought tolerance in chickpea (*Cicer arietinum* L.) induced by PGPR and PGRs. Sci Rep 9:1–19. https://doi.org/10.1038/s41598-019-38702-8

Khan N, Bano A, Ali S, Babar MA (2020) Crosstalk amongst phytohormones from planta and PGPR under biotic and abiotic stresses. Plant Growth Regul 90:189–203. https://doi.org/10.1007/S10725-020-00571-X

Kizheva Y, Georgiev G, Donchev D, Dimitrova M, Pandova M, Rasheva I, Hristova P (2022) Cross-over pathogenic bacteria detected in infected tomatoes (*Solanum lycopersicum* L.) and peppers (*Capsicum annuum* L.) in Bulgaria. Pathogens 11:1507. https://doi.org/10.3390/pathogens11121507

Kumar A, Munder A, Aravind R, Eapen SJ, Tümmler B, Raaijmakers JM (2013) Friend or foe: genetic and functional characterization of plant endophytic *Pseudomonas aeruginosa*. Environ Microbiol 15:764–779. https://doi.org/10.1111/1462-2920.12031

Lee S, Chakma N, Joung S, Lee JM, Lee J (2022) QTL mapping for resistance to bacterial wilt caused by two isolates of *Ralstonia solanacearum* in chili pepper (*Capsicum annuum* l.). Plan Theory 11:1551. https://doi.org/10.3390/plants11121551

Li Q, Hou Z, Zhou D, Jia M, Lu S, Yu J (2022a) Antifungal activity and possible mechanism of *Bacillus amyloliquefaciens* FX2 against the postharvest apple ring rot pathogen. Phytopathology 112:2486–2494. https://doi.org/10.1094/PHYTO-02-22-0047-R

Li D, Zhao C, Li P, Liu C, Gong D, Liu S, Yuan Z, Chen Y (2022b) Macro-and micro-physical characteristics of different parts of mixed convective-stratiform clouds and differences in their responses to seeding. Adv Atmos Sci 39:2040–2055. https://doi.org/10.1007/S00376-022-2003-8

Lopes MJ, Dias-Filho MB, Castro TH, Silva GB (2018) Light and plant growth-promoting rhizobacteria effects on *Brachiaria brizantha* growth and phenotypic plasticity to shade. Grass Forage Sci 73:493–499. https://doi.org/10.1111/GFS.12336

Lopes MJS, Dias-Filho MB, Gurgel ESC (2021) Successful plant growth-promoting microbes: inoculation methods and abiotic factors. Front Sustain Food Sys 5:606454. https://doi.org/10.3389/FSUFS.2021.606454/BIBTEX

Ma Y, Rajkumar M, Zhang C, Freitas H (2016) Beneficial role of bacterial endophytes in heavy metal phytoremediation. J Environ Manag 174:14–25. https://doi.org/10.1016/J.JENVMAN.2016.02.047

MacLaren C, Mead A, van Balen D, Claessens L, Etana A, de Haan J, Haagsma W, Jäck O, Keller T, Labuschagne J, Myrbeck A (2022) Long-term evidence for ecological intensification as a pathway to sustainable agriculture. Nat Sustain 5:770–779. https://doi.org/10.1038/s41893-022-00911-x

Mageshwaran V, Gupta R, Singh S, Sahu PK, Singh UB, Chakdar H, Bagul SY, Paul S, Singh HV (2022) Endophytic *Bacillus subtilis* antagonize soil-borne fungal pathogens and suppress wilt complex disease in chickpea plants (*Cicer arietinum* L.). Front Microbiol 13:994847. https://doi.org/10.3389/fmicb.2022.994847

Malgioglio G, Rizzo GF, Nigro S, Lefebvre du Prey V, Herforth-Rahmé J, Catara V, Branca F (2022) Plant-microbe interaction in sustainable agriculture: the factors that may influence the efficacy of PGPM application. Sustain For 14:2253. https://doi.org/10.3390/su14042253

Masmoudi F, Tounsi S, Dunlap CA, Trigui M (2021) Halotolerant Bacillus spizizenii FMH45 promoting growth, physiological, and antioxidant parameters of tomato plants exposed to salt stress. Plant Cell Rep 40:1199–1213. https://doi.org/10.1007/s00299-021-02702-8

Mayak S, Tirosh T, Glick BR (2004) Plant growth-promoting bacteria confer resistance in tomato plants to salt stress. Plant Physiol Biochem 42:565–572. https://doi.org/10.1016/J.PLAPHY.2004.05.009

Mirghasempour SA, Studholme DJ, Chen W, Zhu W, Mao B (2022) Molecular and pathogenic characterization of Fusarium species associated with corm rot disease in saffron from China. J Fungi 8:515. https://doi.org/10.3390/jof8050515

Mitter B, Pfaffenbichler N, Flavell R, Compant S, Antonielli L, Petric A, Berninger T, Naveed M, Sheibani-Tezerji R, von Maltzahn G, Sessitsch A (2017) A new approach to modify plant microbiomes and traits by introducing beneficial bacteria at flowering into progeny seeds. Front Microbiol 8:230405. https://doi.org/10.3389/FMICB.2017.00011/BIBTEX

Monaghan J, Zipfel C (2012) Plant pattern recognition receptor complexes at the plasma membrane. Curr Opin Plant Biol 15:349–357. https://doi.org/10.1016/J.PBI.2012.05.006

Moore FC, Baldos U, Hertel T, Diaz D (2017) New science of climate change impacts on agriculture implies higher social cost of carbon. Nat Commun 8:1607. https://doi.org/10.1038/S41467-017-01792-X

Mullineaux P, Ball L, Escobar C, Karpinska B, Creissen G, Karpinski S (2000) Are diverse signalling pathways integrated in the regulation of Arabidopsis antioxidant defence gene expression in response to excess excitation energy? Philos Trans R Soc B Biol 355:1531–1540. https://doi.org/10.1098/RSTB.2000.0713

Munir N, Hanif M, Abideen Z, Sohail M, El-Keblawy A, Radicetti E, Mancinelli R, Haider G (2022) Mechanisms and strategies of plant microbiome interactions to mitigate abiotic stresses. Agronomy 12:2069. https://doi.org/10.3390/agronomy12092069

Muthuraja R, Muthukumar T (2022) Co-inoculation of halotolerant potassium solubilizing *Bacillus licheniformis* and *Aspergillus violaceofuscus* improves tomato growth and potassium uptake in different soil types under salinity. Chemosphere 294:133718. https://doi.org/10.1016/j.chemosphere.2022.133718

Nadarajah K, Abdul Rahman NSN (2021) Plant–microbe interaction: aboveground to belowground, from the good to the bad. Int J Mol Sci 22:10388. https://doi.org/10.3390/ijms221910388

Nayana AR, Joseph BJ, Jose A, Radhakrishnan EK (2020) Nanotechnological advances with PGPR applications. Sustainable agriculture reviews 41: nanotechnology for plant growth and development. Springer, Cham, pp 163–180. https://doi.org/10.1007/978-3-030-33996-8_9

Nchu F, Macuphe N, Rhoda I, Niekerk LA, Basson G, Keyster M, Etsassala NG (2022) Endophytic *Beauveria bassiana* induces oxidative stress and enhances the growth of Fusarium oxysporum-infected tomato plants. Plan Theory 11:3182. https://doi.org/10.3390/plants11223182

Palmer CL, Skinner W (2002) *Mycosphaerella graminicola*: latent infection, crop devastation and genomics. Mol Plant Pathol 3:63–70. https://doi.org/10.1046/j.1464-6722.2002.00100.x

Pandey G (2018) Challenges and future prospects of Agri-nanotechnology for sustainable agriculture in India. Environ Technol Innov 11:299–307. https://doi.org/10.1016/j.eti.2018.06.012

Parray JA, Jan S, Kamili AN, Qadri RA, Egamberdieva D, Ahmad P (2016) Current perspectives on plant growth-promoting Rhizobacteria. J Plant Growth Regul 35:877–902. https://doi.org/10.1007/S00344-016-9583-4

Pii Y, Penn A, Terzano R, Crecchio C, Mimmo T, Cesco S (2015) Plant-microorganism-soil interactions influence the Fe availability in the rhizosphere of cucumber plants. Plant Physiol Biochem 87:45–52. https://doi.org/10.1016/J.PLAPHY.2014.12.014

Pollak S, Cordero OX (2020) Rhizobiome shields plants from infection. Nat Microbiol 5:978–979. https://doi.org/10.1038/s41564-020-0766-1

Radhakrishnan R, Hashem A, Abd Allah EF (2017) Bacillus: a biological tool for crop improvement through bio-molecular changes in adverse environments. Front Physiol 8:293128. https://doi.org/10.3389/FPHYS.2017.00667/BIBTEX

Ramakrishna A, Ravishankar GA (2011) Influence of abiotic o loop stress signals on secondary metabolites in plants. Plant Signal Behav 6:1720–1731. https://doi.org/10.4161/PSB.6.11.17613

Reddy AS, Ali GS, Celesnik H, Day IS (2011) Coping with stresses: roles of calcium-and calcium/calmodulin-regulated gene expression. Plant Cell 23:2010–2032. https://doi.org/10.1105/TPC.111.084988

Reinecke T, Kindl H (1994) Inducible enzymes of the 9, 10-dihydro-phenanthrene pathway. Sterile orchid plants responding to fungal infection. Mol Plant Microbe Interact 7:449–454. https://doi.org/10.1094/MPMI-7-0449

Ripa FA, Cao WD, Tong S, Sun JG (2019) Assessment of plant growth promoting and abiotic stress tolerance properties of wheat endophytic fungi. Bio Med Res Int 2019:6105865. https://doi.org/10.1155/2019/6105865

Romero FM, Marina M, Pieckenstain FL (2014) The communities of tomato (*Solanum lycopersicum* L.) leaf endophytic bacteria, analyzed by 16S-ribosomal RNA gene pyrosequencing. FEMS Microbiol Lett 351:187–194. https://doi.org/10.1111/1574-6968.12377

Saeed Q, Xiukang W, Haider FU, Kučerik J, Mumtaz MZ, Holatko J, Naseem M, Kintl A, Ejaz M, Naveed M, Brtnicky M (2021) Rhizosphere bacteria in plant growth promotion, biocontrol, and bioremediation of contaminated sites: a comprehensive review of effects and mechanisms. Int J Mol Sci 22:10529. https://doi.org/10.3390/IJMS221910529

Saha P, Majumder P, Dutta I, Ray T, Roy SC, Das S (2006) Transgenic rice expressing *Allium sativum* leaf lectin with enhanced resistance against sap-sucking insect pests. Planta 223:1329–1343. https://doi.org/10.1007/S00425-005-0182-Z

Salwan R, Sharma A, Sharma V (2019) Microbes mediated plant stress tolerance in saline agricultural ecosystem. Plant Soil 442:1–22. https://doi.org/10.1007/S11104-019-04202-X

Santoyo G, Moreno-Hagelsieb G, del Carmen O-MM, Glick BR (2016) Plant growth-promoting bacterial endophytes. Microbiol Res 183:92–99. https://doi.org/10.1016/J.MICRES.2015.11.008

Saravanakumar D (2012) Rhizobacterial ACC deaminase in plant growth and stress amelioration. In: Bacteria in agrobiology: stress management. Springer, Cham, pp 187–204. https://doi.org/10.1007/978-3-642-23465-1_9

Schoebitz M, López M, Roldán A, López MD (2013) Bioencapsulation of microbial inoculants for better soil-plant fertilization. A review. Agron Sustain Dev 33:751–765. https://doi.org/10.1007/s13593-013-0142-0

Synek L, Rawat A, L'Haridon F, Weisskopf L, Saad MM, Hirt H (2021) Multiple strategies of plant colonization by beneficial endophytic *Enterobacter* sp. SA187. Environ Microbiol 23:6223–6240. https://doi.org/10.1111/1462-2920.15747

Ullah MA, Mahmood IA, Ali A, Nawaz Q, Sultan T (2017) Effect of inoculation methods of Biozotemax (plant growth promoting Rhizobacteria-Pgpr) on growth and yield of Rice under naturally salt-affected soil. Res Plant Biol 7:24–26. https://doi.org/10.25081/RIPB.2017.V7.3602

Van Oosten MJ, Pepe O, De Pascale S, Silletti S, Maggio A (2017) The role of biostimulants and bioeffectors as alleviators of abiotic stress in crop plants. Chem Biol Technol Agric 4:1–12. https://doi.org/10.1186/S40538-017-0089-5

Vishwakarma K, Kumar N, Shandilya C, Mohapatra S, Bhayana S, Varma A (2020) Revisiting plant–microbe interactions and microbial consortia application for enhancing sustainable agriculture: a Review. Front Microbiol 11:560406. https://doi.org/10.3389/FMICB.2020.560406/BIBTEX

Wang L, Hu J, Li D, Reymick OO, Tan X, Tao N (2022) Isolation and control of *Botrytis cinerea* in postharvest green pepper fruit. Sci Hortic 302:111159. https://doi.org/10.1016/j.scienta.2022.111159

Worley JN, Russell AB, Wexler AG, Bronstein PA, Kvitko BH, Krasnoff SB, Munkvold KR, Swingle B, Gibson DM, Collmer A (2013) *Pseudomonas syringae* pv. Tomato DC3000 CmaL (PSPTO 4723), a DUF1330 family member, is needed to produce L-Allo-isoleucine, a precursor for the phytotoxin coronatine. J Bacteriol 195:287–296

Yuan Y, Brunel C, van Kleunen M, Li J, Jin Z (2019) Salinity-induced changes in the rhizosphere microbiome improve salt tolerance of hibiscus hamabo. Plant Soil 443:525–537. https://doi.org/10.1007/s11104-019-04258-9

Zhang H, Zhu J, Gong Z, Zhu JK (2022) Abiotic stress responses in plants. Nat Rev Genet 23:104–119. https://doi.org/10.1038/s41576-021-00413-0

Zhao Y, Mao W, Tang W, Soares MA, Li H (2023) Wild Rosa endophyte M7SB41-mediated host Plant's powdery mildew resistance. J Fungi 9:620. https://doi.org/10.3390/jof9060620

Zhu Y, Zhu S, Zhang F, Zhao Z, Christensen MJ, Nan Z, Zhang X (2022) Transcriptomic analyses reveals molecular regulation of photosynthesis by Epichloe endophyte in A*chnatherum inebrians* under *Blumeria graminis* infection. J Fungi 8:1201. https://doi.org/10.3390/jof8111201

Chapter 2
An Outline of Plant–Microbe Interactions for Stress Management

Shalini Tiwari

Abstract Climate change and environmental stresses, including both biotic and abiotic stress, directly influence plant growth imparting losses in crop yield and productivity worldwide. Plants develop intricate physiological and molecular responses to these unfavorable environments to improve their tolerance, adaptability, and survival. These days, many agronomists and researchers support a variety of strategies, such as conventional breeding and transgenic technology, for the development of stress-resistant varieties. The use of plant growth-promoting microbes (PGPM) as an alternative technology to increase crop resistance to stress is becoming more popular. Also, the use of PGPM to mitigate environmental changes is becoming increasingly important given the current climate condition and overuse of fertilizers in agricultural soils. The use of these beneficial microorganisms is considered one of the most promising methods for safe crop management practices. In the presence of these beneficial microbes, several events at the molecular and cellular levels take place from the perception of environmental stimuli to metabolic changes for better adaption in plants. Thus, this chapter takes into account our current state of knowledge about plant stress management in the presence of PGPM under various adverse environmental conditions and also considers the mechanism involved behind the plant–PGPM interaction and its application for sustainable agriculture.

Keywords Climate change · Plant growth-promoting microbe · Mechanism · Stress

---

S. Tiwari (✉)
Department of Biochemistry and Molecular Biology, Oklahoma State University, Stillwater, OK, USA

P. Singh Chauhan et al. (eds.), *Plant-Microbe Interaction and Stress Management*, Rhizosphere Biology,
https://doi.org/10.1007/978-981-97-4239-4_2

2.1 Introduction

Environmental stress, including both biotic and abiotic stress, poses a serious threat to sustainable agriculture production and productivity, and subsequently global food security by causing major yield loss or compromising the nutritional value of cereal crops. To cope with these adverse conditions, plants adopt various stress tolerance mechanisms for their better development. Plant stress tolerance involves changes in whole plant at cellular, physiological, and molecular levels for stress tolerance or avoidance (Tiwari et al. 2017). The expression of several thousand genes is also known to be altered during various individual and multiple stresses. The sensitivity of plants toward a range of stresses is a prime challenge for their overall growth and productivity.

This global climate crisis has heightened the need for innovative strategies to enhance plant stress tolerance and productivity. Several crop improvement programs are directed toward developing an improved variety of crops; however, the application of plant growth-promoting microorganisms (PGPM) is gaining momentum these days as it is a cost and time-effective technique to save plants from adverse environmental conditions, without changing their genotype or developing a new variety. Plant–microbe interactions have emerged as key players in this quest, with a myriad of microbial partners influencing plant stress responses. These beneficial microbes can be either bacteria or fungus. These microorganisms have several plant growth-promoting (PGP) traits that aid plants in combating the stress condition. These traits mainly include phytohormone secretion, nutrient stabilization, nitrogen fixation, siderophore production, biocontrol activity, exo-polysaccharides production, and biofilm formation. Earlier studies reported that microorganisms having these traits promote the growth and development of plants and also can increase production and yield in crop plants. Altogether we can say that plant–microbe interactions play a pivotal role in the stress management of plants, influencing their growth, development, and overall health. Therefore, in the present review, we shed light on the recent advancements in our understanding of the intricate relationships between plants and microbes, focusing on their collaborative mechanisms in mitigating various environmental stresses. The symbiotic associations between plants and beneficial microbes, such as rhizobacteria and mycorrhizal fungi, are explored for their potential applications in sustainable agriculture and stress resilience. This section provides an overview of the importance of stress management in plants and the critical role of microbes in this context.

2.2 Beneficial Plant–Microbe Interactions

2.2.1 Plant Growth-Promoting Bacteria

Plant growth-promoting bacteria (PGPB) encompass a diverse group of bacteria with the ability to positively influence plant growth and development. These bacteria belong to various genera, including *Azospirillum*, *Pseudomonas*, *Bacillus*, and *Rhizobium* (Tiwari et al. 2019). Each genus and species of PGPB possesses unique mechanisms that contribute to their plant growth-promoting capabilities. This group includes both exophytic and endophytic bacteria. The best examples of exophytic bacteria are mainly rhizobacteria, which incorporate those soil-dwelling bacteria that inhabit the rhizosphere, the zone around plant roots. Unlike pathogenic soil microbes, plant growth-promoting rhizobacteria (PGPR) engage in mutualistic associations with plants, conferring various benefits to their host. PGPR have been extensively studied for their ability to enhance plant growth and stress tolerance. Plants release root exudates containing sugars, amino acids, and organic acids that serve as chemical signals, attracting rhizobacteria to the rhizosphere. The commonality between these rhizobacteria is that they are known to exhibit several PGP traits, thereby modulating the overall plant's tolerance capability and its stress response. Several previous studies reported the role of rhizobacteria in various plant stress management in different crops, including rice, chickpea, and maize (Kar et al. 2024; Tiwari et al. 2016, 2021a; Mishra et al. 2020; Chauhan et al. 2019).

Conversely, endophytic microbes inhabit plant tissues and contribute to stress management by producing bioactive compounds and promoting systemic resistance. Actinobacteria, Firmicutes, and Proteobacteria are among the diverse bacterial phyla found as endophytes, each contributing unique capabilities to the plant host. Endophytic bacteria are prolific producers of secondary metabolites, including alkaloids, terpenoids, and polyketides, which play a role in stress management and defense against pathogens. Some endophytes produce phytohormones, influencing plant growth and stress responses, such as auxins, cytokinins, and gibberellins. These bacteria also contribute to the activation of antioxidant enzymes in plants, mitigating oxidative stress associated with environmental challenges. Apart from nutrient solubilization and making essential minerals available to the plant host, a few endophytic bacteria are known to fix atmospheric nitrogen, providing a supplementary nitrogen source for the plant. These microbes aid in osmotic adjustment in plants, helping maintain cell turgor and hydration during periods of drought stress. Endophytes also assist in ion homeostasis, preventing the toxic accumulation of ions and enhancing the plant's ability to tolerate saline conditions. In the context of biotic stress, endophytic microbes serve as biological control agents, protecting plants from pathogenic infections and contributing to disease management.

2.2.2 *Mycorrhizal Fungi*

Mycorrhizal fungi establish mutualistic relationships with plants, facilitating nutrient exchange and enhancing resistance to various stresses. The symbiotic relationship between plants and mycorrhizal fungi is characterized by a harmonious exchange of resources, leading to enhanced plant growth and stress resilience. The most prevalent type of mycorrhizal association, arbuscular mycorrhiza (AM) fungi, forms arbuscules within plant root cells, aiding in nutrient transfer. However, ectomycorrhizae (ECM) fungi envelop plant roots, primarily benefiting trees and woody plants by enhancing nutrient uptake and providing protection against stress. These fungi establish extensive underground networks that facilitate nutrient exchange not only between plants and fungi but also among neighboring plants, creating a "common mycorrhizal network" (Figueiredo et al. 2021). Apart from nutrient acquisition and mobilization, mycorrhizal associations enhance the plant's ability to extract water from the soil, a crucial advantage during periods of water scarcity (Huey et al. 2020). Mycorrhizal fungi contribute to osmotic adjustment in plants, aiding in the maintenance of turgor pressure under drought conditions. However, under saline conditions, mycorrhizal fungi contribute to ion homeostasis in plants, preventing the toxic accumulation of sodium while enhancing the uptake of essential nutrients (Dastogeer et al. 2020). These fungi stimulate the activity of antioxidant enzymes, mitigating oxidative stress induced by salinity. Mycorrhizal associations also induce systemic resistance in plants, activating defense mechanisms that protect against various stresses, including pathogens and pests (Boyno et al. 2023). Mycorrhizal associations trigger the production of signaling molecules, such as jasmonic acid and salicylic acid that play a central role in stress responses. Interestingly, it is also observed that rhizobacteria often form symbiotic relationships with mycorrhizal fungi, creating a tripartite interaction that further enhances nutrient uptake and plant resilience (Wilkes et al. 2021).

2.3 Mechanism Behind Plant–Microbe Interactions and Stress Management

The intricate interplay between plants and beneficial microbes during stress conditions involves a complex web of molecular interactions that orchestrate adaptive responses in both partners. This interaction involves the key signaling pathways that are known for the recognition of microbial partners and the activation of stress-related responses, leading to genetic responses, and physiological changes that contribute to enhanced stress tolerance in plants.

In that instance, rhizobacteria release a repertoire of signaling molecules, including quorum-sensing compounds, lipopolysaccharides (LPS), and volatile organic compounds (VOCs), in response to stress conditions. These signals act as elicitors, triggering plant responses to combat stress. Plants, in turn, possess pattern recognition receptors (PRRs) that detect these rhizobacterial signals. Further, the recognition process initiates a cascade of molecular events, activating stress-related signaling pathways in the host plant.

Particularly in the context of hormonal regulation, rhizobacteria influence ABA signaling in plants, modulating stomatal closure, and promoting water-use efficiency under stress conditions. Rhizobacterial interactions induce the synthesis of SA and JA, signaling molecules associated with systemic acquired resistance (SAR) and induced systemic resistance (ISR), respectively, enhancing the plant's defense against stress. Interestingly, studies reported the rhizobacteria-induced modulation in genes of ABA, SA, and JA signaling cascade under different abiotic stresses (Tiwari et al. 2016, 2017). Rhizobacteria are also known to induce ISR by priming the plant's immune system through the activation of mitogen-activated protein kinase (MAPK) cascades and the expression of pathogenesis-related (PR) proteins (Zhu et al. 2022). This heightened state of readiness enables the plant to mount a quicker and more robust defense against stressors. Additionally, rhizobacteria produce enzymes and organic acids that modify the rhizosphere, enhancing nutrient availability for plants. Plants, in response to rhizobacterial signals, release specific root exudates that attract beneficial rhizobacteria, forming a positive feedback loop that promotes nutrient mobilization and uptake (Pantigoso et al. 2022).

The molecular basis behind the abovementioned mechanisms is due to changes in gene expression patterns in both plants and rhizobacteria in response to stress. Transcriptomic analyses reveal the activation of stress-responsive genes involved in defense mechanisms, antioxidant pathways, and nutrient transport (Chauhan et al. 2019). Emerging evidence suggests the existence of cross-kingdom signaling, where genetic responses in rhizobacteria influence gene expression in plants and vice versa, fostering a harmonized response to stress. Advanced omics technologies provide a holistic understanding of the molecular intricacies in plant–rhizobacteria interactions under various stresses. Integration of these data sets allows for the identification of key regulatory networks and potential targets for crop improvement. A study by Tiwari et al. (2021a) in rice reported that *OsNAM* gene plays important roles in root rhizobacteria interaction through abiotic stress and phytohormone crosstalk. Similarly, *OsJAZ* gene deciphers rhizobacteria-mediated nutrient stress modulation in rice (Bisht et al. 2023). Interestingly, earlier studies reported that the whole GRAM-domain-containing gene family showed a response to PGPR treatment under different abiotic stress conditions (Tiwari et al. 2020, 2021b). Overall, it can be said that numerous rhizobacteria-induced genes have involvement in the betterment of plant growth as well as stress management.

2.4 Applications in Sustainable Agriculture

Harnessing the plant growth-promoting abilities of microbes has led to the development of biofertilizers, providing a sustainable alternative to chemical fertilizers. Microbe-based inoculants are being explored for their potential to improve crop yield, reduce dependency on synthetic inputs, and enhance soil health. The application of endophytic microbial inoculants has shown promise in promoting crop growth, yield, and stress resilience in various agricultural systems. Understanding the host specificity of endophytes and their adaptability to different plant species is crucial for effective applications. A study by Agarwal et al. (2019) reported PGPR *Bacillus amyloliquefaciens*-induced *OsASR6* improves plant growth and yield by altering root auxin sensitivity and the xylem structure in transgenic *Arabidopsis thaliana*. Drought tolerance induction and growth promotion by indole acetic acid-producing *Pseudomonas aeruginosa* in *Vigna radiata* was also reported (Uzma et al. 2022). However, the impact of endophytic microbes on ecosystem dynamics and their interactions with other microorganisms warrants further exploration. Similarly, mycorrhizal fungi alter the expression of stress-responsive genes in plants, orchestrating a finely tuned molecular response. The application of mycorrhizal inoculants in agriculture has shown promise in enhancing crop yield, particularly in nutrient-poor soils and under stress conditions. Harnessing mycorrhizal symbiosis aligns with the principles of sustainable agriculture, reducing the reliance on chemical inputs and fostering ecosystem health.

2.5 Conclusion and Future Perspectives

Increasing global population, urbanization, and industrialization are increasing the rate of conversion of arable land into wasteland. Supplying food to an ever-increasing population is one of the biggest challenges that agriculturalists and plant scientists are currently confronting. Environmental stresses make this situation even graver. Despite the induction of several tolerance mechanisms, sensitive plants often fail to survive under environmental extremes. Therefore, addressing current challenges and understanding and harnessing plant–microbe interactions for stress management is the need of the hour. Understanding the molecular mechanisms behind plant–rhizobacteria interactions during stress is crucial for harnessing these partnerships to develop sustainable agricultural practices. The insights gained from such investigations offer promising avenues for engineering crops with enhanced stress tolerance, ultimately contributing to global food security in the face of changing environmental conditions. Similarly, the effectiveness of mycorrhizal interactions can be influenced by environmental factors, necessitating a deeper understanding of optimized applications. Future research should aim to tailor rhizobacterial interventions based on specific plant–microbe interactions, soil types, and environmental conditions, paving the way for precision agriculture.

References

Agarwal P, Singh PC, Chaudhry V, Shirke PA, Chakrabarty D, Farooqui A, Nautiyal CS, Sane AP, Sane VA (2019) PGPR-induced OsASR6 improves plant growth and yield by altering root auxin sensitivity and the xylem structure in transgenic *Arabidopsis thaliana*. J Plant Physiol 240:153010. https://doi.org/10.1016/j.jplph.2019.153010

Bisht N, Anshu A, Singh PC, Chauhan PS (2023) Comprehensive analysis of OsJAZ gene family deciphers rhizobacteria-mediated nutrient stress modulation in rice. Int J Biol Macromol 253:126832. https://doi.org/10.1016/j.ijbiomac.2023.126832

Boyno G, Rezaee Danesh Y, Demir S, Teniz N, Mulet JM, Porcel R (2023) The complex interplay between arbuscular mycorrhizal fungi and Strigolactone: mechanisms, Sinergies, applications and future directions. Int J Mol Sci 24:16774. https://doi.org/10.3390/ijms242316774

Chauhan PS, Lata C, Tiwari S, Chauhan AS, Mishra SK, Agrawal L, Chakrabarty D, Nautiyal CS (2019) Transcriptional alterations reveal *Bacillus amyloliquefaciens*-rice cooperation under salt stress. Sci Rep 9:11912. https://doi.org/10.1038/s41598-019-48309-8

Dastogeer KM, Zahan MI, Tahjib-Ul-Arif M, Akter MA, Okazaki S (2020) Plant salinity tolerance conferred by arbuscular mycorrhizal fungi and associated mechanisms: a meta-analysis. Front Plant Sci 11:1927. https://doi.org/10.3389/fpls.2020.588550

Figueiredo AF, Boy J, Guggenberger G (2021) Common mycorrhizae network: a review of the theories and mechanisms behind underground interactions. Front Fungal Biol 2:48. https://doi.org/10.3389/ffunb.2021.735299

Huey CJ, Gopinath SC, Uda MNA, Zulhaimi HI, Jaafar MN, Kasim FH, Yaakub ARW (2020) Mycorrhiza: a natural resource assists plant growth under varied soil conditions. 3 Biotech 10:1–9

Kar S, Mishra SK, Misra S, Agarwal R, Kumar S, Chauhan PS (2024) Endophytic Alkalotolerant plant growth-promoting bacteria render maize (*Zea mays* L.) growth under alkaline stress. Curr Microbiol 81:43. https://doi.org/10.1007/s00284-023-03557-w

Mishra SK, Khan MH, Misra S, Dixit VK, Gupta S, Tiwari S, Gupta SC, Chauhan PS (2020) Drought tolerant *Ochrobactrum* sp. inoculation performs multiple roles in maintaining the homeostasis in *Zea mays* L. subjected to deficit water stress. Plant Physiol Biochem 150:1–4. https://doi.org/10.1016/j.plaphy.2020.02.025

Pantigoso HA, Newberger D, Vivanco JM (2022) The rhizosphere microbiome: plant–microbial interactions for resource acquisition. J Appl Microbiol 133:2864–2876. https://doi.org/10.1111/jam.15686

Tiwari S, Lata C, Chauhan PS, Nautiyal CS (2016) *Pseudomonas putida* attunes morphophysiological, biochemical and molecular responses in *Cicer arietinum* L. during drought stress and recovery. Plant Physiol Biochem 99:108–117. https://doi.org/10.1016/j.plaphy.2015.11.001

Tiwari S, Prasad V, Chauhan PS, Lata C (2017) Bacillus amyloliquefaciens confers tolerance to various abiotic stresses and modulates plant response to phytohormones through osmoprotection and gene expression regulation in rice. Front Plant Sci 8:1510. https://doi.org/10.3389/fpls.2017.01510

Tiwari S, Prasad V, Lata C (2019) Bacillus plant growth promoting bacteria for sustainable agriculture and environment. In: New and future developments in microbial biotechnology and bioengineering 1. Elsevier, Amsterdam, pp 43–55. https://doi.org/10.1016/B978-0-444-64191-5.00003-1

Tiwari S, Shweta S, Prasad M, Lata C (2020) Genome-wide investigation of GRAM-domain containing genes in rice reveals their role in plant-rhizobacteria interactions and abiotic stress responses. Int J Biol Macromol 156:1243–1257. https://doi.org/10.1016/j.ijbiomac.2019.11.162

Tiwari S, Gupta SC, Chauhan PS, Lata C (2021a) An OsNAM gene plays important role in root rhizobacteria interaction in transgenic Arabidopsis through abiotic stress and phytohormone crosstalk. Plant Cell Rep 40:143–155. https://doi.org/10.1007/s00299-020-02620-1

Tiwari S, Muthamilarasan M, Lata C (2021b) Genome-wide identification and expression analysis of *Arabidopsis* GRAM-domain containing gene family in response to abiotic stresses and PGPR treatment. J Biotechnol 325:7–14. https://doi.org/10.1016/j.jbiotec.2020.11.021
Uzma M, Iqbal A, Hasnain S (2022) Drought tolerance induction and growth promotion by indole acetic acid producing Pseudomonas aeruginosa in *Vigna radiata*. PLoS One 17:e0262932. https://doi.org/10.1371/journal.pone.0262932
Wilkes TI, Warner DJ, Edmonds-Brown V, Davies KG, Denholm (2021) The tripartite Rhizobacteria-AM fungal-host plant relationship in winter wheat: impact of multi-species inoculation, tillage regime and naturally occurring Rhizobacteria species. Plan Theory 10:1357. https://doi.org/10.3390/plants10071357
Zhu L, Huang J, Lu X, Zhou C (2022) Development of plant systemic resistance by beneficial rhizobacteria: recognition, initiation, elicitation and regulation. Front Plant Sci 13:952397. https://doi.org/10.3389/fpls.2022.952397

Chapter 3
Exploring the Phyllosphere: Microbial Diversity, Interactions, and Ecological Significance in Plant Health

Ariyan Manikandan, Rangasamy Anandham, Sivakumar Madhan, Rajasekaran Raghu, Ramasamy Krishnamoorthy, and Murugaiyan Senthilkumar

Abstract The phyllosphere, which refers to the surfaces of plant leaves, is a unique biological niche that harbors a diverse microbial community. The bacteria, fungus, yeasts, cyanobacteria, and algae present in this specific niche play a significant role in the intricate interactions within the phyllosphere. The resilience of the microbiome and its distinctive survival strategies demonstrate its capacity to endure UV radiation, limited access to nutrients, and the impacts of climate change.

The presence of fungi and yeasts in the phyllosphere has a significant impact on the nutrient cycle and serves as a link between terrestrial and aquatic ecosystems. Genetic characteristics of host plants, analyzing seasonal variations, examining geographical locations, and environmental factors, also influence the composition of phyllosphere microbiota. The microbial community composition and function are intricately influenced by these factors, which reflect the dynamic nature of the phyllosphere. The phyllosphere microbiome provides plants with benefits that extend beyond nutrient uptake. These microorganisms also produce various plant growth-promoting substances (PGPS). These activities contribute to the enhance-

A. Manikandan
Department of Agricultural Microbiology, Tamil Nadu Agricultural University, Coimbatore, Tamil Nadu, India

The Institute of Ecology and Earth Sciences, University of Tartu, Tartu, Estonia

R. Anandham (✉) · S. Madhan · M. Senthilkumar
Department of Agricultural Microbiology, Tamil Nadu Agricultural University, Coimbatore, Tamil Nadu, India

R. Raghu
Department of Plant Biotechnology, Tamil Nadu Agricultural University, Coimbatore, Tamil Nadu, India

R. Krishnamoorthy
Department of Crop Management, Vanavarayar Institute of Agriculture, Pollachi, Tamil Nadu, India

P. Singh Chauhan et al. (eds.), *Plant-Microbe Interaction and Stress Management*, Rhizosphere Biology,
https://doi.org/10.1007/978-981-97-4239-4_3

ment of plant growth. Furthermore, the phyllosphere microbiome exerts control over plant diseases and contributes to the plants' ability to withstand biotic stressors. This chapter highlights the ecological significance of the phyllosphere microbiome and their potential applications in agricultural and environmental management. The comprehension of plant–microbe interactions within the phyllosphere facilitates the utilization of microbial resources to enhance plant resilience and promote sustainability within ecosystems.

Keywords Phylloplane · Plant growth-promoting bacteria · *Methylobacterium* · Plant growth-promoting bacteria

3.1 Introduction

The phyllosphere refers to the ecological niche formed by the surfaces of plant leaves. The phyllosphere encompasses the aboveground components of plants, such as leaves, flowers, stems, fruits, and pollens, each possessing distinct morphological and physical characteristics. The microbiota, which inhabits the surface of the phyllosphere, refers to epiphytes. At the same time, endophytes are the microbiota members that reside within the tissues of the phyllosphere, either in the intercellular spaces (apoplast) or within the plant cell. The phyllosphere constitutes approximately 60% of the total biomass across all species on Earth, thus serving as a crucial habitat for microbial life. Leaves are the extensively studied components within the phyllosphere (Koskella 2020).

Leaves are highly structured and layered plant organs. They are distinguished by a thin, protective layer of wax called the cuticle, which surrounds the top and lower epidermis. This cuticle is a physical barrier, shielding the leaf from harmful factors such as living organisms and nonliving environmental stresses. Consequently, leaves provide an inhospitable environment for microbial life. This diverse ecosystem is commonly subjected to various climatic factors, including temperature changes, UV radiation, nutrient scarcity, humidity, and dryness. Despite these challenging conditions, it supports a wide array of microbial communities that are highly adapted to survive in such harsh surroundings. The survival of these microorganisms relies on their ability to develop specific mechanisms to withstand the unfavorable environmental conditions found in the phyllosphere. These mechanisms include producing pigments to protect against UV radiation, generating extracellular polysaccharides (EPS) and biosurfactants to enhance surface attachment and protect against desiccation, and producing chemical compounds to compete for resources and space (Sundin 2002; Helfrich et al. 2018; Gal et al. 2003; Schreiber et al. 2005).

However, the phyllosphere contains many bacteria, fungi, yeast, archaea, and other microorganisms that interact with the plant host in various ways, including commensal, pathogenic, and mutualistic relationships. According to estimates, bacteria are the most prevalent inhabitants of leaf surfaces, with densities reaching up to 10^8 cells per cm^2 (Leveau 2006). The average yeast concentration on a leaf typically falls between 10 and 10^4 cells per square centimeter (Shivas and Brown 1984). Filamentous fungi are typically short-lived inhabitants of leaf surfaces, primarily existing as spores. Alongside various plant pathogenic bacteria and oomycetes, they frequently play a significant role in the phyllosphere, negatively impacting plants' health and vitality. Phyllospheric communities play a crucial role in nitrogen fixation, bioremediation of harmful chemicals or pollutants, and as biocontrol agents against foliar plant pathogens. Environmental variables influence the microbial diversity in the phyllosphere and play a vital role in global food webs and nutrient connections (Glawe 2008).

3.2 Diversity of Phyllosphere Microbiome

Plants and their accompanying microbial communities have a deep ecological and evolutionary relationship with each other (Vandenkoornhuyse et al. 2015). The phyllosphere contains three primary categories of microorganisms: bacteria, fungi, and oomycetes. These microorganisms engage with abiotic stimuli and inherit the host's genes, successfully colonizing plants. Therefore, it is crucial to classify the system as a holobiont, in which the plants and their accompanying creatures have a close and interconnected relationship. The colonization of microorganisms in the phyllosphere begins as soon as the seedlings appear. In the case of perennial plants, the microbes establish themselves during bud bursts after having spent the winter in buds or twigs (Osono et al. 2014). The microbial populations reside on the leaf surface as both endophytes and epiphytes, and the ecological characteristics of each compartment exhibit notable distinctions.

3.2.1 Bacteria

Bacteria are the predominant organisms in the phyllosphere, and they can have beneficial, harmful, or neutral effects on the host plants they colonize. The phyllospheric bacteria can exist in either endophytic or epiphytic forms (Newton et al. 2010). They can also be classified as pathogenic or nonpathogenic (Mazinani et al. 2017). The predominant bacterial groups found in the phyllosphere include Proteobacteria (specifically α, β, and γ-proteobacteria), Bacteroidetes, Actinobacteria, and Firmicutes. Even among individuals within the same group, variations might occur due to differences in habitat, geographical location, or even variations in the host plants they rely on. The literature survey on phyllosphere microbial communities has shown that Proteobacteria and Firmicutes are the most prevalent phyla in different plants' phyllosphere. *Bacillus* and *Pseudomonas* are the most prominent genera. Actinobacteria and Bacteriodetes were prevalent in most instances. The predominant species identified were *Bacillus*, *Pantoea*, *Erwinia*, *Pseudomonas*, *Sphingomonas*, *Acinetobacter*, *Xanthomonas*, and *Gluconobacter*.

Yashiro et al. (2011) studied the apple phyllosphere microbiome using culture-dependent and culture-independent methods. A total of 309 isolates were identified using the culture-dependent approach. The most common orders were Sphingomonadales and Actinomycetales, Rhizobiales, Pseudomonadales, and Burkholderiales. Additionally, specific isolates were shown to be closely related to plant pathogens such as *Curtobacterium flaccumfaciens* pv., *flaccumfaciens*, *Pseudomonas syringae*, and *Xylophilus ampelinus*. Among the 317 clones analyzed using a culture-independent approach, the most common sequences originated from bacteria belonging to the Sphingomonadales order. The Pseudomonadales, Burkholderiales, Sphingobacteriales, and Rhizobiales orders followed this. In their study, Kim et al. (2012) utilized a metagenomic technique to explore the bacterial population of six different kinds of tropical trees in the phyllosphere. The bacterial phyla that are most plentiful are Alphaproteobacteria, Acidobacteria, Gammaproteobacteria, Actinobacteria, Bacteroidetes, and Deltaproteobacteria. Several phyla, including Betaproteobacteria, Planctomycetes, Firmicutes, Cyanobacteria, and Chloroflexi, were regularly found in all samples, comprising 1–4% of the total. Despite not being the dominating species in the phyllosphere, they were continuously present.

Methanol, an evaporative organic molecule, is emitted from the plant surface due to pectin demethylation. Due to its vast size, the phyllosphere serves as the most significant interface on a global scale, resulting in a substantial release of methanol. The genus *Methylobacterium* includes several bacteria frequently encountered as members of the pink-pigmented facultative methylotrophs (PPFM). These bacteria can use methanol as their exclusive source of carbon and energy. Metaproteomic analyses have confirmed the presence of related proteins involved in this process (Delmotte et al. 2009). Research conducted on the metaproteogenomics of various plant species has revealed that *Methylobacterium* sp. is the predominant microorganism found in the phyllosphere of plants. This finding is reasonable, considering that leaf surfaces are known to have the highest levels of methanol emissions (Delmotte et al. 2009). Environmental factors influence the population of PPFM in the phyllosphere. It tends to decrease from spring to summer and then gradually increases toward the end of the cropping season in red clover and wheat crops, as observed in a study conducted by Omer et al. (2004). According to Mizuno (2013), *M. mesophilicum*, *M. radiotolerans*, and *M. fujisawanse* are efficient colonizers of perilla plant species. PPFM produces auxins, gibberellins, N_2 fixation, siderophore production, cytokinin, and phosphorous solubilization reported (Yadav et al. 2018; Madhaiyan et al. 2010; Krishnamoorthy et al. 2018; Verma et al. 2014). These substances play a role in plant growth, development, and seed germination, leading to increased crop yields (Madhaiyan et al. 2005; Meena et al. 2012). The ability of aerobic methylobacteria to withstand heavy metals such as nickel, cadmium, copper, zinc, chromium, mercury, lead, and arsenic has been extensively documented (Idris et al. 2006; Dourado et al. 2012).

Pseudomonas syringae, a bacterial pathogen, has been the subject of extensive research in numerous studies. The pathogen induces illnesses in various commercially significant plant species (Burch et al. 2014). The pathogen proliferates on the leaf surface but can also infiltrate the inner leaf tissue through stomata or wounds, leading to illness in the intercellular space. This plant pathogen is capable of inducing ice nucleation, resulting in the formation of precipitate (Morris et al. 2014a, b). Ice nucleation is believed to enhance nutrient uptake in injured leaves and encourage bioprecipitation, which involves the transfer of microbes from nutrient-deficient clouds to nutrient-abundant environments (Morris et al. 2014a, b).

3.2.2 Fungi and Yeasts

The fungal community on the surface of plant leaves can be classified as either epiphytic or endophytic. The population of filamentous fungi can vary from 10^2 to 10^8 colony-forming units (CFU) per gram of leaf, while the population of yeast can range from 10 to 10^{10} CFU per leaf (Thompson et al. 1993). The leaves regularly harbor fungal genera such as *Cladosporium*, *Acremonium*, *Mucor*, *Alternaria*, *Penicillium*, and *Aspergillus*. Additionally, yeasts such as *Cryptococcus*, *Sporobolomyces*, and *Rhodotorula* are frequently found. In their study, Sadaka and Ponge (2003) investigated the analysis of fungal communities in the leaves of *Quercus rotundifolia*. *Trichothecium roseum* and *Ulocladium chartarum* were observed to demonstrate a significant level of detection on green leaves. *Aureobasidium pullulans*, *Cladosporium cladosporioides*, *Alternaria alternata*, and *Epicoccum purpurascens* were detected on both living and senescent leaves, with the highest rates of isolation observed on senescent leaves. Venkatachalam et al. (2016) isolated *Curvularia lunata* and *Bipolaris ravenelii* from the surface of rice plants. These microorganisms are frequently found to cause plant diseases and have been linked to many leaf-related ailments. Fungi also experience seasonal fluctuations (Rastogi et al. 2012). The fungi found in the phyllosphere can exhibit many behaviors, such as acting as saprophytes or phytopathogens or forming associations with algae to form lichens (Frey-Klett et al. 2011). The phyllosphere fungal communities have a crucial role in the recycling of nutrients and in connecting terrestrial and aquatic ecosystems. In their study, Fu et al. (2016) isolated yeast from the phyllosphere of the *Drosera spatulate*. The isolated yeast strains were tested for multiple plant growth-promoting (PGP) traits. These traits included the ability to produce indole-3-acetic acid (IAA), ammonia, and polyamines, as well as the capacity to solubilize calcium phosphate and zinc oxide. Additionally, the yeast strains were evaluated for their siderophore production. A total of 32 yeasts and yeast-like fungi have been found, including *Aureobasidium pullulans*, *Aureobasidium* sp., *Candida akabanensis*, *Dothideomycetes* sp., *Hanseniaspora uvarum*, *Meyerozyma Caribbean*, *Saccharomyces cerevisiae*, *Barnettozyma California*, and *Cryptococcus laurentii*.

3.2.3 Cyanobacteria and Algae

Algae and cyanobacteria are the foundation of most ecosystems' food chains. The extensive presence and variety of algae in soil, including the rhizosphere, have been thoroughly recorded (Prasanna et al. 2009; Lin and Wu 2013), especially in rice fields. Scant reports have shown the range of cyanobacteria and algal variety in the phyllosphere of several plants. Most published material on algae or cyanobacteria documents their prevalence in tropical woods due to the prevailing optimal humidity conditions in a study by Zhu et al. (2018), who analyzed phyllosphere algal communities in tropical regions. A total of 2922 operational taxonomic units (OTUs) have been identified, which are distributed among five distinct classes: Ulvophyceae, Trebouxiophyceae, Chlorophyceae, Dinophyceae, and Eustigmatophyceae. Rigonato et al. (2012) examined the range of cyanobacterial species in a pristine mangrove ecosystem in Brazil by employing metagenomic methods. The findings indicate that the habitat and plant species influence the diversity of cyanobacteria. The orders of *Nostocales* and *Oscillatoriales* were the most abundant. Venkatachalam et al. (2016) documented the occurrence of cyanobacteria in the phyllosphere of rice for the initial instance, employing both culture enrichment and culture-independent methods (PCR-DGGE). Bright-field microscopy identified unicellular and filamentous morphotypes, including *Lyngbya*, *Plectonema*, and *Nostoc*. Metagenomic analysis detected 4 and 12 unique species (phylotypes) in samples obtained from different cultivation strategies. Kim et al. (2012) discovered the presence of cyanobacteria in the phyllosphere of the tropical rainforest in Malaysia. They were consistently found in all the samples but did not constitute most of the population. In a similar study, Lin et al. (2013) examined the diversity, with a particular focus on the epiphyllous algae, in the rainforest of Taiwan. The researchers observed a total of six species of green algae, specifically *Chloroidium saccharophilum*, *Ettlia pseudoalveolaris*, *Klebsormidium flaccidum*, *Prasiococcus calcarius*, *Rosenvingiella radicans*, and *Trebouxia* sp., as well as one species of cyanobacteria, *Leptolyngbya* sp. The algae play a crucial role in nutrient cycling by mobilizing nitrogen and affecting water loss through evaporation on the host plant. These structures are also believed to play a role in producing photosynthates utilized by heterotrophs, such as bacteria and fungi, in the phyllosphere.

3.3 Factors Affecting Phyllosphere Microbiota

3.3.1 Host Plant

The host plant's genetic makeup and physical characteristics significantly influence microbial communities' composition in the phyllosphere (Bodenhausen et al. 2014; Wagner et al. 2016). Plants with varying genetic compositions exhibit distinct physical characteristics, which subsequently impact the community of microorganisms that inhabit their leaf surfaces, known as the phyllosphere microbiota (Li et al. 2018). The leaves' chemical and physical characteristics and growth stages greatly influence the composition of the phyllosphere microbiome. For example, a change in the process of producing a protective layer on the surface of leaves called cuticular wax impacts the types of bacteria that live on the leaves of *Arabidopsis thaliana* plants. Bodenhausen et al. (2014) studied developing plant host genotype-dependent communities using the SynCom technique. Their findings showed that the synthesis of cuticle and the perception of ethylene impacted the bacterial communities in the phyllosphere of *A. thaliana*. Kembel and Mueller (2014) conducted a study to estimate the fungal communities on the leaves of 51 different tree species in a lowland tropical rainforest in Panama. The purpose was to investigate how the taxonomy and characteristics of the host plants affect the fungi that colonize the phyllosphere. The taxonomic identity of the host plant accounted for almost 50% of the diversity in fungal community composition among trees. Additionally, several functional aspects of the host plant, such as leaf morphology, leaf chemistry, and plant growth and mortality, were significantly linked to the structure of the fungal community.

3.3.2 Seasonal Variation

The composition of the phyllosphere microbiome also varies with seasonal changes. Temporal and seasonal fluctuations cause significant shifts in temperature, humidity, and solar radiation levels, leading to changes in the structure and composition of the foliar microbiota (Beattie 2011). Various studies have reported the seasonal changes greatly influence the bacterial and fungal populations (Li et al. 2022). Li et al. (2022) studied the phyllosphere and fungal communities in 45 tree species in forest plot in wet and dry season. They found significant clustering of the phyllosphere microbial assemblage among trees based on seasonality. Furthermore, they found that the effects of seasonality have a greater impact on the assembly of bacterial communities compared to fungal communities. Bao et al. (2020) demonstrated comparable seasonal changes in the makeup of phyllosphere microbial communities in three tree species: *Ginkgo biloba*, *Pinus bungeana*, and *Sabina chinensis*. In their study, Ibekwe et al. (2021) investigated the impact of seasonal variations on the bacterial community composition in *Spinacia oleracea* L., cv. Racoon. The researchers identified the four primary phyla present in the phyllosphere, namely Proteobacteria, Firmicutes, Bacteroidetes, and Actinobacteria. Additionally, the researchers have investigated the impact of seasonality on the phyllosphere microbiomes. Variations in the seasonal climate cause long-term fluctuations in these communities over a year, the process of microbial succession, and the age of the plant host (Jackson and Denney 2011; Redford and Fierer 2009). The fluctuation in seasonal conditions emphasizes the importance of considering the timing of sampling and the host plant in studies of phyllosphere microbial ecology.

3.3.3 Geographical Location

Like plant and animal biogeography, microbial biogeography examines microbial diversity across different geographical locations and periods (Andrews and Harris 2000). The study undertaken by Finkel et al. (2011) focus on Tamarix trees. A study has shown that the microbial phyllosphere communities found on several Tamarix species are highly homogeneous within a certain geographical region. Nevertheless, trees of the identical species that thrive in distinct climatic regions possess distinct microbial communities. Redford et al. (2010) conducted a study to estimate the phyllosphere communities in *Pinus ponderosa* from various locations all over the world. They found that the variety in bacterial community composition between different species was greater than the variability within the same species. Also, there are not much differences even across continents. The study revealed that the geographical origin of the host is the primary factor influencing the composition of epiphytic fungal communities, while the host plant mainly shapes bacterial communities. Therefore, fungi demonstrated a greater degree of endemism than bacteria, potentially influenced by distinct limitations related to climate and dispersal (Bonito et al. 2014). Qian et al. (2018) observed significant differences in the taxonomic diversity of fungi associated with *Mussaenda shikokiana* across a small range of altitudes. The researchers showed that phyllosphere fungal communities associated with a single host exhibited a wide range of taxonomic diversity and various trophic guilds. Elevation has a major impact on the richness of OTUs.

3.3.4 Environmental Factors

Geographical location, climatic circumstances, temperature, moisture, and CO_2 levels significantly impact the microbial community found on the surface of plants, known as the phyllosphere (Xu et al. 2020). For example, studies have shown that temperature and CO_2 levels impact molecular pathways and can considerably influence the variety and composition of microorganisms in the phyllosphere (Aydogan et al. 2020). The effects mentioned contribute to the significance of the phyllosphere as a crucial environment for thermophilic bacteria (Godon et al. 2020). A recent study demonstrated that exposure to O_3 and water-deficit stress can decrease the diversity of microorganisms in the phyllosphere, as well as a reduction in the abundance of Betaproteobacteria and an increase in the abundance of gamma proteobacteria. These changes in microbial composition can indicate the risk of disease in poplar trees (Li et al. 2023).

3.4 Benefits of Phyllosphere Microbiome over Plants

In the stimulation of plant growth when it comes to the acquisition of nutrients by plants, the phyllosphere is regarded as less active compared to the rhizosphere, which is widely known for its crucial function in making nutrients soluble and mobile. Nitrogen-fixing bacteria have recently been found in the phyllosphere, as Delmotte et al. (2009) and Holland (2011) described. However, the mechanisms involved in nitrogen fixation have not been thoroughly investigated, as noted by Stone et al. (2018). Nitrogen-fixing organisms provide fixed nitrogen to leaves, stimulating plant growth and augmenting the nitrogen concentration in plants (Giri and Pati 2004). Chemoautotrophic nitrifiers, specifically Archaea, have been discovered in the phyllosphere, suggesting that the ability to reduce ammonia pollution is ubiquitous and not limited to bacteria (Stone et al. 2018). In their study, Abadi et al. (2020) investigated nitrogen-fixing microorganisms present in the phyllosphere of maize. Seventy-seven isolates of N-fixing bacteria were obtained from the maize phyllosphere in various geographical regions. *Bacillus* spp. was the most prevalent among these bacteria, accounting for 37 isolates.

The rhizosphere microbiome has significantly enhanced plant development by synthesizing hormones, but the phyllosphere microbiome has made relatively minor contributions (Liu et al. 2013). IAA-producing bacteria help promote plant development and productivity (Romero et al. 2016; Kushwaha et al. 2020). *Methylobacterium* sp., obtained from the phyllosphere of pepper plants, can synthesize IAA (Tani et al. 2015). Similarly, in a different study conducted by Krishnamoorthy et al. (2018) on groundnut plants, *M. populi* TMV7-4, *M. thiocyanatum* VRI7-A4, and *M. radiotolerans* VRI8-A4 were shown to generate IAA at levels ranging from 3 to 4 μg/mL. These strains were obtained from the surface of groundnut plants.

A crucial mechanism for enhancing plants' ability to withstand drought is the closing of stomata mediated by abscisic acid (ABA) as ABA primarily regulates stomatal closure (Sussmilch et al. 2017). Phyllosphere bacteria have been discovered to synthesize ABA and modify the plant's immunological response (Cao et al. 2011) during periods of drought. Remarkably, it has been noted that bacteria residing in the phyllosphere generate more IAA and cytokinins under typical circumstances, which counteract the stomatal closure regulated by ABA. However, the regulation of these different activities in regulating the plant stress response is not well investigated (Stone et al. 2018). Also, various studies reported that the phyllosphere microorganism can be able to solubilize the plant nutrients' P solubilization (Krishnamoorthy et al. 2018), Verma et al. 2014); sulfur oxidation (Anandham et al. 2007, 2014). Various microbes involved in the PGPG activities are listed in Table 3.1.

Table 3.1 Diverse phylloplane bacteria and their plant growth-promoting traits

S. No.	Organisms	Isolation source	Characters	Reference
1.	*Methylobacterium populi* TMV7–4 *M. thiocyanatum* VRI7-A4 *M. radiotolerans* VRI8-A4 *M. thiocyanatum* SM11490T *Xanthomonas sacchari* TMV2–3 *Rhizobium pusense* TMV2–6 *Cronobacter malonaticus* TMV2–7 *Psuedomonas psychrotolerans* TMV2-A1	Ground nut	Auxins, ACC, P solubilization, sulfur oxidation Plant growth promotion	Krishnamoorthy et al. (2018)
2.	Delftia *lacustris* PPO-1	Onion	Biocontrol ability	Janahiraman et al. (2016)
3.	*Bacillus subtilis* PPT-1	Tomato		
4.	*Bacillus cereus* PPB-1	Bhendi		
5.	*Pantoea agglomerans, Klebsiella terrigena* *Erwinia rhapontici, Rahnella aquatilis*	*Rosa rugosa*	–	Hashidoko et al. (2002)
6.	*Roseomonas musae*	Banana	–	Nutaratat et al. (2013)
7.	*Acinetobacter*	Tomato		Dong et al. (2019)
8.	*B. subtilis, Bacillus megaterium, Bacillus paralicheniformis, Bacillus pumilus, Microbacterium arborescens, Pseudarthrobacter oxydans, Psuedomonas putida, Stenotrophomonas maltophilia, Enterobacter hormaechei, Kocuria polaris*	Maize	IAA, nitrogenase activity, , EPS, inorganic P solubilization and siderophore	Abadi et al. (2020)
9.	*Psuedomonas stutzeri,* *Acinetobacter* sp., *B. mojavensis, Pseudomonas chlororaphis,* *Enterobacter asburia*	Cotton	ACC deaminase production, IAA, siderophore, and nutrient solubilization	Sharath et al. (2021)
10.	*B. pumilus, B. amyloliquefaciens, B. subtilis, B. paralicheniformis, Bacillus licheniformis, Bacillus niabensis, B. megaterium*	Maize	N fixation	Abadi et al. (2021)
11.	*Rhizobium* b1 and *B. subtilis* b2	Tomato	Biocontrol of *P. syringae* pv. tomato and *Alternaria solani*	Shao et al. (2023)

(continued)

Table 3.1 (continued)

S. No.	Organisms	Isolation source	Characters	Reference
12.	*Aureobasidium pullulans* JYC101, *Aureobasidium* sp. JYC378, *Dothideomycetes* sp. JYC359, *Galactomyces candidum* JYC360, *Kazachstania jiainicus* JYC361, *S. cerevisiae* JYC382, *Sporidiobolus ruineniae* JYC099, *Ustilago esculenta* JYC368, *Pseudozyma* sp. JYC372	*Drosera spatula*	IAA, NH3, P solubilization, and ACC deaminase production	Fu et al. (2016)
13.	*Chromelosporium*, *Tricharina*, *Fusarium*, *Neofabraea*, and *Pyronema*, *Hyalodendriella*, *Phoma*, *Biscogniiauxia*, and *Coniozyma*, *Peinicillium*, *Lewia*, Nemainia, and *Phaeosphaeria*, *Copirinopsis*, *Coprinellus*, *Ulocladium*	Olive tree	–	Gomes et al. (2018)
14.	*Colletotrichum gloeosporioides*, *Colletotrichum acutatum*, *Pestalotiopsis* sp.1 *Aureobasidium pullulans*, *Phoma* sp. and *Ramichloridium* sp., *Geniculosporium* sp.	*Camellia japonica*	–	Osono (2008)
15.	*Aspergillus* spp., *Chaetomium* spp., *Colletotrichum* spp., *Curvularia* spp., *Fusarium* spp., *Penicillium* spp., *Phytophthora* spp., *Rhizopus* spp., *Trichoderma* spp.	Rice		Leewijit et al. (2016)
16.	*Aureobasidium pullulans* and *Rhodotorula glutinis*	Wheat	Biodegradation	Kucharska et al. (2020)
17.	*Cryptococcus* sp. OTU 9, *Cryptococcus* sp. OTU 4, *Leucosporidium* sp. OTU 26, *Taphrina* sp. OTU 3, *Protomyces* sp. OTU 1, *Dioszegia* sp. OTU 23 strain 1, *Dioszegia* sp. OTU, *Cryptococcus* sp. OTU 17, *Cystobasidiomycetes* sp. OTU 32, *Microbotryozyma* sp. OTU 28, *Leucosporidium* sp. OTU 27	*Arabidopsis thaliana*		Wang et al. (2021)

3.4.1 Abiotic Stress Tolerance

Multiple studies have demonstrated the significance of phyllosphere bacteria in enhancing plants' ability to withstand drought, extreme temperatures (Grover et al. 2011; Meena et al. 2015), and saline environments (Sharma et al. 2019). Only a limited number of research have investigated the impact of phyllosphere microorganisms in enhancing plants' capacity to withstand different environmental conditions. The extracellular polymeric substances (EPS) generated by bacteria offer defense not only to the bacteria itself but also, to some degree, to plants against desiccation and UV radiation (Stone et al. 2018). *Pseudomonas* spp., commonly found in the phyllosphere, naturally create biofilms that reduce water loss and protect against UV radiation. The presence of pigmented bacteria enhances their resistance to UV radiation, increasing UV tolerance across the phyllosphere microbiome (Jacobs and Sundin 2001). The phyllosphere microbiome likely has a crucial impact on the heat tolerance of plants. The hypothesis suggests that tolerance is achieved through two mechanisms: the release of heat by the cell walls of fungi and the activation of heat stress-responsive genes (Redman et al. 2002). Heat shock proteins have conserved functions throughout several organisms, including bacteria, fungi, and plants (Queitsch et al. 2000). Gene- and functional-level homology is a promising approach for investigating plant–microbe interactions and enhancing plant heat resistance, especially in the phyllosphere (Stone et al. 2018). The advantages of phyllosphere bacteria in frost production have been well demonstrated, in addition to their ability to tolerate high temperatures. Phyllosphere bacteria, such as *P. syringae* and *Pantoea agglomerans*, have a cell surface protein complex that increases the temperature required for ice crystal formation (Attard et al. 2012).

Ethylene is a prominent phytohormone that is predominantly present in the gaseous state. The hormone is endogenous, meaning it is spontaneously produced within the plant. It exhibits efficacy in regulating diverse processes, such as growth, cellular metabolism, and senescence, even when administered in small quantities. Methylotrophic bacteria possess an enzyme called ACC deaminase, which converts ACC, the direct precursor of ethylene, into α-ketobutyrate and ammonia. This process decreases ethylene concentration in the surroundings when plants are under stress, stimulating plant growth. The strain *M. oryzae* CBMB20, obtained from the surface disinfected rice stem, was found to utilize ACC and produce ACC deaminase (Kim et al. 2012). The strains *M. populi* TMV7-4 and *M. thiocyanatum* VRI7-A4 have been found to produce ACC deaminase activity, resulting in an enhanced yield in groundnut (Krishnamoorthy et al. 2018). In the study conducted by Indiragandhi et al. (2008), the stress ethylene level in tomato plants was reduced by *Methylobacterium oryzae* CBMB20 by producing ACC deaminase. This reduction occurred due to the challenge of inoculation of *Pseudomonas syringae* pv. Tomato.

3.4.2 Biotic Stress Tolerance

The biological control of a plant pathogen can be achieved through various methods. One approach is through the direct competition between pathogenic microorganisms and antagonistic microorganisms for resources and habitat. Another method involves the production of antibiotics that specifically target the pathogen (Kashyap et al. 2020; Jamali et al. 2018; Stone et al. 2018). Microorganisms trigger plants' defense against diseases in two ways. Firstly, synthesizing harmless gene products that activate the immune response genes in plants leads to local and global reactions (Conrath et al. 2002). Another mechanism involves stimulating the development of degradative enzymes in plants, such as β-1, 3-glucanase, chitinase, and peroxidase. This leads to increased resistance against infections (Fernando et al. 2007).

In a study conducted by Janahiraman et al. (2016), tomato plants were inoculated with bacteria isolated from the phylloplane, including methylotrophic *Delftia lacustris* PPO-1 and various *Bacillus* strains. These plants were then exposed to pathogens such as *F. oxysporum,* f. sp. *lycopersici*, *S. rolfsii*, *Pythium ultimum*, and *Rhizoctonia solani*. As a result, the levels of pathogenesis-related proteins (specifically beta-1, 3-glucanase, and chitinase) and defensive enzymes (phenylalanine ammonia lyase, peroxidase, polyphenol oxidase, and catalase) were observed to increase (Janahiraman et al. 2016). Some dominant methylotrophs that have been isolated from mangrove sediment have been found to demonstrate biocontrol properties against *Macrophomina phaseolina*, a pathogen responsible for root rot (Kumar et al. 2016). The research conducted by Madhaiyan et al. (2006) found that *Methylobacterium* sp., obtained from groundnut, exhibited strong biocontrol properties against *A. niger* and *Sclerotium rolfsii* in groundnuts when tested in pot culture conditions. In a study conducted by Shao et al. (2023), two bacterial strains from the phyllosphere, namely *Rhizobium* b1 and *B. subtilis* b2, were discovered to possess inhibitory properties against 14 tomato pathogens. These pathogens include *P. syringae* pv. tomato and *Alternaria solani*.

3.4.3 Phyllosphere Microbiota Acquisition by Plants

The phyllosphere actively acquires its microbiome through vertical transmission from the mother plant (such as through seeds), horizontal environmental transmission, or mixed mechanisms from neighboring microbial reservoirs (Fig. 3.1). The colonization of microorganisms in the phyllosphere begins early during seedling emergence. In perennial plants, the microbes establish themselves during budburst, having overwintered in buds or twigs (Osono et al. 2014). The microbial populations reside on the leaf surface as both endophytes and epiphytes, and the ecological characteristics of both compartments exhibit notable distinctions. Leaf epiphytes are mostly populated through internal colonization, specifically by seed tissue and the environment in which germination occurs. Following the appearance of seedlings, it is probable that the bacteria are transported by wind, precipitation, irrigated water (Morris 2001), insects (Osono et al. 2014; Vacher et al. 2016), or bioaerosols (Bulgarelli et al. 2013). After being deposited, the establishment and survival on the leaf surface will be influenced by crucial microbial properties such as the acquisition of foliar nutrients, the capacity to cling to the cuticle, and growth in the leaf microclimate. The host leaf's morphology, physiology, and chemistry will also play a role. Once the microbes are fully established, their contact becomes active.

Phyllosphere endophytes are transferred systemically through the xylem (Compant et al. 2010). Alternatively, they can enter the plant through openings on

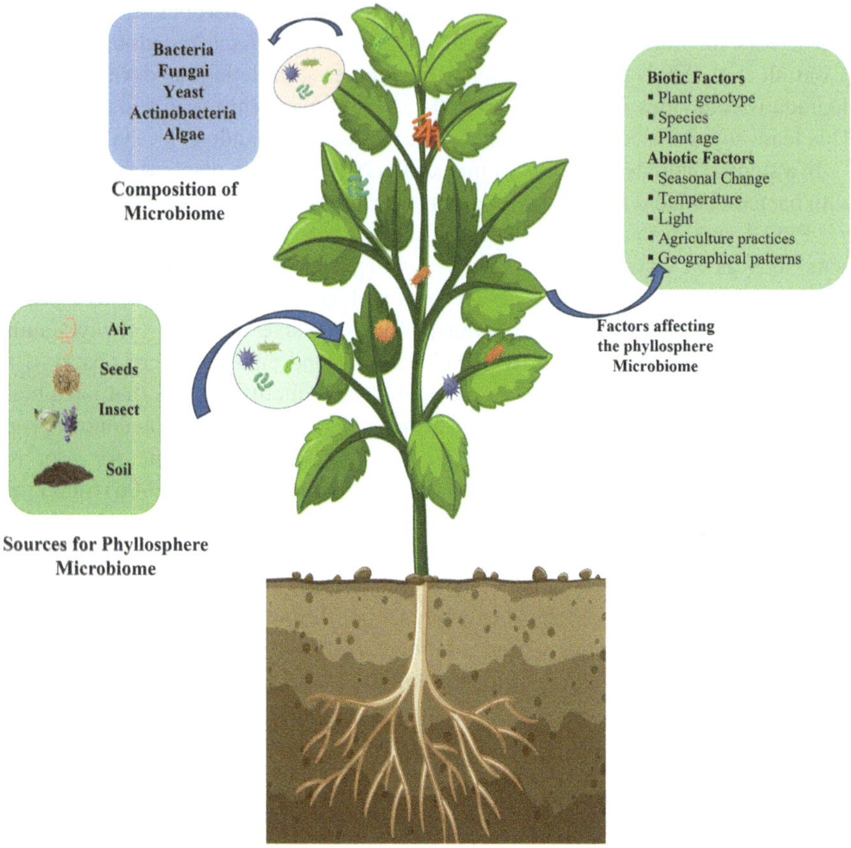

Fig. 3.1 Various factors influencing the phyllosphere microbiome

the leaf surface, such as stomata, lenticels, and hydathodes. While bacteria can enter through several openings, stomata are regarded as the primary entry point to the internal tissues of leaf surfaces (Melotto et al. 2008). Nevertheless, it is essential to note that not every bacterium gains admission through stomata. The access is regulated by the guard cells that surround the stomatal orifice. For instance, most disease-causing microorganisms are prevented from entering through pattern-triggered immunity (PTI). This occurs when the pattern recognition receptors on guard cells detect microbe-associated molecular patterns (MAMPs), leading to the closure of stomatal pores. Additionally, these pathogens employ effectors secreted by type III secretion systems. These effectors either prevent the closure of the pore or induce its opening, facilitating their entry (Hurley et al. 2014). In addition, nucleotide-binding leucine-rich repeat proteins (NLRs) act as intercellular receptors inside the leaf surface to restrict the presence of harmful organisms. These NLRs specifically recognize effector molecules linked with a type of plant immunity known as effector-triggered immunity (ETI) (Jacob et al. 2023).

3.5 Conclusion

In summary, the phyllosphere microbiome is a dynamic and intricate ecosystem that serves a vital function in plant growth, development, and adaptability. Despite the challenging conditions found on leaf surfaces, the phyllosphere is inhabited by many microorganisms, such as bacteria, fungi, yeasts, cyanobacteria, and algae. The microorganisms involved in these interactions exhibit many relationships with the host plant, including commensal, pathogenic, and mutualistic associations. Various factors, including host plant characteristics, seasonal fluctuations, geographical locations, and environmental conditions, determine the phyllosphere microbiome's diversity. The adaptability of the microbiome to various environmental stressors, such as fluctuations in temperature, exposure to UV radiation, and limited nutrient availability, underscores the remarkable resilience and specialized mechanisms evolved by these microorganisms to ensure their survival. The phyllosphere microbiome offers a range of advantages to plants, such as facilitating nutrient acquisition, promoting plant growth by synthesizing hormones, and improving tolerance to abiotic and biotic stresses. In addition, specific microorganisms present in the phyllosphere play a role in nitrogen fixation and bioremediation and act as biocontrol agents to combat plant pathogens. Comprehending the phyllosphere microbiome dynamics is crucial for effectively utilizing its potential in sustainable agriculture, plant protection, and environmental management. Further investigation is warranted to understand better the precise mechanisms that govern plant–microbe interactions in the phyllosphere. The research should focus on exploring the potential agricultural and ecosystem health benefits that can be derived from harnessing the capabilities of these microorganisms.

Ethical Approval and Consent to Participate Not applicable.

Acknowledgments This work was supported by the Tamil Nadu Agricultural University and Biostimulants Efficacy Testing Structure (BETS) (BMGF/BETS/DR/CBE2023/R001), Bill and Melinda Gates Foundation, USA.

Conflict of Interest The authors declare that there are no conflicts of interest related to this chapter.

References

Abadi VA, Sepehri M, Rahmani HA, Zarei M, Ronaghi A, Taghavi SM, Shamshiripour M (2020) Role of dominant phyllosphere bacteria with plant growth–promoting characteristics on growth and nutrition of maize (*Zea mays* L.). J Soil Sci Plant Nutr 20:2348–2363. https://doi.org/10.1007/s42729-020-00302-1

Abadi VA, Sepehri M, Rahmani HA, Dolatabad HK, Shamshiripour M, Khatabi B (2021) Diversity and abundance of culturable nitrogen-fixing bacteria in the phyllosphere of maize. J Appl Microbiol 131:898–912. https://doi.org/10.1111/jam.14975

Anandham R, Indiragandhi P, Madhaiyan M, Kim K, Yim W, Saravanan VS, Chung J, Sa T (2007) Thiosulfate oxidation and mixotrophic growth of *Methylobacterium oryzae*. Can J Microbiol 53:869–876. https://doi.org/10.1139/W07-057

Anandham R, Janahiraman V, Gandhi PI, Kwon S, Chung K, Han GH, Choi JH, Sa TM (2014) Early plant growth promotion of maize by various sulfur oxidizing bacteria that uses different thiosulfate oxidation pathway. Afr J Microbiol Res 8:19–27

Andrews JH, Harris RF (2000) The ecology and biogeography of microorganisms on plant surfaces. Annu Rev Phytopathol 38:145–180. https://doi.org/10.1146/annurev.phyto.38.1.145

Attard E, Yang H, Delort AM, Amato P, Pöschl U, Glaux C, Koop T, Morris CE (2012) Effects of atmospheric conditions on ice nucleation activity of Pseudomonas. Atmospheric Chem Phys 12:10667–10677. https://doi.org/10.5194/acp-12-10667-2012

Aydogan EL, Budich O, Hardt M, Choi YH, Jansen-Willems AB, Moser G, Müller C, Kämpfer P, Glaeser SP (2020) Global warming shifts the composition of the abundant bacterial phyllosphere microbiota as indicated by a cultivation-dependent and-independent study of the grassland phyllosphere of a long-term warming field experiment. FEMS Microbiol Ecol 96:fiaa087. https://doi.org/10.1093/femsec/fiaa087

Bao L, Gu L, Sun B, Cai W, Zhang S, Zhuang G, Bai Z, Zhuang X (2020) Seasonal variation of epiphytic bacteria in the phyllosphere of *Gingko biloba*, *Pinus bungeana* and *Sabina chinensis*. FEMS Microbiol Ecol 96:fiaa017. https://doi.org/10.1093/femsec/fiaa017

Beattie GA (2011) Water relations in the interaction of foliar bacterial pathogens with plants. Annu Rev Phytopathol 49:533–555. https://doi.org/10.1146/annurev-phyto-073009-114436

Bodenhausen N, Bortfeld-Miller M, Ackermann M, Vorholt JA (2014) A synthetic community approach reveals plant genotypes affecting the phyllosphere microbiota. PLoS Genet 10:e1004283. https://doi.org/10.1371/journal.pgen.1004283

Bonito G, Reynolds H, Robeson MS, Nelson J, Hodkinson BP, Tuskan G, Schadt CW, Vilgalys R (2014) Plant host and soil origin influence fungal and bacterial assemblages in the roots of woody plants. Mol Ecol 23:3356–3370. https://doi.org/10.1111/mec.12821

Bulgarelli D, Schlaeppi K, Spaepen S, Van Themaat EV, Schulze-Lefert P (2013) Structure and functions of the bacterial microbiota of plants. Annu Rev Plant Biol 64:807–838. https://doi.org/10.1146/annurev-arplant-050312-120106

Burch AY, Zeisler V, Yokota K, Schreiber L, Lindow SE (2014) The hygroscopic biosurfactant syringafactin produced by *Pseudomonas syringae* enhances fitness on leaf surfaces during fluctuating humidity. Environ Microbiol 16:2086–2098. https://doi.org/10.1111/1462-2920.12437

Cao FY, Yoshioka K, Desveaux D (2011) The roles of ABA in plant–pathogen interactions. J Plant Res 12:489–499. https://doi.org/10.1007/s10265-011-0409-y

Compant S, Clément C, Sessitsch A (2010) Plant growth-promoting bacteria in the rhizo-and endosphere of plants: their role, colonization, mechanisms involved and prospects for utilization. Soil Biol Biochem 42:669–678. https://doi.org/10.1016/j.soilbio.2009.11.024

Conrath U, Pieterse CM, Mauch-Mani B (2002) Priming in plant–pathogen interactions. Trends Plant Sci 7(5):210–216

Delmotte N, Knief C, Chaffron S, Innerebner G, Roschitzki B, Schlapbach R, von Mering C, Vorholt JA (2009) Community proteogenomics reveals insights into the physiology of phyllosphere bacteria. Proc Natl Acad Sci USA 106:16428–16433. https://doi.org/10.1073/pnas.0905240106

Dong CJ, Wang LL, Li Q, Shang QM (2019) Bacterial communities in the rhizosphere, phyllosphere and endosphere of tomato plants. PLoS One 14:e0223847. https://doi.org/10.1371/journal.pone.0223847

Dourado MN, Andreote FD, Dini-Andreote F, Conti R, Araújo JM, Araújo WL (2012) Analysis of 16S rRNA and mxaF genes reveling insights into Methylobacterium niche-specific plant association. Genet Mol Biol 35:142–148. https://doi.org/10.1590/S1415-47572012005000017

Fernando WG, Nakkeeran S, Zhang Y, Savchuk S (2007) Biological control of *Sclerotinia sclerotiorum* (Lib.) de Bary by *Pseudomonas* and *Bacillus* species on canola petals. J Crop Prot 26:100–107. https://doi.org/10.1016/j.cropro.2006.04.007

Finkel OM, Burch AY, Lindow SE, Post AF, Belkin S (2011) Geographical location determines the population structure in phyllosphere microbial communities of a salt-excreting desert tree. 7. Appl Environ Microbiol 7:7647–7655. https://doi.org/10.1128/AEM.05565-11

Frey-Klett P, Burlinson P, Deveau A, Barret M, Tarkka M, Sarniguet A (2011) Bacterial-fungal interactions: hyphens between agricultural, clinical, environmental, and food microbiologists. Microbiol Mol Biol Rev 75:583–609. https://doi.org/10.1128/mmbr.00020-11

Fu SF, Sun PF, Lu HY, Wei JY, Xiao HS, Fang WT, Cheng BY, Chou JY (2016) Plant growth-promoting traits of yeasts isolated from the phyllosphere and rhizosphere of *Drosera spatulata* Lab. Fungal Biol 120:433–448. https://doi.org/10.1016/j.funbio.12.006

Gal M, Preston GM, Massey RC, Spiers AJ, Rainey PB (2003) Genes encoding a cellulosic polymer contribute toward the ecological success of *Pseudomonas fluorescens* SBW25 on plant surfaces. Mol Ecol 12:3109–3121. https://doi.org/10.1046/j.1365-294x.2003.01953.x

Giri S, Pati B (2004) A comparative study on phyllosphere nitrogen fixation by newly isolated *Corynebacterium* sp. & *Flavobacterium* sp. and their potentialities as biofertilizer. Acta Microbiol Immunol Hung 51:47–56. https://doi.org/10.1556/amicr.51.2004.1-2.3

Glawe DA (2008) The powdery mildews: a review of the world's most familiar (yet poorly known) plant pathogens. Annu Rev Phytopathol 46:27–51. https://doi.org/10.1146/annurev.phyto.46.081407.104740

Godon JJ, Galès A, Latrille E, Ouichanpagdee P, Seyer JP (2020) An "overlooked" habitat for thermophilic bacteria: the phyllosphere. BioDiscovery 30:23. https://doi.org/10.3897/biodiscovery.23.e47033

Gomes T, Pereira JA, Benhadi J, Lino-Neto T, Baptista P (2018) Endophytic and epiphytic phyllosphere fungal communities are shaped by different environmental factors in a Mediterranean ecosystem. Microb Ecol 76:668–679. https://doi.org/10.1007/s00248-018-1161-9

Grover M, Ali SZ, Sandhya V, Rasul A, Venkateswarlu B (2011) Role of microorganisms in adaptation of agriculture crops to abiotic stresses. World J Microbiol Biotechnol 27:1231–1240. https://doi.org/10.1007/s11274-010-0572-7

Hashidoko Y, Itoh E, Yokota K, Yoshida T, Tahara S (2002) Characterization of five phyllosphere bacteria isolated from Rosa rugosa leaves, and their phenotypic and metabolic properties. Biosci Biotech Biochem 66:2474–2478. https://doi.org/10.1271/bbb.66.2474

Helfrich EJ, Vogel CM, Ueoka R, Schäfer M, Ryffel F, Müller DB, Probst S, Kreuzer M, Piel J, Vorholt JA (2018) Bipartite interactions, antibiotic production and biosynthetic potential of the Arabidopsis leaf microbiome. Nat Microbiol 3:909–919. https://doi.org/10.1038/s41564-018-0200-0

Holland MA (2011) Nitrogen: give and take from phylloplane microbes. In: Ecological aspects of nitrogen metabolism in plants. Wiley, New York, pp 215–230. https://doi.org/10.1002/9780470959404

Hurley B, Lee D, Mott A, Wilton M, Liu J, Liu YC, Angers S, Coaker G, Guttman DS, Desveaux D (2014) The *Pseudomonas syringae* type III effector HopF2 suppresses Arabidopsis stomatal immunity. PLoS One 9:e114921. https://doi.org/10.1371/journal.pone.0114921

Ibekwe AM, Ors S, Ferreira JF, Liu X, Suarez DL (2021) Influence of seasonal changes and salinity on spinach phyllosphere bacterial functional assemblage. PLoS One 16:e0252242. https://doi.org/10.1371/journal.pone.0252242

Idris R, Kuffner M, Bodrossy L, Puschenreiter M, Monchy S, Wenzel WW, Sessitsch A (2006) Characterization of Ni-tolerant methylobacteria associated with the hyperaccumulating plant *Thlaspi goesingense* and description of *Methylobacterium goesingense* sp. nov. Syst Appl Microbiol 29:634–644. https://doi.org/10.1016/j.syapm.2006.01.011

Indiragandhi P, Anandham R, Kim K, Yim W, Madhaiyan M, Sa T (2008) Induction of defense responses in tomato against *Pseudomonas syringae* pv. Tomato by regulating the stress ethylene level with *Methylobacterium oryzae* CBMB20 containing 1-aminocyclopropane-1-carboxylate deaminase. World J Microbiol Biotechnol 24:1037–1045. https://doi.org/10.1007/s11274-007-9572-7

Jackson CR, Denney WC (2011) Annual and seasonal variation in the phyllosphere bacterial community associated with leaves of the Southern Magnolia (*Magnolia grandiflora*). Microb Ecol 61:113–122. https://doi.org/10.1007/s00248-010-9742-2

Jacob P, Hige J, Dangl JL (2023) Is localized acquired resistance the mechanism for effector-triggered disease resistance in plants. Nat Plants 9:1184–1190. https://doi.org/10.1038/s41477-023-01466-1

Jacobs JL, Sundin GW (2001) Effect of solar UV-B radiation on a phyllosphere bacterial community. Appl Environ Microbiol 67:5488–5496. https://doi.org/10.1128/AEM.67.12.5488-5496.2001

Jamali JH, Sharma A, Kushwaha P, Roohi PL, Srivastava AK (2018) Exploitation of multifarious abiotic stresses, antagonistic activity and plant growth promoting attributes of *Bacillus amyloliquefaciens* AH53 for sustainable agriculture production. Int J Curr Microbiol App Sci 7:751–763. https://doi.org/10.20546/ijcmas.2018.710.083

Janahiraman V, Anandham R, Kwon SW, Sundaram S, Karthik Pandi V, Krishnamoorthy R, Kim K, Samaddar S, Sa T (2016) Control of wilt and rot pathogens of tomato by antagonistic pink pigmented facultative methylotrophic *Delftia lacustris* and *Bacillus* spp. Front Plant Sci 7:1626. https://doi.org/10.3389/fpls.2016.01626

Kashyap PL, Solanki MK, Kushwaha P, Kumar S, Srivastava AK (2020) Biocontrol potential of salt-tolerant Trichoderma and Hypocrea isolates for the management of tomato root rot under saline environment. J Soil Sci Plant Nutr 20:160–176. https://doi.org/10.1007/s42729-019-00114-y

Kembel SW, Mueller RC (2014) Plant traits and taxonomy drive host associations in tropical phyllosphere fungal communities. Botany 92:303–311. https://doi.org/10.1139/cjb-2013-0194

Kim M, Singh D, Lai-Hoe A, Go R, Abdul Rahim R, An A, Chun J, Adams JM (2012) Distinctive phyllosphere bacterial communities in tropical trees. Microb Ecol 63:674–681. https://doi.org/10.1007/s00248-011-9953-1

Koskella B (2020) The phyllosphere. Curr Biol 30:R1143–R1146

Krishnamoorthy R, Kwon SW, Kumutha K, Senthilkumar M, Ahmed S, Sa T, Anandham R (2018) Diversity of culturable methylotrophic bacteria in different genotypes of groundnut and their potential for plant growth promotion. 3 Biotech 8:1. https://doi.org/10.1007/s13205-018-1291-2

Kucharska K, Wachowska U, Czaplicki S (2020) Wheat phyllosphere yeasts degrade propiconazole. BMC Microbiol 20:1–4

Kumar GD, Natarajan N, Nakkeeran S (2016) Antifungal activity of nanofungicide Trifloxystrobin 25%+ Tebuconazole 50% against *Macrophomina phaseolina*. Afr Microbiol Res 10:100–105. https://doi.org/10.5897/AJMR2015.7692

Kushwaha P, Kashyap PL, Kuppusamy P, Srivastava AK, Tiwari RK (2020) Functional characterization of endophytic bacilli from pearl millet (*Pennisetum glaucum*) and their possible role in multiple stress tolerance. Plant Biosyst 154:503–514. https://doi.org/10.1080/11263504.2019.1651773

Leewijit T, Pongnak W, Soytong K, Poeaim S (2016) Isolation of soil and endophytic fungi from rice (*Oryza sativa* L.). Int J Agric Technol 12:2191–2202

Leveau JH (2006) 11 Microbial communities in the phyllosphere. In: Annual plant reviews, biology of the plant cuticle, vol 23. Wiley, New York, p 334

Li Y, Wu X, Chen T, Wang W, Liu G, Zhang W, Li S, Wang M, Zhao C, Zhou H, Zhang G (2018) Plant phenotypic traits eventually shape its microbiota: a common garden test. Front Microbiol 9:2479. https://doi.org/10.3389/fmicb.2018.02479

Li M, Hong L, Ye W et al (2022) Phyllosphere bacterial and fungal communities vary with host species identity, plant traits and seasonality in a subtropical forest. Environ Microbiome 17:29. https://doi.org/10.1186/s40793-022-00423-3

Li P, Wu X, Gao F (2023) Ozone pollution, water deficit stress and time drive poplar phyllospheric bacterial community structure. Ecotoxicol Environ Saf 262:115148. https://doi.org/10.1016/j.ecoenv.2023.115148

Lin C, Wu J (2013) Tolerance of soil algae and cyanobacteria to drought stress. J Phycol 50:131–139. https://doi.org/10.1111/jpy.12141

Lin CS, Chou TL, Wu JT (2013) Biodiversity of soil algae in the farmlands of mid-Taiwan. Bot Stud 54:41. https://doi.org/10.1186/1999-3110-54-41

Liu F, Xing S, Ma H, Du Z, Ma B (2013) Cytokinin-producing, plant growth-promoting rhizobacteria that confer resistance to drought stress in *Platycladus orientalis* container seedlings. Appl Microbiol Biotechnol 97:9155–9164. https://doi.org/10.1007/s00253-013-5193-2

Madhaiyan M, Poonguzhali S, Lee HS, Hari K, Sundaram SP, Sa TM (2005) Pink-pigmented facultative methylotrophic bacteria accelerate germination, growth and yield of sugarcane clone Co86032 (*Saccharum officinarum* L.). Biol Fertil Soils 41:350–358. https://doi.org/10.1007/s00374-005-0838-7

Madhaiyan M, Suresh Reddy BV, Anandham R, Senthilkumar M, Poonguzhali S, Sundaram SP, Sa T (2006) Plant growth–promoting Methylobacterium induces defense responses in groundnut (*Arachis hypogaea* L.) compared with rot pathogens. Curr Microbiol 53:270–276. https://doi.org/10.1007/s00284-005-0452-9

Madhaiyan M, Poonguzhali S, Kang BG, Lee YJ, Chung JB, Sa TM (2010) Effect of co-inoculation of methylotrophic *Methylobacterium oryzae* with *Azospirillum brasilense* and *Burkholderia pyrrocinia* on the growth and nutrient uptake of tomato, red pepper and rice. Plant Soil 328:71–82. https://doi.org/10.1007/s11104-009-0083-1

Mazinani Z, Zamani M, Sardari S (2017) Isolation and identification of phyllospheric bacteria possessing antimicrobial activity from *Astragalus obtusifolius*, *Prosopis juliflora*, *Xanthium strumarium* and *Hippocrepis unisiliqousa*. Avicenna J Med Biotechnol 9:31

Meena KK, Kumar M, Kalyuzhnaya MG, Yandigeri MS, Singh DP, Saxena AK, Arora DK (2012) Epiphytic pink-pigmented methylotrophic bacteria enhance germination and seedling growth of wheat (*Triticum aestivum*) by producing phytohormone. Antonie Van Leeuwenhoek 101:777–786. https://doi.org/10.1007/s10482-011-9692-9

Meena RK, Singh RK, Singh NP, Meena SK, Meena VS (2015) Isolation of low temperature surviving plant growth–promoting rhizobacteria (PGPR) from pea (*Pisum sativum* L.) and documentation of their plant growth promoting traits. Biocatal Agric Biotechnol 4:806–811. https://doi.org/10.1016/j.bcab.2015.08.006

Melotto M, Underwood W, He SY (2008) Role of stomata in plant innate immunity and foliar bacterial diseases. Annu Rev Phytopathol 46:101–122

Mizuno M (2013) Studies on distribution and colonization of facultative methylotrophic bacteria *Methylobacterium* spp. On the perilla plant. Dissertation, Kyoto University

Morris CE (2001) Phyllosphere. https://doi.org/10.1038/npg.els.0000400

Morris CE, Conen F, Alex Huffman J, Phillips V, Pöschl U, Sands DC (2014a) Bioprecipitation: a feedback cycle linking earth history, ecosystem dynamics and land use through biological ice nucleators in the atmosphere. Glob Chang Bio 20:341–351. https://doi.org/10.1111/gcb.12447

Morris CE, Georgakopoulos DG, Sands DC (2014b) Ice nucleation active bacteria and their potential role in precipitation. J Phys IV 121:87–103. https://doi.org/10.1051/jp4:2004121004

Newton AC, Gravouil C, Fountaine JM (2010) Managing the ecology of foliar pathogens: ecological tolerance in crops. Ann Appl Biol 157:343–359. https://doi.org/10.1111/j.1744-7348.2010.00437.x

Nutaratat P, Srisuk N, Duangmal K, Yurimoto H, Sakai Y, Muramatsu Y, Nakagawa Y (2013) *Roseomonas musae* sp. a new bacterium isolated from a banana phyllosphere. Antonie Van Leeuwenhoek 103:617–624. https://doi.org/10.1007/s10482-012-9845-5

Omer ZS, Tombolini R, Gerhardson B (2004) Plant colonization by pink-pigmented facultative methylotrophic bacteria (PPFMs). FEMS Microbiol Ecol 47:319–326. https://doi.org/10.1016/S0168-6496(04)00003-0

Osono T (2008) Endophytic and epiphytic phyllosphere fungi of *Camellia japonica*: seasonal and leaf age-dependent variations. Mycologia 100:387–391. https://doi.org/10.3852/07-110R1

Osono T, Azuma JI, Hirose D (2014) Plant species effect on the decomposition and chemical changes of leaf litter in grassland and pine and oak forest soils. Plant Soil 376:411–421. https://doi.org/10.1007/s11104-013-1993-5

Prasanna R, Jaiswal P, Nayak S, Sood A, Kaushik BD (2009) Cyanobacterial diversity in the rhizosphere of rice and its ecological significance. Indian J Microbiol 49:89–97. https://doi.org/10.1007/s12088-009-0009-x

Qian X, Chen L, Guo X, He D, Shi M, Zhang D (2018) Shifts in community composition and co-occurrence patterns of phyllosphere fungi inhabiting *Mussaenda shikokiana* along an elevation gradient. Peer J 6:e5767. https://doi.org/10.7717/peerj.5767/supp-7

Queitsch C, Hong SW, Vierling E, Lindquist S (2000) Heat shock protein 101 plays a crucial role in thermotolerance in Arabidopsis. Plant Cell 12:479–492. https://doi.org/10.1105/tpc.12.4.479

Rastogi G, Sbodio A, Tech JJ, Suslow TV, Coaker GL, Leveau JH (2012) Leaf microbiota in an agroecosystem: spatiotemporal variation in bacterial community composition on field-grown lettuce. ISME J 6(10):1812–1822

Redford AJ, Fierer N (2009) Bacterial succession on the leaf surface: a novel system for studying successional dynamics. Microbial Ecol 58:189–198. https://doi.org/10.1007/s00248-009-9495-y

Redford AJ, Bowers RM, Knight R, Linhart Y, Fierer N (2010) The ecology of the phyllosphere: geographic and phylogenetic variability in the distribution of bacteria on tree leaves. Environ Microbiol 12:2885–2893. https://doi.org/10.1111/j.1462-2920.2010.02258.x

Redman RS, Sheehan KB, Stout RG, Rodriguez RJ, Henson JM (2002) Thermotolerance generated by plant/fungal symbiosis. Science 12:2885–2893. https://doi.org/10.1126/science.1078055

Rigonato J, Alvarenga DO, Andreote FD, Dias ACF, Melo IS, Kent A, Fiore MF (2012) Cyanobacterial diversity in the phyllosphere of a mangrove forest. FEMS Microbiol Ecol 80(2):312–322. https://doi.org/10.1111/j.1574-6941.2012.01299.x

Romero FM, Marina M, Pieckenstain FL (2016) Novel components of leaf bacterial communities of field-grown tomato plants and their potential for plant growth promotion and biocontrol of tomato diseases. Res Microbiol 167:222–233. https://doi.org/10.1016/j.resmic.2015.11.001

Sadaka N, Ponge JF (2003) Fungal colonization of phyllosphere and litter of *Quercus rotundifolia* Lam. In a holm oak forest (high atlas, Morocco). Biol Fertil Soils 39:30–36. https://doi.org/10.1007/s00374-003-0666-6

Schreiber L, Krimm U, Knoll D, Sayed M, Auling G, Kroppenstedt RM (2005) Plant–microbe interactions: identification of epiphytic bacteria and their ability to alter leaf surface permeability. New Phytol 166:589–594. https://doi.org/10.1111/j.1469-8137.2005.01343.x

Shao Z, Schenk PM, Dart P (2023) Phyllosphere bacterial strains *Rhizobium* b1 and *Bacillus subtilis* b2 control tomato leaf diseases caused by *Pseudomonas syringae* pv. Tomato and *Alternaria solani*. Appl Microbiol 134:lxad139. https://doi.org/10.1093/jambio/lxad139

Sharath S, Triveni S, Nagaraju Y, Latha PC, Vidyasagar B (2021) The role of phyllosphere bacteria in improving cotton growth and yield under drought conditions. Front Agron 3:680466. https://doi.org/10.3389/fagro.2021.680466

Sharma A, Kashyap PL, Srivastava AK, Bansal YK, Kaushik R (2019) Isolation and characterization of halotolerant bacilli from chickpea (*Cicer arietinum* L.) rhizosphere for plant growth promotion and biocontrol traits. Eur J Plant Pathol 153:787–800. https://doi.org/10.1007/s10658-018-1592-7

Shivas RG, Brown JF (1984) Identification and enumeration of yeasts on *Banksia collina* and *Callistemon viminalis* leaves. Trans Br Mycol Soc 83:687–689. https://doi.org/10.1016/S0007-1536(84)80188-6

Stone BW, Weingarten EA, Jackson CR (2018) The role of the phyllosphere microbiome in plant health and function. Annu Plant Rev Online 1:533–556. https://doi.org/10.1002/9781119312994.apr0614

Sundin GW (2002) Ultraviolet radiation on leaves: its influence on microbial communities and their adaptations. In: Phyllosphere microbiology. CABI, London, pp 27–41

Sussmilch FC, Brodribb TJ, McAdam SA (2017) Up-regulation of NCED3 and ABA biosynthesis occur within minutes of a decrease in leaf turgor but AHK1 is not required. J Exp Bot 68:2913–2918. https://doi.org/10.1093/jxb/erx124

Tani A, Sahin N, Fujitani Y, Kato A, Sato K, Kimbara K (2015) Methylobacterium species promoting rice and barley growth and interaction specificity revealed with whole-cell matrix-assisted laser desorption/ionization-time-of-flight mass spectrometry (MALDI-TOF/MS) analysis. PLoS One 10:e0129509. https://doi.org/10.1371/journal.pone.0129509

Thompson IP, Bailey MJ, Fenlon JS, Fermor TR, Lilley AK, Lynch JM, McCormack PJ, McQuilken MP, Purdy KJ, Rainey PB, Whipps JM (1993) Quantitative and qualitative seasonal changes in the microbial community from the phyllosphere of sugar beet (*Beta vulgaris*). Plant Soil 150:177–191. https://doi.org/10.1007/BF00013015

Vacher C, Hampe A, Porté AJ, Sauer U, Compant S, Morris CE (2016) The phyllosphere: microbial jungle at the plant–climate interface. Ann Rev Ecol Evol Syst 47:1–24

Vandenkoornhuyse P, Quaiser A, Duhamel M, Le Van A, Dufresne A (2015) The importance of the microbiome of the plant holobiont. New Phytol 206:1196–1206. https://doi.org/10.1111/nph.13312

Venkatachalam S, Ranjan K, Prasanna R, Ramakrishnan B, Thapa S, Kanchan A (2016) Diversity and functional traits of culturable microbiome members, including cyanobacteria in the rice phyllosphere. Plant Biol 18:627–637. https://doi.org/10.1111/plb.12441

Verma P, Yadav AN, Kazy SK, Saxena AK, Suman A (2014) Evaluating the diversity and phylogeny of plant growth promoting bacteria associated with wheat (*Triticum aestivum*) growing in central zone of India. Int J Curr Microbiol Appl Sci 3:432–447

Wagner MR, Lundberg DS, Del Rio TG, Tringe SG, Dangl JL, Mitchell-Olds T (2016) Host genotype and age shape the leaf and root microbiomes of a wild perennial plant. Nat Commun 7:12151. https://doi.org/10.1038/ncomms12151

Wang K, Chen H, Ortega-Perez M, Miao Y, Ma Y, Henschen A, Lohmann JU, Laubinger S, Bayer M (2021) Independent parental contributions initiate zygote polarization in *Arabidopsis thaliana*. Curr Biol 31:4810–4816

Xu N, Zhao Q, Zhang Z, Zhang Q, Wang Y, Qin G, Ke M, Qiu D, Peijnenburg WJ, Lu T, Qian H (2020) Phyllosphere microorganisms: sources, drivers, and their interactions with plant hosts. J Agric Food Chem 70:4860–4870. https://doi.org/10.1021/acs.jafc.2c01113

Yadav AN, Kumar V, Dhaliwal HS, Prasad R, Saxena AK (2018) Microbiome in crops: diversity, distribution, and potential role in crop improvement. In: Crop improvement through microbial biotechnology. Elsevier, Amsterdam, pp 305–332. https://doi.org/10.1016/B978-0-444-63987-5.00015-3

Yashiro E, Spear R, McManus P (2011) Culture-dependent and culture-independent assessment of bacteria in the apple phyllosphere. J Appl Microbiol 110:1284–1296. https://doi.org/10.1111/j.1365-2672.2011.04975.x

Zhu H, Li S, Hu Z, Liu G (2018) Molecular characterization of eukaryotic algal communitiesin the tropical phyllosphere based on real-time sequencing of the 18S rDNA gene. BMC Plant Biol 18:1–4. https://doi.org/10.1186/s12870-018-1588-7

Chapter 4
Drought and Waterlogging Stress Responses in Crops

Priyam Vandana, Anmol Gupta, and Manoj Kumar

Abstract Around 70% of the freshwater is used globally for agriculture. Due to worldwide agricultural expansion and climate change, water demand is rising day by day. Therefore, creating crop cultivars that require less water is necessary for achieving sustainable agricultural output. As the climate continues to change, extreme soil conditions and soil water availability, such as severe drought and waterlogging from high rains, are predicted to result. In the absence of novel resistance variations to these abiotic stressors, both waterlogging and drought are harmful to plant species and might pose a danger to global food security. Agricultural production is severely hampered by drought and waterlogging, both of which have become more frequent and severe over time as a result of climate change. Both waterlogging and drought stress may cause changes in a plant's shape and photosynthetic activity, affecting leaf shape (such as crimping and wilting), antioxidant enzyme system, and hormone levels. In particular, microbes with the potential to promote plant development might be useful instruments to achieve the objective of sustainable agriculture. As a viable long-term solution, particular focus has been placed on the use of microorganisms in agriculture to boost production and tolerance to shocks. This chapter will concentrate on the effects of drought and waterlogging on physiology, plant responses to climatic changes, and molecular pathways. Further, the processes behind these alterations are described. Additionally, it will

Priyam Vandana and Anmol Gupta contributed equally with all other contributors.

P. Vandana
Division of Crop Protection, ICAR-Indian Institute of Sugarcane Research, Lucknow, Uttar Pradesh, India

A. Gupta
Integral University, Lucknow, Uttar Pradesh, India

M. Kumar (✉)
Institute of Plant Sciences, Agricultural Research Organization, Volcani Center, Rishon LeZion, Israel

P. Singh Chauhan et al. (eds.), *Plant-Microbe Interaction and Stress Management*, Rhizosphere Biology,
https://doi.org/10.1007/978-981-97-4239-4_4

discuss the methods by which microorganisms that promote plant development might increase a plant's ability to withstand drought and waterlogging stress.

Keywords Drought · Waterlogging · PGPR · Stress tolerance · Crop improvement

4.1 Introduction

Water stress is one of the abiotic stresses brought on by climate change on plants. It covers both water surplus (drought) and water shortage (flooding). The plant's growth, development, and eventually yield are constrained by these pressures. Water stress is a problem for about 60% of crops. Water stress affects a plant's physiological processes (stomata closure, reduced photosynthesis, and transpiration), as well as its biochemistry [protein denaturing, reactive oxygen species (ROS) formation, rise in osmoprotectants, and antioxidant enzymes] that affect crop yield. Tolerant genotypes have been developed to deal with these challenges, but more work is needed to modify plant genomes so that we can sustain plant growth and enhance production to satisfy the demands of a growing global population.

According to Sasidharan et al. (2017), flooding is characterized as waterlogging (WL), where the root system is submerged, or submersion, when the aerial plant tissues are submerged. Flooding and waterlogging stress (WLS) can be brought on by external environmental factors such as heavy rainfall, storms, river overflow, and excessive irrigation. The availability of soil water for agricultural development and growth is reduced during a drought because the projected rainfall is lower than usual over a protracted period (Ceccarelli et al. 2007; Blum 2011). Drought stress (DS) is brought on by several elements, including salt buildup in the soil, large temperature swings, decreased rainfall, and excessive light levels. When there is WL, plants are primarily harmed by a sharp decline in the oxygen content of the rhizosphere.

Long-term low oxygen levels frequently impede plant development and cause premature plant mortality. During WL, leaves as well as roots frequently endure some amount of stress. Stomatal closure and decreases in photosynthetic rate are two of the early leaf responses to WLS. Waterlogging stress largely limits aerobic cellular respiration in roots and activates fermentative metabolism, allowing plants to continue glycolysis for ATP generation.

Roots and leaves are the primary organs of plants that respond to water stress, and their internal anatomy and morphology are the greatest indicators of their capacity to adapt to harsh circumstances (Cal et al. 2019; Liu et al. 2020; Lozano et al. 2020; Pedersen et al. 2021). When it comes to long-term environmental adaptability, leaves are a plant's most changeable organ. Both DS and WLS cause them to behave similarly, exhibiting curling, senescence atrophy, abscission, and even etiolation (Patharkar and Walker 2019; Bhusal et al. 2020). Stress has been shown to lower the quantity of leaves and inhibit leaf growth in some instances as a result of reduction in productivity (Wei et al. 2013; Fang and Xiong 2015; Nadal et al.

2020). The ability of microorganisms to encourage plant growth may make them extremely useful tools for sustaining sustainable agriculture.

4.2 Morphological Responses of Plants to Waterlogging and Drought Stress

When leaves are stressed by WL, they may wilt, yellow, fall off, curl, rot, and other signs. In response to flooding stress, leaves can adjust in two ways: either by becoming thinner or by becoming thicker. In the case of the former, spongy and palisade tissue growth, as well as a reduction in the size of stomata and leaves, increase the plant's ability to retain water and minimize water loss (Yin and Luo 2012; Tahira et al. 2014; Zuniga et al. 2017). To enhance the inorganic nutrients and ability of CO_2 to infiltrate into the leaves, certain plants thin or form special leaves (Colmer and Pedersen 2008; Brodersen et al. 2017). These plants also enhance gas exchange to sustain and restore respiration in conditions of WLS (Lawson et al. 2017; Mommer and Visser 2005). One of the most noticeable adaptations to WLS is the formation of aerenchyma in adventitious roots. As a result, various plant species under WLS have distinct methods of developing adventitious roots (Table 4.1). By generating air holes in the aerenchyma to increase storage capacity and prevent soil pollutants from entering plants, roots help plants become more adaptable.

In addition to causing complex changes in leaf thickness, spongy tissue, and palisade tissue, during adaptation, DS can also limit plant growth by preventing the cell division of leaf meristematic tissue and cell expansion in elongation areas (Binks et al. 2016; Nelissen et al. 2018; Meng et al. 2021). According to Rueda et al. (2017), in drought-prone areas, plants can retain more water by thickening their leaves and spongy tissue and lowering the thickness of their palisade tissue. Zheng et al. (2010) however, discovered that *Lycium barbarum* enhanced the thickness of palisade tissue, preserving tissue from extreme dryness, and decreased the thickness of spongy tissue, reducing transpiration (Table 4.1). Transpiration was reduced and the photosynthetic rate increased as a result of alterations in the internal structure of the leaf.

Under DS conditions, there is a reduction in vessel diameter, secondary root cortex cells, and stele area, while the number of vessels in the stele increases to facilitate water movement (De Bauw et al. 2019; Thangthong et al. 2019; Hazman and Brown 2018). To affect root spatial distribution, nutrient absorption, soil fixation, and plants can alter their root structure, including root hair and density. In drought-stressed environments, they can also alter their root length and internal structure to increase their capacity for water absorption. Plants extend their root systems by growing more functional roots, but they also boost the root sheath's ability to absorb water to improve DS tolerance and water retention (Lee et al. 2016; Pierret et al. 2016). Stress has been proven to reduce the number of leaves and, in certain cases, impede leaf development, which leads to a decline in crop productivity (Wei et al. 2013; Fang and Xiong 2015; Nadal et al. 2020). According to Farooq

Table 4.1 Characteristics of leaves and roots under DS and WLS

Stress	Leaf	Reference	Root	Reference
Drought stress	Crimping, wilting, stomatal closure	Song et al. (2018), Yang et al. (2020)	Functional root number increases, root system lengthens, distribution and breadth increase	Gupta et al. (2020), Djanaguiraman et al. (2019)
	Vascular bundle increases, thickness of spongy tissue decreases	Zhang et al. (2018), Ouyang et al. (2017)	Several vascular bundles increase but their diameter reduces, area of the stele reduces	Gupta et al. (2020), Canales et al. (2021), Xiao et al. (2020)
Waterlogging stress	Wilting, abscission, stomatal closure, etiolation	Yin et al. (2010), Ploschuk et al. (2018), Fukao et al. (2006)	Adventitious roots are generated, the number of roots decreases, and root activity decreases	Palta et al. (2010), Domisch et al. (2020), Dresbøll and Thorup-Kristensen (2012), Pan et al. (2019)
	The thickness of the blades decreases together with the quantity and area of leaves	Yin et al. (2010), Mommer et al. (2005), Challabathula et al. (2018)	The size of the stele reduces, and Aerenchyma is formed in adventitious roots	Celedonio et al. (2017), Zhang et al. (2015), Sundgren et al. (2018)

et al. (2009), DS has a negative impact on phenology, leaf size, stem extension, root division, seed germination, and plant growth. Additionally, it restricts the availability of nutrients for plants, and plants are crucial for the underlying mechanisms of tolerance (Fig. 4.1).

4.3 Physiological Responses of Crop Plants to Waterlogging Stress

Waterlogging can lead to physiological problems in plants, such as imbalanced hormones, reduced photosynthetic rate, and insufficient intake of water, nutrients, and minerals (Fig. 4.2). Waterlogging and flooding result in stomata closure, which is linked to a plant's photosynthetic efficiency. This disrupts gas exchange, which in turn lowers productivity and yield (Yu et al. 2015; Zhu et al. 2016). Waterlogging stress damages the photosynthesis reaction centers, modifies the structure of chloroplasts, and influences the activity of photosynthetic enzymes (Ren et al. 2016). Plants cultivated in water have been shown to have lower chlorophyll concentrations, according to Bansal and Srivastava (2015). As a result, the photosynthetic rate generally declines, which in turn lowers agricultural yield and output (Zhang et al. 2019). Many ROS were generated as a result of oxidative damage under WLS

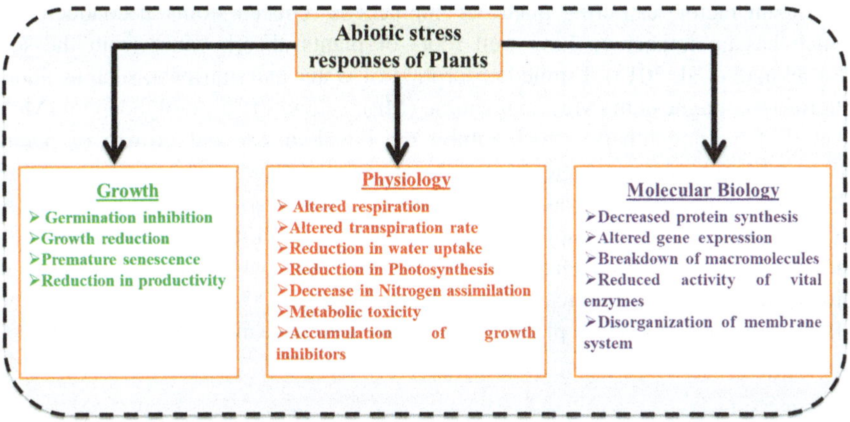

Fig. 4.1 Adaptive mechanisms of plants in response to abiotic stressors

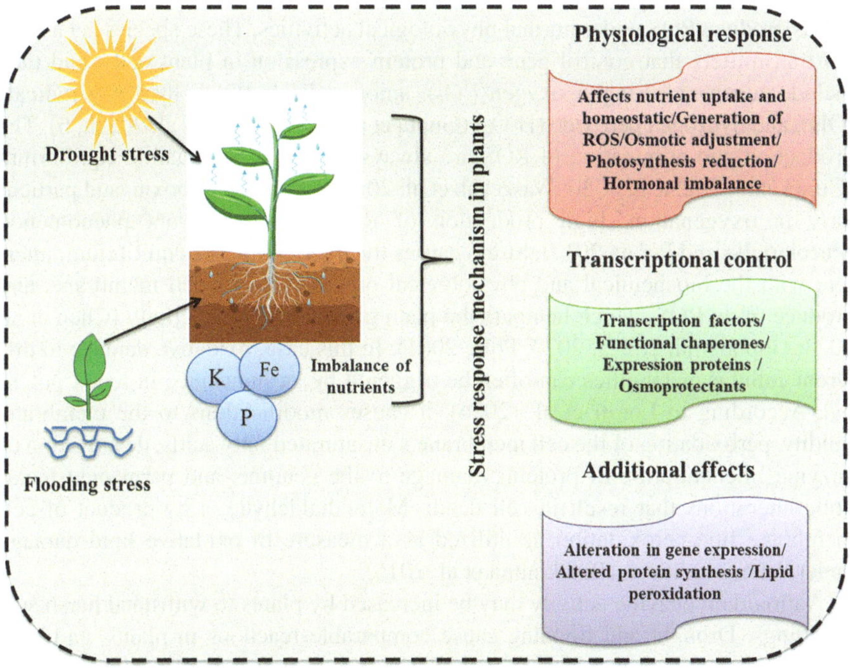

Fig. 4.2 Plant response cascades to drought and flooding stress

circumstances because of an excessive decrease in the electron transport chain (ETC) (Lal et al. 2019). Bermuda grass exhibits a decreased transpiration rate, decreased stomatal conductance, decreased leaf photosynthesis, and decreased root fresh weight when under WL conditions (Xiao and Jespersen 2019). The most significant factor restricting plant development is reduced stomatal conductivity, which has an impact on how well roots of plants absorb water from the soil (Barickman et al. 2019). Limited absorption of water and nutrients, such as phosphorus (P), magnesium (Mg), manganese (Mn), calcium (Ca), molybdenum (Mo), iron (Fe), among others, which inhibit the development and growth of plants (Fig. 4.2) and organic matter formation, results in decreased net CO_2 buildup (Ploschuk et al. 2018; Ye et al. 2018). Stress from flooding and droughts negatively impacts photosynthesis, respiration, and nutrient availability in plants. It also disturbs water relations and plant water usage efficiency, which causes a variety of biochemical and physiological reactions (Farooq et al. 2009). Figure 4.2 shows the effects of floods and DS on plants as well as the response mechanisms involved.

4.4 Understanding the Antioxidant Mechanism for Mitigating Waterlogging and Drought-Induced Stress

Plants produce ROS under normal physiological activities. These species act as signal transmitters that control gene and protein expression in plant cells, and they include superoxide singlet oxygen ($^1O_2$), anion radicals (O_2^-), hydroxyl radicals (OH), and hydrogen peroxide (H_2O_2) (Gupta et al. 2020; Gupta et al. 2022a, b). The production and elimination of ROS are always in a state of dynamic equilibrium (Gupta et al. 2022a, b, 2020; Waszczak et al. 2018). In anoxia/hypoxia, and particularly in oxygenation, high production of ROS is a significant phenomenon (Pucciariello and Perata 2017). Stress causes the plant to lose its equilibrium, interfere with the biochemical and physiological processes of its cell membrane, and produce more ROS, which hampers the plant growth and productivity (Chen et al. 2023; Gupta et al. 2022a, 2020; Polle 2001). In this case, oxidative damage to different cellular metabolites can often be triggered by an imbalance in redox potential. According to Loreti et al. (2016), it causes modifications to the membrane fluidity, peroxidation of the cell membrane's unsaturated fatty acids, deactivation of enzymes, denaturation of proteins, damage to the genome, and permanent metabolic alterations that result in cell death. Malondialdehyde, a by-product of cell membrane lipo-peroxidation, is utilized as a measure of oxidative lipid damage under WLS conditions (Balakhnina et al. 2015).

Antioxidant enzyme activity may be increased by plants to withstand harsh surroundings. Drought and flooding cause comparable reactions in plants, and both trigger the plant's antioxidant defense mechanism to prevent cell damage (Kumar et al. 2024). Malondialdehyde and antioxidant activity have an inverse relationship; a high level of malondialdehyde indicates a low level of antioxidant capacity, which lowers resistance under conditions of WL. Enzymatic and nonenzymatic antioxidants make up the antioxidant defense system. Peroxidases (POD), glutathione reductase (GR), monodehydro ascorbic acid reductase (MDHAR), ascorbate peroxidase (APX), and dehydroascorbate reductase glutathione (DHAR) are the enzymatic antioxidants. Ascorbic acid (AA), carotenoids, tocopherols, and glutathione (GSH) are examples of nonenzymatic antioxidants (Gupta et al. 2022a; Bano et al. 2021; Gupta et al. 2020). Both elements mitigate the damage brought forth by ROS (Miller et al. 2010; Laxa et al. 2019). Tolerance and stress threshold play a major role in antioxidant enzyme responses in plants to water stress. As the degree of water stress increases, SOD activity rises in the roots and leaves of the same species. Moreover, there is a reduction in the concentration of O_2^- and a rise in the disproportionation conversion of O_2^- to H_2O_2. According to Laishram et al. (2024), catalase (CAT) and peroxidase (POD) break down H_2O_2 into H_2O, efficiently prevent H_2O_2 from building up, shield plants from oxidative stress, and lessen the harmful effects of water stress on plants.

4.5 Nutritional Aspects of Plant Growth

Plant nutrition concerning minerals can be affected differently by drought and flooding. Waterlogging prevents the majority of vital nutrients from being absorbed by the soil, which results in deficits in Ca, Mg, P, N, and K. Nutrient shortages may result from the rivalry of Na^+ and Cl^- with nutrients like K, Ca, and NO. Certain nutrients' translocation and absorption can be impacted by drought (Hu and Schmidhalter 2003). The decrease in oxygen supply in the root zone may be the cause of the nutritional imbalance since it lowers root respiration and, consequently, the capacity to absorb nutrients. Thus, one factor contributing to yield loss is nutritional imbalance. The energy held in roots is decreased in waterlogged conditions, which prevents these nutrients from being actively transported to other organs. One of the most vital minerals, potassium plays a crucial function in a plant's ability to withstand stress (Baque et al. 2006). SOD concentrations are increased by macronutrients like N, K, and Ca and micronutrients like Mg, Si, and Zn, which also help plants tolerate drought better by reducing the toxicity of ROS. Iron toxicity is significant when anaerobic conditions are met. Oxidized polyphenols are produced as a result of an increase in polyphenol oxidase activity. In addition, it results in decreased root oxidation power and leaf bronzing (Fig. 4.3).

4.6 Influence of PGPR on Crop Resilience and Yield in Challenging Conditions

Plant growth-promoting rhizobacteria (PGPR) are free-living, plant-associated soil-borne bacteria that naturally dwell in the rhizospheric region of the plant roots. They can support plant growth in several ways. Based on their inherent traits, such as (a) root colonization, (b) plant growth promotion, and (c) the capacity to adapt, survive, multiply, and engage in competition with other microbial flora to manifest their

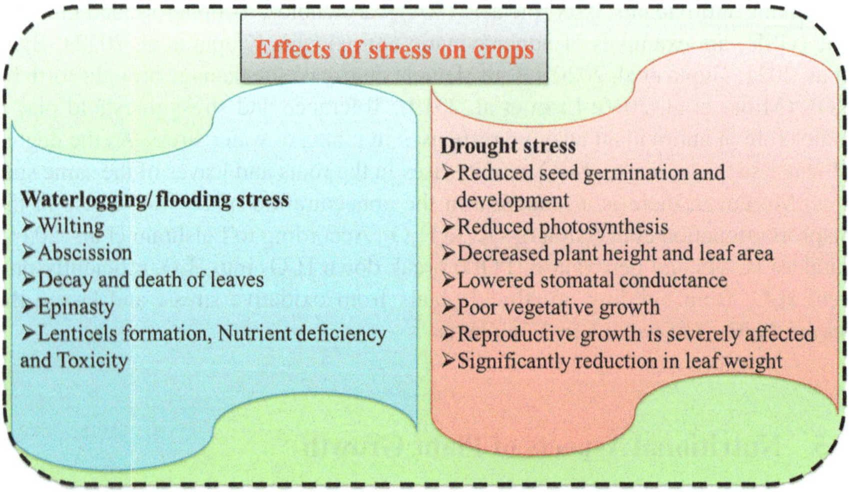

Fig. 4.3 Impact of waterlogging and drought stress on crop physiology

plant growth promotion and protection activities, Kloepper (1996) classed PGPR. These PGPRs are mostly Firmicutes, Bacteroidetes, Proteobacteria, and Actinobacteria, but they also represent a wide range of taxa. Many bacteria are consistently included in the PGPR-diversified taxa, including *Bacillus subtilis*, *B. megaterium*, *Rahnella aquatilis*, *Azotobacter salinestris*, *Pseudomonas fluorescens*, *P. stutzeri*, *Burkholderia phytofirmans*, and *Paenibacillus favisporus* (Gupta et al. 2020; Ferchichi et al. 2019). Numerous plant advantages are offered by these PGPRs, including increased plant hormone stimulation, nutrient absorption, plant development, pathogenic activity inhibition, and soil health restoration (Artyszak and Gozdowski 2020; Kour et al. 2019). Plant development is stimulated by PGPR strains through many mechanisms, including root growth promotion, enhanced nutrient availability in the soil, elongation, cell division, and metabolic stability. PGPR strains produce numerous metabolites and substances that promote plant growth, including phytohormones, hydrogen cyanide, siderophores, and ammonia. These compounds improve soil nutrient availability by P-solubilization, N_2 fixation, and organic compound mineralization, all of which promote plant growth (Gupta and Vandana 2019).

Numerous stressors, including drought, heavy metals, floods, infections, temperatures, and so on, can impede the development of plants. These stresses can also be exacerbated by anthropogenic activities (Gupta et al. 2022c). Plant growth regulators, such as ABA (Fibach-Paldi et al. 2012), nitric oxide (Cohen et al. 2008), polyamines like spermine and spermidine (Cassan et al. 2009), synthesis and exudation of IAA, ethylene, gibberellins, cytokinin (Meza et al. 2015), and ACC (1-amin ocyclopropane-1-carboxylate)deaminase (Gupta et al. 2022a; Gupta et al. 2020), increased solubilization and availability of nutrients (Rosa et al. 2020; Jalal et al. 2023a, b), and nitrate nutrient use efficiency and reductase activity (Galindo et al. 2022). By employing several strategies such as phosphate solubilization, nitrogen fixation, ACC deaminase (ACCD) synthesis, and siderophore synthesis, stress-tolerant PGPR can reduce these stressors. Plants gain greatly from bacterial strains' capacity to withstand stress. In reaction to environmental stressors such as cold, drought, floods, pathogen infection, and heavy metals, plants produce ACC (Glick 2012).

Plant roots colonized by PGPR with ACCD activity showed greater resistance to environmental stressors (Gupta et al. 2022a, 2020). It has been demonstrated that a variety of ACCD-producing bacteria stimulate plant development and lessen the negative effects of waterlogging stress (Nascimento et al. 2012). According to reports, *Pseudomonas* spp. Generate ACCD and lessen agricultural plant damage caused by flooding stress (Shaharoona et al. 2006; Khandelwal and Sindhu 2013). When exposed to *Pseudomonas veronii* KJ, which produces ACCD, sesame plants demonstrated markedly greater resistance to WLS in comparison to untreated flooded plants. Root and branch lengths, chlorophyll content, and plant biomass all significantly differed in flooded plants treated with *P. veronii* KJ bacterium that produces ACCD (Ali et al. 2018). The decline in electron transport rate (ETR) might be a sign of degradation in PS-II's ETC. *Pseudomonas veronii* KJ, a bacterium that produces ACCD, is used to treat sesame plants, which lessens damage and promotes plant development and growth (Ali et al. 2018). According to Grichko and Glick's (2001) research, flooded plants with ACCD-producing bacteria had a far greater concentration of chlorophyll than control plants, whose leaf chlorophyll content had dropped. Furthermore, Rahman et al. (2023) employ the strains of *Aspergillus nomiae* and *A. fumigatus*, MA1 and MA4, to mitigate the hypoxic stress caused by WLS in *Zea mays* L. through ethylene metabolism.

In addition to regularizing and transforming soluble sugars, amino acids, proline, and mineral nutrients in the soil above plant parts, these bacteria activate several processes that change soil organic matter to an instantly available form (Jalal et al. 2023a, b). This improves nutrient accumulation in nutrient-deficient soils (Singh et al. 2019). Through the synthesis of certain phytohormones like gibberellins, cytokinin, auxins, ethylene, and salicylic acid, the PGPRs enhanced the development and performance of plants (Vejan et al. 2016; Zaheer et al. 2019). Positive interactions between several vegetable crops and several species of PGPR, such as *Bacillus*, *Pseudomonas*, *Azospirillum*, *Bradyrhizobium*, and *Rhizobium*, were observed (Zeffa et al. 2018; Jalal et al. 2021; Kumar et al. 2021) (Table 4.2). Numerous research works demonstrate the ability of various rhizobacteria that promote plant growth to fix nitrogen biologically (Teixeira Filho and Galindo 2019; Galindo et al. 2020), as well as to increase the availability of P, Fe (Sivasakthi et al. 2014), and Zn for transportation and solubilization (Jalal et al. 2022). To improve the resistance of tomato and potato crops to DS, PGPR such as *Achromobacter*, *Citrobacter*, *Bacillus*, *Pseudomonas*, *Mesorhizobium*, and *Variovorax* can be employed (Belimov et al. 2015; Ullah et al. 2016). Research by Smethurst and Shabala (2003) found that WLS often causes decreased photochemical efficiency in leaves and can have a substantial impact on plant photosynthesis. According to a study by Forghani et al. (2018), sweet sorghum leaves increase their total chlorophyll and carotenoid levels

Table 4.2 Plant growth-promoting rhizobacteria (PGPR) strains mediated beneficial features related to stress tolerance in crops

PGPR strains	Plant species	Type of stress	Mechanisms	Beneficial features	Reference
Bacillus amyloliquefaciens	Alfalfa	Drought	EPS, QST713, and IAA	• Induction of systemic resistance • Reduction in H_2O_2, O_{2-} content • Increase in plant growth and biomass (dry, fresh weight of shoots and roots), RWC, antioxidant enzyme activities (APX, SOD, POD, CAT), photosynthesis, chlorophyll a, chlb, and carotenoid content	Han et al. (2022)
Bacillus spp.	Rice	Drought	IAA, EPS, and ACCD	• Improved germination, yield, and plant biomass • Reducing membrane and oxidative damage	Joshi et al. (2020)
Azospirillum brasilense, Bacillus sp.; *Azospirillum lipoferum*	Wheat	Drought	ABA, IAA, and CK	• Increases seed germination and root growth • Improved plant biomass, nutrient absorption, photosynthetic pigment concentration (carotenoids and Chl a, chlb), and plant yield • Increase the amount of antioxidant enzymes (SOD, CAT, APX) and osmolytes (sugar and proline) • Decrease in EL while maintaining membrane stability	Akhtar et al. (2021)

(continued)

Table 4.2 (continued)

PGPR strains	Plant species	Type of stress	Mechanisms	Beneficial features	Reference
Pseudomonas SP2	*Arabidopsis thaliana*	Drought	IAA, ABA, and ACCD	• Improved SP2 colonization and survival • Improved chlorophyll levels, plant biomass, RWC, and reduction in MDA • Increase in glycine-betaine, osmolyte, and proline accumulation	Yasmin et al. (2022)
Enterobacter cloacae, *Achromobacter xylosoxidans*	Maize	Drought	IAA and ACCD	• Increase in plant height, size, shoot dry weight, grain production, transpiration rate, photosynthetic rate, stomatal conductance, and chlorophyll a, b, and a+b	Danish et al. (2020)
Pseudomonas otitidis AKAD A1-2, *Pseudomonas* sp. AKAD A1–16, *B. cereus* AKAD A1–1	Soybean	Drought	IAA, ACCD, and EPS	• Improves chlorophyll content, plant biomass, and RWC • Increase in soluble carbohydrates, protein, and proline content • Decrease in MDA and H_2O_2 levels	Dubey et al. (2021)

(continued)

Table 4.2 (continued)

PGPR strains	Plant species	Type of stress	Mechanisms	Beneficial features	Reference
Pseudomonas sp. N66 L.	*Sorghum bicolour*	Drought	ACCD	• Reduces the level of ethylene • Improves root system architecture growth and via regulation of phytohormones such as IAA, GA, and CK • Increase in osmolyte accumulation (proline, glutamic acid, choline)and antioxidant activity • Activates induced systemic resistance by the signaling hormones SA, and JA and brassinolides, the signaling molecules psychosine and sphingosine	Carlson et al. (2020)
B. amyloliquefaciens GB03, *Pseudomonas fluorescens* WCS417	Peppermint	Drought	ACCD and IAA	• Increase in both enzymatic (SOD, APX) and nonenzymatic (ascorbic acid, phenol) antioxidant activities • Decrease in MDA levels • Increase in growth of shoots and roots	Chiappero et al. (2019)
Pseudomonas sp., *bacillus* sp.	Tomato	Drought	EPS, ACCD, and IAA	• Increase in rate of germination, plant biomass, and photosynthesis • Improved resilience to drought and recovery	Astorga-Eló et al. (2021)

(continued)

Table 4.2 (continued)

PGPR strains	Plant species	Type of stress	Mechanisms	Beneficial features	Reference
Bacillus sp. AGS84, *B. amyloliquefaciens* LMG9814, *B. amyloliquefaciens* RHF6	Spinach	Drought	Phosphate solubilization, IAA, and NH_3	• Increases in germination rate, the content of photosynthetic pigments (Chl a, Chl b, total Chl, and carotenoids), and plant biomass	Petrillo et al. (2022)
Pseudomonas sp. UW4	*Brassica napus*	Flooding	ACCD	• Increased canola germination and survival, and root and shoot biomass	Farwell et al. (2007)
B. mojavensis, *Mesorhizobium Cicero*, *B. subtilis*	Chickpea	Drought	ACCD, and IAA	• Improved plant growth, nutrient accumulation in seeds and leaves, and root colonization • Improved seed yield	Mahmood Aulakh et al. (2020)
Pseudomonas sp. UW4	*Lycopersicon esculentum*	Flooding	ACCD	• Increased dry and shoot fresh weights, and chlorophyll contents	Grichko and Glick (2001)
B. subtilis	Okra	Drought	Antioxidant enzymes	• Increase in photosynthesis, enzymatic antioxidants (GR, SOD, APX, CAT, DHAR, and MDHAR), osmolyte (proline and sugar) accumulation, and nonenzymatic antioxidants (ascorbate AsA, glutathione GSH) • Reduction in MDA and O_2^- levels	Puthiyottil and Akkara (2021)

(continued)

Table 4.2 (continued)

PGPR strains	Plant species	Type of stress	Mechanisms	Beneficial features	Reference
B. amyloliquefaciens	Candle millet	Drought	ACCD	• Increase in germination rate and seedling vigor, proline, APX, and SOD levels • Enhanced plant growth, SOD1 and APX1 gene expression, total chlorophyll, and RWC • Reduction in MDA content, relative expression of drought response, and ethylene response factor (ERF-1B)	Murali et al. (2021)
B. subtilis RJ46, *Pseudomonas* sp. RJ15, *Ochrobactrum pseudogrignonense* RJ12	Black gram, pea	Drought	ACCD	• Improvement in seed germination, shoot and root length, and plant biomass • Increase in the accumulation of osmolytes (phenol, proline), antioxidant enzyme activity like POD and CAT, RWC, and chlorophyll content • Decrease in ACC oxidase gene production and ACC accumulation in plants	Saikia et al. (2018)

(continued)

Table 4.2 (continued)

PGPR strains	Plant species	Type of stress	Mechanisms	Beneficial features	Reference
Planomicrobium chinense P1; *B. cereus* P2	Sunflower	Drought	Production of phytohormone (GA, IAA), and phosphate solubilization	• Reduction in MDA levels • Increase in quantity of leaf proteins, sugar, phenolic substances, chlorophyll, and carotenoids • Improved root length, shoot length, and plant biomass	Khan et al. (2020)
Pseudomonas palleroniana DPB15, *Ochrobactrum anthropi* DPB13, *Pseudomonas palleroniana*, *Pseudomonas fluorescens* DPB16	Finger millet	Drought	IAA and ACCD	• Increase in growth metrics, proline, phenol, total chlorophyll, antioxidant enzyme activity accumulation (GPX, SOD, APX), and nutrient supply (Na^+, Ca^{2+}, N, P, K) • Reduction in MDA levels and H_2O_2, improves membrane stability	Chandra and Sharma (2021)
Pseudomonas veronii KJ	Sesame	Flooding	ACCD	• Increase in root length, shoot length, biomass, and chlorophyll content	Ali et al. (2018)

EPS exopolysaccharides, *RWC* relative water content, *EL* electrolyte leakage, *APX* ascorbate peroxidase, *SOD* superoxide dismutase, *POD* peroxidase, *CAT* catalase, *GPX* glutathione peroxidase, *MDA* malondialdehyde, H_2O_2 hydrogen peroxide, *ACCD* 1-aminocyclopropane-1-carboxylate deaminase, *MDHAR* monodehydroascorbate reductase, *DHAR* dehydroascorbate reductase, *IAA* indole acetic acid, *GA* gibberellic acid, *CK* creatine kinase, *JA* jasmonic acid, *SA* salicylic acid, *AsA* ascorbate, *Na* sodium, *Ca* calcium, *N* nitrogen, *P* phosphorus, *K* potassium when stress conditions are imposed, indicating an adaptive mechanism to enhance photosynthetic efficiency and protect against oxidative damage (Table 4.2).

4.7 Shotgun Approach for Enhancing Waterlogging Tolerance in Plants

Through the exogenous administration of plant hormones and nutrients, scientists from many parts of the world are actively working to make plants resistant to drought and flood/WL stress. According to Ashraf et al. (2011), potassium applied exogenously to the soil and as foliar spray helped cotton plants recover from the damaging effects of WLS. Potassium plays an important role in osmoregulation, stomatal regulation, and in several activities of enzymes involved in the mechanism of photosynthesis (Benlloch-González et al. 2008; Taiz and Zeiger 2010) increased root shoot growth due to exogenous application of potassium through all modes in waterlogging-stressed as well as nonstressed cotton plants may have been due to K-induced enhanced uptake and accumulation of essential nutrients, photosynthetic capacity, and maintenance of water relation components. According to Ashraf and Rehman (1999), adding nitrate to the soil helped lessen the negative impacts of waterlogging on the various physiological characteristics of maize. When onions are subjected to flooding stress, the exogenous injection of spermine and spermidine causes a number of physiological and biochemical changes (Yiu et al. 2009). Exogenous uniconazole injection has been found to help wheat plants and oil seed rape resist the negative effects of waterlogging (Webb and Fletcher 1996; Zhou et al. 1997). As a result, using these organic and inorganic materials offers a solid basis for developing resilience to the stress caused by WLS.

4.8 Utilizing Transgenic Techniques for Enhancing DS and WLS Resilience in Crops

Conventional breeding technologies have failed to develop many stress-tolerant high-yielding crop varieties (Raina et al. 2022). Transgenic approaches can be used in understanding the stress tolerance mechanisms in plants. Transgenic technologies can be used to incorporate stress-responsive genes from wild varieties into major crops (Li et al. 2018). It offers novel opportunities to develop stress-tolerant crop varieties by introducing genes implicated in stress defense into species. Plants change their gene expression in response to various stressors (He et al. 2017). Regulating stress-responsive genes is important for plants to cope with abiotic challenges and build stress tolerance (Ullah et al. 2020; Basu et al. 2022). Plants adapt stress-like drought and activate different molecular mechanism including drought-responsive genes, transcription factors, secondary messengers, and signaling pathways (Nakashima et al. 2014). In rice, *Arabidopsis*, and other crops, several drought-inducible genes encoding important enzymes in the ABA production pathway, signaling proteins, transcription factors (TFs), and cellular protection enzymes have been found (Manna et al. 2021). Stress-inducible protein late embryogenic abundant (LEA) has an important role in cell membrane protection under water stress conditions (Chen et al. 2021a, b). Various genes and TFs have been engineered through overexpression or silencing in model plants to develop waterlogging tolerance in barley (Luan et al. 2020), kiwifruit (Pan et al. 2021), and wheat (Wei et al. 2019) and drought tolerance in chickpea (Das et al. 2021) and rice (Shanmugam et al. 2021). Moreover, there is a need to identify novel stress-responsive genes and TFs to develop new crop stress-tolerant varieties WL and DS.

4.9 Utilizing QTL Manipulation to Breed Waterlogging-Tolerant Varieties

Abiotic stressors like flooding damage rice crops worldwide, particularly in low-lying regions. The heredity of abiotic stress resistance is a complex phenomenon, and quantitative trait loci (QTL) mapping is the most popular method of discovering QTL, as well as genetic and linkage mapping of genomic regions related to tolerance. Quantitative trait loci analysis allows researchers to examine the genetic structure of a characteristic. QTLs can identify genomic areas linked to the expression of the trait being researched (Kushwah et al. 2021). Markers are successful in determining the quantitative trait loci for specific traits from various regions of chromosomes for the identification of genotypes against the different multifunctional and biotic stresses. QTLs were previously identified from the many legume crops by using linked markers (Luo et al. 2021; Zate 2021). Scientists can regress this gene via MAB into the widely used high-yielding crop lines due to the identification of the SUB1 QTL. Single-nucleotide polymorphisms (SNP) inside Sub1A and Sub1C are revealed by mapping the Sub1 allele, which is helpful for breeders using precision breeding and molecular markers. Several kinds, including CR1009, BR11, Samba Mahsuri IR64, and Swarna, have been created by the usage of MAB by inserting Sub1A. In comparison to their parents, newly created Sub1 lines have a greater degree of WL tolerance (Manzanilla et al. 2011). These studies demonstrate the potential for producing tolerant varieties with a variety of genetic origins by introducing the Sub1 region from existing tolerant varieties by MAB. When compared to their original parents, the submergence-resistant cultivars created by the insertion of the Sub1 gene do not change field performances, including grain quality and yield under normal growing conditions (Minhas and Grover 1999). The recently created cultivars exhibit superior yield and Sub1 performance in the event of rice submersion. It has been established that soybeans are genetically viable for flood resistance, and crops with genetically enhanced QTLs have been successfully bred using marker-assisted selection (MAS) (Boru et al. 2001). Advancements in genomics and phenomics now allowed us to better characterize the QTLs that influence a certain trait, admired as QTLome (Salvi and Tuberosa 2015). Breeders can use QTLome's immense knowledge effectively because of its vastness. The identification of these QTLs against abiotic stress in various crops is helpful for breeders during breeding activities.

4.10 Enhancing Crop Stress Resilience Through Genetic Engineering Techniques

Biotechnological and molecular biology techniques have advanced to the point that they are now regarded as important crop breeding strategies. Before, the main technique for transforming genes was the protoplast-based approach, which involved using a microprojectile gun. However, with the advancement of technology, it is now possible to transfer genes using an *Agrobacterium*. Transgenes can now be produced regularly or artificially in response to a given stimulus due to the availability of regulatory sequences controlling transgene expression in a variety of crops. Under anaerobiosis, the respiration route changes from oxidation to the fermentation process. It is thought that one of the key mechanisms that plants use to respond to WLS is the ethanolic fermentation pathway. The two enzymes involved in ethanol fermentation are pyruvate decarboxylase (PDC) and alcohol dehydrogenase (ADH). The process is quite straightforward. Because cloned adh and *pdc* are now available, molecular scientists are interested in using these in transgenic research. PDC and ADH levels were changed in rice at CISRO, Australia. This led to the discovery of three distinct pdc genes in rice (*pdc1*, *pdc2*, and *pdc3*), which have been cloned and sequenced. The International Rice Research Institute has been using one of the aforementioned gene constructions (*actin 1-pdc1*, *sense*) to produce transgenic rice that has a higher metabolic capacity under anaerobic circumstances, which results in a water submergence tolerance. Plants that underwent transgenic modification exhibited higher ethanol production and increased *pdc* enzyme activity compared to nontransformed plants. This suggests a direct relationship between the ethanol yield of transgenic plants and their capacity to endure submersion. Researchers are working to modify the amount of ADH in rice (Dadrwal et al. 2020). Furthermore, prospective crops' genetic resources differ significantly in terms of their ability to withstand WL and DS. Plant species such as cucumber (Yeboah et al. 2008), maize (Silva et al. 2005), and oats (Lemsons e Silva et al. 2003) have also been documented to exhibit genetic diversity.

4.11 Conclusion and Future Prospects

The prior debate suggests that the two main obstacles to sustainable agriculture are drought and waterlogging. Both at the cellular and whole plant levels, its impacts are apparent. In order to make other possible crops resistant as well, it is necessary to screen the existing germplasm for tolerance to drought and waterlogging and to employ the genes that cause tolerance in other potential crops. Drought and waterlogging lead to nutritional deficiencies in various areas. Consequently, exogenous treatment of these nutrients or other plant hormones may be employed to mitigate the negative consequences associated with waterlogging and drought. The creation of stress-tolerant crop varieties through genetic engineering and plant breeding is crucial, but at the same time inoculation with PGPR to relieve abiotic stresses like DS and WLS in plants opens a new chapter in the use of microorganisms in agriculture. PGPB-based bioformulations can be used as biocontrol and biofertilizers to increase crop yields and productivity through IAA production, siderophore production, ACCD production, phosphate solubilization, and other beneficial traits. Therefore, to accomplish sustainable agriculture, the right microorganisms must be identified. In addition, advances in biotechnology tools such as genetic engineering and genome editing have opened the door to improve yields and stress tolerance in crop plants. Many stress-responsive genes and TFs have been identified for engineering (overexpression/silencing) and CRISPR-Cas9-mediated editing (knock-out/knock-in) in crop plants to develop drought and WLS-tolerant crop varieties. Thus, there is a need to identify more stress-responsive genes and TFs for application in crop improvement and sustainable agriculture.

References

Akhtar N, Ilyas N, Hayat R, Yasmin H, Noureldeen A, Ahmad P (2021) Synergistic effects of plant growth promoting rhizobacteria and silicon dioxide nano-particles for amelioration of drought stress in wheat. Plant Physiol Biochem 166:160–176

Ali S, Khan MA, Kim WC (2018) *Pseudomonas veronii* KJ mitigates flood stress-associated damage in *Sesamum indicum* L. Appl Biol Chem 61:575–585. https://doi.org/10.1007/s13765-018-0392-2

Artyszak A, Gozdowski D (2020) The effect of growth activators and plant growth-promoting rhizobacteria (PGPR) on the soil properties, root yield, and technological quality of sugar beet. Agronomy 10:1262. https://doi.org/10.3390/agronomy10091262

Ashraf M, Rehman H (1999) Mineral nutrient status of corn in relation to nitrate and long term waterlogging. J Plant Nutr 22:1253–1268. https://doi.org/10.1080/01904169909365710

Ashraf MA, Ahmad MSA, Ashraf M, Al-Qurainy F, Ashraf MY (2011) Alleviation of waterlogging stress in upland cotton (*Gossypium hirsutum* L.) by exogenous application of potassium in soil and as a foliar spray. Crop Pasture Sci 62:25–38. https://doi.org/10.1071/CP09225

Astorga-Eló M, Gonzalez S, Acuña JJ, Sadowsky MJ, Jorquera MA (2021) Rhizobacteria from 'flowering desert'events contribute to the mitigation of water scarcity stress during tomato seedling germination and growth. Sci Rep 11(1):13745

Balakhnina TI, Bulak P, Matichenkov VV, Kosobryukhov AA, Włodarczyk TM (2015) The influence of Si-rich mineral zeolite on the growth processes and adaptive potential of barley plants under cadmium stress. J Plant Growth Regul 75:557–565. https://doi.org/10.1007/s10725-014-0021-y

Bano A, Gupta A, Rai S, Fatima T (2021) Mechanistic role of reactive oxygen species and its regulation via the antioxidant system under environmental stress. In: Plant stress physiology—perspectives in agriculture. IntechOpen, London, UK, pp 1–18

Bansal R, Srivastava JP (2015) Effect of waterlogging on photosynthetic and biochemical parameters in pigeonpea. Russ J Plant Physiol 62:322–327. https://doi.org/10.1134/S1021443715030036

Baque MA, Karim MA, Hamid A, Tetsushi H (2006) Effects of fertilizer potassium on growth, yield and nutrient uptake of wheat (*Triticum aestivum*) under water stress conditions. South Pac Stud 27:25–35

Barickman TC, Simpson CR, Sams CE (2019) Waterlogging causes early modification in the physiological performance, carotenoids, chlorophylls, proline, and soluble sugars of cucumber plants. Plan Theory 8:160. https://doi.org/10.3390/8060160

Basu S, Sarkar S, Prasad R, Kumar G (2022) Current approaches for alleviating abiotic stress tolerance in crops: role of beneficial microorganisms. Benef Microorg Agric:231–248. https://doi.org/10.1007/978-981-19-0733-3_9

Belimov AA, Dodd IC, Safronova VI, Shaposhnikov AI, Azarova TS, Makarova NM, Davies WJ, Tikhonovich IA (2015) Rhizobacteria that produce auxins and contain 1-amino-cyclopropane-1-carboxylic acid deaminase decrease amino acid concentrations in the rhizosphere and improve growth and yield of well-watered and water-limited potato (*Solanum tuberosum*). Ann Appl Biol 167:11–25. https://doi.org/10.1111/aab.12203

Benlloch-González M, Arquero O, Fournier JM, Barranco D, Benlloch M (2008) K^+ starvation inhibits water-stress-induced stomatal closure. J Plant Physiol 165:623–630. https://doi.org/10.1016/j.jplph.2007.05.010

Bhusal N, Kim HS, Han SG, Yoon TM (2020) Photosynthetic traits and plant–water relations of two apple cultivars grown as bi-leader trees under long-term waterlogging conditions. Environ Exp Bot 176:104111. https://doi.org/10.1016/j.envexpbot.2020.104111

Binks O, Meir P, Rowland L, da Costa AC, Vasconcelos SS, de Oliveira AA, Ferreira L, Mencuccini M (2016) Limited acclimation in leaf anatomy to experimental drought in tropical rainforest trees. Tree Physiol 36:1550–1561. https://doi.org/10.1093/treephys/tpw078

Blum A (2011) Plant breeding for water-limited environments. Springer Science and Business Media, LLC, New York, pp 217–234

Boru G, Van Ginkel M, Kronstad WE, Boersma L (2001) Expression and inheritance of tolerance to waterlogging stress in wheat. Euphytica 117:91–98. https://doi.org/10.1023/A:1003929803920

Brodersen KE, Hammer KJ, Schrameyer V, Floytrup A, Rasheed MA, Ralph PJ, Kühl M, Pedersen O (2017) Sediment resuspension and deposition on seagrass leaves impedes internal plant aeration and promotes phytotoxic H_2S intrusion. Front Plant Sci 8:657. https://doi.org/10.3389/fpls.2017.00657

Cal AJ, Sanciangco M, Rebolledo MC, Luquet D, Torres RO, McNally KL, Henry A (2019) Leaf morphology, rather than plant water status, underlies genetic variation of rice leaf rolling under drought. Plant Cell Environ 42:1532–1544. https://doi.org/10.1111/pce.13514

Canales FJ, Rispail N, García-Tejera O, Arbona V, Pérez-de-Luque A, Prats E (2021) Drought resistance in oat involves ABA-mediated modulation of transpiration and root hydraulic conductivity. Environ Exp Bot 182:104333. https://doi.org/10.1016/j.envexpbot.2020.104333

Carlson R, Tugizimana F, Steenkamp PA, Dubery IA, Hassen AI, Labuschagne N (2020) Rhizobacteria- induced systemic tolerance against drought stress in Sorghum bicolor (L.) Moench. Microbiol Res 232:126388

Cassan F, Maiale S, Masciarelli O, Vidal A, Luna V, Ruiz O (2009) Cadaverine production by *Azospirillum brasilense* and its possible role in plant growth promotion and osmotic stress mitigation. Eur J Soil Biol 45:12–19. https://doi.org/10.1016/j.ejsobi.2008.08.003

Ceccarelli S, Grando S, Baum M (2007) Participatory plant breeding in water-limited environments. Exp Agric 43:411–435. https://doi.org/10.1017/S0014479707005327

Celedonio SRP, Abeledo LG, Mantese AI, Miralles DJ (2017) Differential root and shoot biomass recovery in wheat and barley with transient waterlogging during preflowering. Plant Soil 417:481–498. https://doi.org/10.1007/s11104-017-3274-1

Challabathula D, Zhang Q, Bartels D (2018) Protection of photosynthesis in desiccation-tolerant resurrection plants. J Plant Physiol 227:84–92. https://doi.org/10.1016/j.jplph.2018.05.002

Chandra D, Sharma AK (2021) Field evaluation of consortium of bacterial inoculants producing ACC deaminase on growth, nutrients and yield components of rice and wheat. J Crop Sci Biotechnol 24(3):293–305

Chen J, Li N, Wang X, Meng X, Cui X, Chen Z, Ren H, Ma J, Liu H (2021a) Late embryogenesis abundant (LEA) gene family in salvia miltiorrhiza: identification, expression analysis, and response to drought stress. Plant Signal Behav 16:1891769. https://doi.org/10.1080/15592324.2021.1891769

Chen M, Zhu X, Zhang Y, Du Z, Chen X, Kong X, Sun W, Chen C (2021b) Drought stress modify cuticle of tender tea leaf and mature leaf for transpiration barrier enhancement through common and distinct modes. Sci Rep 10:6696. https://doi.org/10.1038/s41598-020-63683-4

Chen S, Ten Tusscher KH, Sasidharan R, Dekker SC, de Boer HJ (2023) Parallels between drought and flooding: an integrated framework for plant eco-physiological responses to water stress. Plant Environ Interact 4:175–187. https://doi.org/10.1002/pei3.10117

Chiappero J, del Rosario CL, Alderete LGS, Palermo TB, Banchio E (2019) Plant growth promoting rhizobacteria improve the antioxidant status in Mentha piperita grown under drought stress leading to an enhancement of plant growth and total phenolic content. Ind Crop Prod 139:111553

Cohen AC, Bottini R, Piccoli PN (2008) *Azospirillum brasilense* Sp 245 produces ABA in chemically-defined culture medium and increases ABA content in arabidopsis plants. Plant Growth Regul 54:97–103. https://doi.org/10.1007/s10725-007-9232-9

Colmer TD, Pedersen O (2008) Underwater photosynthesis and respiration in leaves of submerged wetland plants: gas films improve CO_2 and O_2 exchange. New Phytol 177:918–926. https://doi.org/10.1111/j.1469-8137.2007.02318.x

Dadrwal BK, Singh UP, Chauhan J, Bhudev RY, Vijai P (2020) Management of abiotic stress in crop plants. Innovative Publication Chapter 9. pp.145–162

Danish S, Zafar-Ul-Hye M, Hussain S, Riaz M, Qayyum MF (2020) Mitigation of drought stress in maize through inoculation with drought tolerant ACC deaminase containing PGPR under axenic conditions. Pak J Bot 52(1):49–60

Das A, Basu PS, Kumar M, Ansari J, Shukla A, Thakur S, Singh P, Datta S, Chaturvedi SK, Sheshshayee MS, Bansal KC (2021) Transgenic chickpea (*Cicer arietinum* L.) harbouring AtDREB1a are physiologically better adapted to water deficit. Plant Biol 21:1–7. https://doi.org/10.1186/s12870-020-02815-4

De Bauw P, Vandamme E, Lupembe A, Mwakasege L, Senthilkumar K, Dramé KN, Merckx R (2019) Anatomical root responses of rice to combined phosphorus and water stress–relations to tolerance and breeding opportunities. Funct Plant Biol 46:1009–1022. https://doi.org/10.1071/FP19002

Djanaguiraman M, Prasad PV, Kumari J, Rengel Z (2019) Root length and root lipid composition contribute to drought tolerance of winter and spring wheat. Plant Soil 439:57–73. https://doi.org/10.1007/s11104-018-3794-3

Domisch T, Qian J, Sondej I, Martz F, Lehto T, Piirainen S, Finér L, Silvennoinen R, Repo T (2020) Here comes the flood! Stress effects of continuous and interval waterlogging periods during the growing season on scots pine saplings. Tree Physiol 40:869–885. https://doi.org/10.1093/treephys/tpaa036

Dresbøll DB, Thorup-Kristensen K (2012) Spatial variation in root system activity of tomato (*Solanum lycopersicum* L.) in response to short and long-term waterlogging as determined by 15 N uptake. Plant Soil 357:161–172. https://doi.org/10.1007/s11104-012-1135-5

Dubey A, Saiyam D, Kumar A, Hashem A, Abd_Allah EF, Khan ML (2021) Bacterial root endophytes: characterization of their competence and plant growth promotion in soybean (Glycine max (L.) Merr.) under drought stress. Int J Environ Res Public Health 18(3):931

Fang Y, Xiong L (2015) General mechanisms of drought response and their application in drought resistance improvement in plants. Cell Mol Life Sci 72:673–689. https://doi.org/10.1007/s00018-014-1767-0

Farooq M, Wahid A, Kobayashi N, Fujita D, Basra SMA (2009) Plant drought stress: effects, mechanisms and management. Agron Sustain Dev 29:185–212153-88. https://doi.org/10.1007/978-90-481-2666-8_12

Farwell AJ, Vesely S, Nero V, Rodriguez H, McCormack K, Shah S (2007) Tolerance of transgenic canola plants (*Brassica napus*) amended with plant growth-promoting bacteria to flooding stress at a metal-contaminated field site. Environ Pollut 147:540–545. https://doi.org/10.1016/j.envpol.2006.10.014

Ferchichi N, Toukabri W, Boularess M, Smaoui A, Mhamdi R, Trabelsi D (2019) Isolation, identification and plant growth promotion ability of endophytic bacteria associated with lupine root nodule grown in Tunisian soil. Arch Microbiol 201:1333–1349. https://doi.org/10.1007/s00203-019-01702-3

Fibach-Paldi S, Burdman S, Okon Y (2012) Key physiological properties contributing to rhizosphere adaptation and plant growth promotion abilities of *Azospirillum brasilense*. FEMS Microbiol Lett 326:99–108. https://doi.org/10.1111/j.1574-6968.2011.02407.x

Forghani AH, Almodares A, Ehsanpour AA (2018) Potential objectives for gibberellic acid and paclobutrazol under salt stress in sweet sorghum (*Sorghum bicolor* [L.] Moench cv. Sofra). Appl Biol Chem 61:113–124. https://doi.org/10.1007/s13765-017-0329-1

Fukao T, Xu K, Ronald PC, Bailey-Serres J (2006) A variable cluster of ethylene response factor–like genes regulates metabolic and developmental acclimation responses to submergence in rice. Plant Cell 18:2021–2034. https://doi.org/10.1105/tpc.106.043000

Galindo FS, Buzetti S, Rodrigues WL, Boleta EH, Silva VM, Tavanti RF, Fernandes GC, Biagini AL, Rosa PA, Teixeira Filho MC (2020) Inoculation of *Azospirillum brasilense* associated with silicon as a liming source to improve nitrogen fertilization in wheat crops. Sci Rep 10:6160. https://doi.org/10.1038/s41598-020-63095-4

Galindo FS, Rodrigues WL, Fernandes GC, Boleta EH, Jalal A, Rosa PA, Buzetti S, Lavres J, Teixeira Filho MC (2022) Enhancing agronomic efficiency and maize grain yield with *Azospirillum brasilense* inoculation under Brazilian savannah conditions. Eur J Agron 134:126471. https://doi.org/10.1016/j.eja.2022.126471

Glick BR (2012) Plant growth-promoting bacteria: mechanisms and applications. Scientifica. https://doi.org/10.6064/2012/963401

Grichko VP, Glick B (2001) Amelioration of flooding stress by ACC deaminase-containing plant growth-promoting bacteria. Plant Physiol Biochem 39:11–17. https://doi.org/10.1016/S0981-9428(00)01212-2

Gupta A, Vandana P (2019) Effect of PGPR isolates on plant growth promotion in relation to salinity stress. Bull Environ Pharmacol Life Sci 8:18–26

Gupta A, Rico-Medina A, Caño-Delgado AI (2020) The physiology of plant responses to drought. Science 368:266–269. https://doi.org/10.1126/science.aaz7614

Gupta A, Rai S, Bano A, Sharma S, Kumar M, Binsuwaidan R, Suhail Khan M, Upadhyay TK, Alshammari N, Saeed M, Pathak N (2022a) ACC deaminase produced by PGPR mitigates the adverse effect of osmotic and salinity stresses in *Pisum sativum* through modulating the antioxidants activities. Plan Theory 11:3419

Gupta A, Bano A, Rai S, Mishra R, Singh M, Sharma S, Pathak N (2022b) Mechanistic insights of plant-microbe interaction towards drought and salinity stress in plants for enhancing the agriculture productivity. Plant Stress 4:100073

Gupta A, Dubey P, Kumar M, Roy A, Sharma D, Khan MM, Hasanuzzaman M (2022c) Consequences of arsenic contamination on plants and mycoremediation-mediated arsenic stress tolerance for sustainable agriculture. Plan Theory 11:3220

Han L, Zhang M, Du L, Zhang L, Li B (2022) Effects of Bacillus amyloliquefaciens QST713 on photosynthesis and antioxidant characteristics of Alfalfa (Medicago sativa L.) under drought stress. Agronomy 12(9):2177

Hazman M, Brown KM (2018) Progressive drought alters architectural and anatomical traits of rice roots. Rice 11:1–16

He P, Zhao P, Wang L, Zhang Y, Wang X, Xiao H, Xiao G (2017) The PIN gene family in cotton (*Gossypium hirsutum*): genome-wide identification and gene expression analyses during root development and abiotic stress responses. BMC Genomics 18:1–10

Hu Y, Schmidhalter U (2003) Drought and salinity: a comparison of their effects on mineral nutrition of plants. J Plant Nutr Soil Sci 168:541–549

Jalal A, Galindo FS, Boleta EHM, Oliveira CES, Reis AR, Nogueira TAR, Moretti Neto MJ, Mortinho ES, Fernandes GC, Teixeira Filho MCM (2021) Common bean yield and zinc use efficiency in association with diazotrophic bacteria co-inoculations. Agronomy 11:959

Jalal A, da Silva Oliveira CE, Freitas LA, Galindo FS, Lima BH, Boleta EHM, Da Silva EC, do Nascimento V, Nogueira TAR, Buzetti S (2022) Agronomic biofortification and productivity of wheat with soil zinc and diazotrophic bacteria in tropical savannah. Crop Pasture Sci 73:817–830

Jalal A, da Silva Oliveira CE, Galindo FS, Rosa PAL, Gato IMB, de Lima BH, Teixeira Filho MCM (2023a) Regulatory mechanisms of plant growth-promoting rhizobacteria and plant nutrition against abiotic stresses in Brassicaceae family. Life 13:211

Jalal A, Oliveira CEDS, Bastos ADC, Fernandes GC, de Lima BH, Furlani Junior E, De Carvalho PHG, Galindo FS, Gato IMB, Teixeira Filho MCM (2023b) Nano-zinc and plant growth-promoting bacteria improve biochemical and metabolic attributes of maize in tropical Cerrado. Front Plant Sci 13:5293

Joshi B, Chaudhary A, Singh H, Kumar PA (2020) Prospective evaluation of individual and consortia plant growth promoting rhizobacteria for drought stress amelioration in rice (Oryza sativa L.). Plant Soil 457:225–240

Khan N, Bano A, Curá JA (2020) Role of beneficial microorganisms and salicylic acid in improving rainfed agriculture and future food safety. Microorganisms 8(7):1018

Khandelwal A, Sindhu SS (2013) ACC deaminase containing rhizobacteria enhance nodulation and plant growth in Clusterbean (*Cyamopsis tetragonoloba* L.). J Microbiol Res 3:117–123

Kloepper JW (1996) Host specificity in microbe-microbe interactions. Bioscience 46:406–409

Kour D, Rana KL, Yadav N, Yadav AN, Kumar A, Meena VS, Singh B, Chauhan VS, Dhaliwal HS, Saxena AK (2019) RhizosphericMicrobiomes: biodiversity, mechanisms of plant growth promotion, and biotechnological applications for sustainable agriculture. In: Plant growth promoting Rhizobacteria for agricultural sustainability. Springer, Singapore, pp 19–65

Kumar M, Giri VP, Pandey S, Gupta A, Patel MK, Bajpai AB, Siddique KH (2021) Plant-growth-promoting Rhizobacteria emerging as an effective bioinoculant to improve the growth, production, and stress tolerance of vegetable crops. Int J Mol Sci 22:12245

Kumar M, Gupta A, Vandana P, Tiwari LD, Patel MK, Siddique KH (2024) Recent advances of plant growth-promoting rhizobacteria (PGPR)-mediated drought and waterlogging stress tolerance in plants for sustainable agriculture. Microbial biostimulants for plant growth and abiotic stress amelioration. Elsevier, In, pp 315–344

Kushwah A, Bhatia D, Rani U, Yadav IS, Singh I, Bharadwaj C, Singh S (2021) Molecular mapping of quantitative trait loci for ascochyta blight and botrytis grey mould resistance in an inter-specific cross in chickpea (*Cicer arietinum* L.) using genotyping by sequencing. Breed Sci 71:229–239. https://doi.org/10.1270/jsbbs.20085

Laishram R, Dutta M, Nagesh CR, Sushmitha J, Meena NL (2024) Effect on morphology, physiology, and biochemistry of plants under different stresses. In: Molecular dynamics of plant stress and its management. Springer Nature Singapore, Singapore, pp 159–192

Lal M, Kumari A, Sheokand S (2019) Reactive oxygen species, reactive nitrogen species and oxidative metabolism under water logging stress. In: Hasanuzzaman M, Fotopoulos V, Nahar K,

Fujita M (eds) Reactive oxygen, nitrogen and sulfur species in plants: production, metabolism, signaling and defense mechanisms. John Wiley & Sons Ltd., pp 777–812

Lawson JR, Fryirs KA, Leishman MR (2017) Interactive effects of waterlogging and atmospheric CO_2 concentration on gas exchange, growth and functional traits of Australian riparian tree seedlings. Ecohydrology 10:e1803

Laxa M, Liebthal M, Telman W, Chibani K, Dietz KJ (2019) The role of the plant antioxidant system in drought tolerance. Antioxidants 8:94

Lee DK, Jung H, Jang G, Jeong JS, Kim YS, Ha SH, Do Choi Y, Kim JK (2016) Overexpression of the *OsERF71* transcription factor alters rice root structure and drought resistance. Plant Physiol 172:575–588

Lemsons e Silva CF, de Mattos LAT, de Oliveria AC, de Carvalho FIF, de Freitas FA, Anjos e Silva SD (2003) Flooding tolerance in oats. Crop Breed Appl Biotechnol 5:29–42

Li H, Rasheed A, Hickey LT, He Z (2018) Fast-forwarding genetic gain. Trends Plant Sci 23:184–186

Liu WS, Zheng L, Qi DH (2020) Variation in leaf traits at different altitudes reflects the adaptive strategy of plants to environmental changes. Ecol Evol 10:8166–8175

Loreti E, VanVeen H, Perata P (2016) Plant responses to flooding stress. Curr Opin Plant Biol 33:64–71

Lozano YM, Aguilar-Trigueros CA, Flaig IC, Rillig MC (2020) Root trait responses to drought are more heterogeneous than leaf trait responses. Funct Ecol 34:2224–2235

Luan H, Guo B, Shen H, Pan Y, Hong Y, Lv C, Xu R (2020) Overexpression of barley transcription factor HvERF2. 11 in *Arabidopsis* enhances plant waterlogging tolerance. Int J Mol Sci 21:1982

Luo H, Guo J, Yu B, Chen W, Zhang H, Zhou X, Chen Y, Huang L, Liu N, Ren X, Yan L (2021) Construction of ddRADseq-based high-density genetic map and identification of quantitative trait loci for trans-resveratrol content in Peanut seeds. Front Plant Sci 12:438. https://doi.org/10.3389/fpls.2021.644402

Mahmood Aulakh A, Qadir G, Hassan FU, Hayat R, Sultan T, Billah M, Hussain M, Khan N (2020) Desert soil microbes as a mineral nutrient acquisition tool for chickpea (Cicer arietinum L.) productivity at different moisture regimes. Plan Theory 9:1629. https://doi.org/10.3390/plants9121629

Manna M, Thakur T, Chirom O, Mandlik R, Deshmukh R, Salvi P (2021) Transcription factors as key molecular target to strengthen the drought stress tolerance in plants. Physiol Plant 172:847–868

Manzanilla DO, Paris TR, Vergara GV (2011) Submergence risks and farmers' preferences: implications for breeding Sub1 rice in Southeast Asia. Agric Syst 104:335–347

Meng D, Dong B, Niu L, Song Z, Wang L, Amin R, Cao H, Li H, Yang Q, Fu Y (2021) The pigeon pea CcCIPK14-CcCBL1 pair positively modulates drought tolerance by enhancing flavonoid biosynthesis. Plant J 106:1278–1297

Meza B, Bashan LE, Bashan Y (2015) Involvement of indole-3-acetic acid produced by *Azospirillum brasilense* in accumulating intracellular ammonium in *Chlorella vulgaris*. Res Microbiol 166:72–83

Miller G, Suzuki N, Ciftci-Yilmaz S, Mittler R (2010) Reactive oxygen species homeostasis and signaling during drought and salinity stresses. Plant Cell Environ 33:453–467

Minhas D, Grover A (1999) Towards developing transgenic rice plants tolerant to flooding stress. Proc Indian Natl Sci Acad-Part B 65:33–50

Mommer L, Visser E (2005) Underwater photosynthesis in flooded terrestrial plants: a matter of leaf plasticity. Ann Bot 96:581–589

Mommer L, Pons TL, Wolters-Arts M, Venema JH, Visser EJ (2005) Submergence-induced morphological, anatomical, and biochemical responses in a terrestrial species affect gas diffusion resistance and photosynthetic performance. Plant Physiol 139:497–508

Murali M, Gowtham HG, Singh SB, Shilpa N, Aiyaz M, Niranjana SR, Amruthesh KN (2021) Bioprospecting of ACC deaminase producing Rhizobacteria towards sustainable agriculture: a special emphasis on abiotic stress in plants. Appl Soil Ecol 168:104142. https://doi.org/10.1016/j.apsoil.2021.104142

Nadal M, Roig-Oliver M, Bota J, Flexas J (2020) Leaf age-dependent elastic adjustment and photosynthetic performance under drought stress in Arbutus unedo seedlings. Flora 271:151662

Nakashima K, Yamaguchi-Shinozaki K, Shinozaki K (2014) The molecular responses to drought stress transcriptional regulatory network in the drought response and its crosstalk in abiotic stress responses including drought, cold, and heat. Front Plant Sci 5:170

Nascimento F, Brigido C, Alho L, Glick BR, Oliveira S (2012) Enhanced chickpea growth-promotion ability of a *Mesorhizobium* strain expressing an exogenous ACC deaminase gene. Plant Soil 353:221–230

Nelissen H, Sun XH, Rymen B, Jikumaru Y, Kojima M, Takebayashi Y, Abbeloos R, Demuynck K, Storme V, Vuylsteke M (2018) The reduction in maize leaf growth under mild drought affects the transition between cell division and cell expansion and cannot be restored by elevated gibberellic acid levels. Plant Biotechnol J 16:615–627

Ouyang WJ, Struik PC, Yin XY, Yang JC (2017) Stomatal conductance, mesophyll conductance, and transpiration efficiency in relation to leaf anatomy in rice and wheat genotypes under drought. J Exp Bot 68:5191–5205

Palta JA, Ganjeali A, Turner NC, Siddique KHM (2010) Effects of transient subsurface waterlogging on root growth, plant biomass and yield of chickpea. Agric Water Manag 97:1469–1476

Pan DL, Wang G, Wang T, Jia ZH, Guo ZR, Zhang JY (2019) *AdRAP2.3*, a novel ethylene response factor vii from *Actinidiadeliciosa*, enhances waterlogging resistance in transgenic tobacco through improving expression levels of PDC and ADH genes. Int J Mol Sci 20:1189

Pan J, Sharif R, Xu X, Chen X (2021) Mechanisms of waterlogging tolerance in plants: research progress and prospects. Front Plant Sci 11:627–331

Patharkar O, Walker JC (2019) Connections between abscission, dehiscence, pathogen defense, drought tolerance, and senescence. Plant Sci 284:25–29

Pedersen O, Sauter M, Colmer TD, Nakazono M (2021) Regulation of root adaptive anatomical and morphological traits during low soil oxygen. New Phytol 229:42–49

Petrillo C, Vitale E, Ambrosino P, Arena C, Isticato R (2022) Plant growth-promoting bacterial consortia as a strategy to alleviate drought stress in Spinacia oleracea. Microorganisms 10(9):1798

Pierret A, Maeght JL, Clement C, Montoroi JP, Hartmann C, Gonkhamdee S (2016) Understanding deep roots and their functions in ecosystems: an advocacy for more unconventional research. Ann Bot 118:621–635

Ploschuk RA, Miralles DJ, Colmer TD, Ploschuk EL, Striker GG (2018) Waterlogging of winter crops at early and late stages: impacts on leaf physiology, growth and yield. Front Plant Sci 9:1863

Polle A (2001) Dissecting the superoxide dismutase-ascorbate-glutathione-pathway in chloroplasts by metabolic modeling. Computer simulations as a step towards flux analysis. Plant Physiol 126:445–462

Pucciariello C, Perata P (2017) New insights into reactive oxygen species and nitric oxide signalling under low oxygen in plants. Plant Cell Environ 40:473–482

Puthiyottil P, Akkara Y (2021) Pre treatment with Bacillus subtilis mitigates drought induced photo-oxidative damages in okra by modulating antioxidant system and photochemical activity. Physiol Mol Biol Plants 27(5):945–957

Rahman KU, Ali K, Rauf M, Arif M (2023) *Aspergillus nomiae* and *fumigatus* ameliorating the hypoxic stress induced by waterlogging through ethylene metabolism in *Zea mays* L. Microorganisms 11(8):2025

Raina A, Laskar RA, Wani MR, Khan S (2022) Plant breeding strategies for abiotic stress tolerance in cereals. In: Omics approach to manage abiotic stress in cereals. Springer Nature Singapore, Singapore, pp 151–177

Ren B, Zhu Y, Zhang J, Dong S, Liu P, Zhao B (2016) Effects of spraying exogenous hormone 6-benzyladenine (6-BA) after waterlogging on grain yield and growth of summer maize. Field Crop Res 188:96–104

Rosa PAL, Mortinho ES, Jalal A, Galindo FS, Buzetti S, Fernandes GC, Barco Neto M, Pavinato PS, Teixeira Filho MCM (2020) Inoculation with growth-promoting bacteria associated with the reduction of phosphate fertilization in sugarcane. Front Environ Sci 8:32

Rueda M, Godoy O, Hawkins BA (2017) Spatial and evolutionary parallelism between shade and drought tolerance explains the distributions of conifers in the conterminous United States. Glob Ecol Biogeogr 26:31–42

Saikia J, Sarma RK, Dhandia R, Yadav A, Bharali R, Gupta VK, Saikia R (2018) Alleviation of drought stress in pulse crops with ACC deaminase producing rhizobacteria isolated from acidic soil of Northeast India. Sci Rep 8(1):3560

Salvi S, Tuberosa R (2015) The crop QTLome comes of age. Curr Opin Biotechnol 32:179–185. https://doi.org/10.1016/j.copbio.2015.01.001

Sasidharan R, Bailey-Serres J, Ashikari M, Atwell BJ, Colmer TD, Fagerstedt K (2017) Community recommendations on terminology and procedures used in flooding and low oxygen stress research. New Phytol 214:1403–1407

Shaharoona B, Arshad M, Zahir ZA, Khalid A (2006) Performance of *Pseudomonas* spp. containing ACC-deaminase for improving growth and yield of maize (*Zea mays* L.) in the presence of nitrogenous fertilizer. Soil Biol Biochem 38:2971–2975

Shanmugam S, Boyett VA, Khodakovskaya M (2021) Enhancement of drought tolerance in rice by silencing of the OsSYT-5 gene. PLoS One 16:e0258171

Silva SD, de Melo Sereno MJ, e Silva CF, de Oliveira AC, Neto JF (2005) Genetic parameters and QTL fortolerance to flooded soils in maize. Crop Breed Appl Biotechnol 5:287–293. https://doi.org/10.1626/pps.9.176

Singh AK, Cabral C, Kumar R, Ganguly R, Kumar Rana H, Gupta A, Pandey AK (2019) Beneficial effects of dietary polyphenols on gut microbiota and strategies to improve delivery efficiency. Nutrients 11:2216

Sivasakthi S, Usharani G, Saranraj P (2014) Biocontrol potentiality of plant growth promoting bacteria (PGPR)—*Pseudomonas fluorescens* and *Bacillus subtilis*: a review. Afr J Agric Res 16:1265–1277

Smethurst CF, Shabala S (2003) Screening methods for waterlogging tolerance in lucerne: comparative analysis of waterlogging effects on chlorophyll fluorescence, photosynthesis, biomass and chlorophyll content. Funct Plant Biol 30:335–343

Song H, Li YB, Zhou L, Xu ZZ, Zhou GS (2018) Maize leaf functional responses to drought episode and rewatering. Agric For Meteorol 249:57–70

Sundgren TK, Uhlen AK, Lillemo M, Briese C, Wojciechowski T (2018) Rapid seedling establishment and a narrow root stele promotes waterlogging tolerance in spring wheat. J Plant Physiol 227:45–55

Tahira MH, Muhammad A, Muhammad SAA, Riffat B, Sana F (2014) Anatomical and physiological adaptations in aquatic ecotypes of *Cyperus alopecuroides* Rottb. under saline and waterlogged conditions. Aquat Bot 116:60–68

Taiz L, Zeiger E (2010) Plant physiology. Sinauer Associates, Inc., Sunderland, MA

Teixeira Filho MM, Galindo FS (2019) Inoculation of bacteria with a focus on biological nitrogen fixation and promotion of plant growth. In: Severiano EC, Moraes MF, Paula AM (eds) Topics in soil science—volume X. Brazilian Society of Soil Science, Brasilia, Brazil, pp 577–648

Thangthong N, Jogloy S, Punjansing T, Kvien CK, Kesmala T, Vorasoot N (2019) Changes in root anatomy of peanut (*Arachis hypogaea* L.) under different durations of early season drought. Agronomy 9:215

Ullah U, Ashraf M, Shahzad SM, Siddiqui AR, Piracha MA, Suleman M (2016) Growth behavior of tomato (*Solanum lycopersicum* L.) under drought stress in the presence of silicon and plant growth promoting rhizobacteria. Soil Environ 35:65–75

Ullah A, Ul Qamar MT, Nisar M, Hazrat A, Rahim G, Khan AH, Hayat K, Ahmed S, Ali W, Khan A, Yang X (2020) Characterization of a novel cotton MYB gene, GhMYB108-like responsive to abiotic stresses. Mol Biol Rep 47:1573–1581

Vejan P, Abdullah R, Khadiran T, Ismail S, Nasrulhaq Boyce A (2016) Role of plant growth promoting Rhizobacteria in agricultural sustainability—a review. Molecules 21:573

Waszczak C, Carmody M, Kangasjarvi J (2018) Reactive oxygen species in plant signaling. Annu Rev Plant Biol 69:209–236

Webb J, Fletcher R (1996) Paclobutrazol protects wheat seedlings from injury due to waterlogging. Plant Growth Regul 18:201–206

Wei WL, Li DH, Wang LH, Ding X, Zhang YX, Gao Y, Zhang XR (2013) Morpho-anatomical and physiological responses to waterlogging of sesame (*Sesamum indicum* L.). Plant Sci 208:102–111

Wei X, Xu H, Rong W, Ye X, Zhang Z (2019) Constitutive expression of a stabilized transcription factor group VII ethylene response factor enhances waterlogging tolerance in wheat without penalizing grain yield. Plant Cell Environ 42:1471–1485

Xiao B, Jespersen D (2019) Morphological and physiological responses of seashore paspalum and bermudagrass to waterlogging stress. J Am Soc Hortic Sci 144:305–313

Xiao S, Liu LT, Zhang YJ, Sun HC, Zhang K, Bai ZY, Dong HZ, Li CD (2020) Fine root and root hair morphology of cotton under drought stress revealed with RhizoPot. J Agron Crop Sci 206:679–693

Yang Z, Chi X, Guo F, Jin X, Luo H, Hawar A, Sun B (2020) SbWRKY30 enhances the drought tolerance of plants and regulates a drought stress-responsive gene, SbRD19, in sorghum. J Plant Physiol 246:153142

Yasmin H, Bano A, Wilson NL, Nosheen A, Naz R, Hassan MN et al (2022) Drought-tolerant Pseudomonas sp. showed differential expression of stress-responsive genes and induced drought tolerance in Arabidopsis thaliana. Physiol Plant 174(1):e13497

Ye H, Song L, Chen H, Valliyodan B, Cheng P, Ali L, Nguyen HT (2018) A major natural genetic variation associated with root system architecture and plasticity improves waterlogging tolerance and yield in soybean. Plant Cell Environ 41:2169–2182

Yeboah MA, Xuehao C, Feng CR, Alfandi M, Liang G, Gu M (2008) Mapping quantitative trait loci for waterlogging tolerance in cucumber using SRAP and ISSR markers. Biotechnology 7:157–167

Yin DM, Luo HL (2012) Anatomical responses to waterlogging in *Chrysanthemum zawadskii*. Sci Hortic 146:86–91

Yin DM, Chen SM, Chen FD, Guan ZY, Fang WM (2010) Morpho-anatomical and physiological responses of two *Dendranthema* species to waterlogging. Environ Exp Bot 68:122–130

Yiu JC, Liu CW, Fang DYT, Lai YS (2009) Waterlogging tolerance of Welsh onion (*Allium fistulosum* L.) enhanced by exogenous spermidine and spermine. Plant Physiol Biochem 47:710–716

Yu F, Han X, Geng C, Zhao Y, Zhang Z, Qiu F (2015) Comparative proteomic analysis revealing the complex network associated with waterlogging stress in maize (*Zea mays* L.) seed-ling root cells. Proteomics 15:135–147

Zaheer MS, Raza MAS, Saleem MF, Khan IH, Ahmad S, Iqbal R, Manevski K (2019) Investigating the effect of *Azospirillum brasilense* and *Rhizobium pisi* on agronomic traits of wheat (*Triticum aestivum* L.). Arch Agron Soil Sci 65:1554–1564

Zate ZZ (2021) Quantitative trait loci mapping of agronomic traits in a cowpea (*Vigna unguiculata* L.) bi-parental cross. J Sustain Dev 9:39–58. https://doi.org/10.20372/au.jssd.9.1.2021.0258

Zeffa DM, Fatin LH, Santos OJAP, Oliveira ALM, Canteri MG, Scapim CA, Gonçalves LSA (2018) The influence of topdressing nitrogen on *Azospirillum* spp. inoculation in maize crops through meta-analysis. Bragantia 77:493–500

Zhang XC, Shabala S, Koutoulis A, Shabala L, Johnson P, Hayes D, Nichols DS, Zhou MX (2015) Waterlogging tolerance in barley is associated with faster aerenchyma formation in adventitious roots. Plant Soil 394:355–372

Zhang JS, Zhang H, Srivastava AK, Pan YJ, Bai JJ, Fang JJ, Shi HZ, Zhu JK (2018) Knockdown of rice MicroRNA166 confers drought resistance by causing leaf rolling and altering stem xylem development. Plant Physiol 176:2082–2094

Zhang F, Zhu K, Wang YQ, Zhang ZP, Lu F, Yu HQ, Zou JQ (2019) Changes in photosynthetic and chlorophyll fluorescence characteristics of sorghum under drought and water-logging stress. Photosynthetica 57:1156–1164
Zheng GQ, Zhang L, Zheng GB, Zhang YP, Wang J, Hu ZH (2010) Effects of irrigation amount on leaf structure, photosynthetic physiology, and fruit yield of *Lycium barbarum* in arid area. Ying Yong sheng tai xue bao= the. J Appl Ecol 21:2806–2813
Zhou WJ, Zhao DS, Lin XQ (1997) Effects of waterlogging on nitrogen accumulation and alleviation of waterlogging damage by application of nitrogen fertilizer and mixtalol in winter rape (*Brassica napus* L.). J Plant Growth Regul 16:47–53
Zhu M, Li FH, Shi ZS (2016) Morphological and photosynthetic response of waxy corn inbred line to waterlogging. Photosynthetica 54:636–640
Zuniga FA, Bustos SA, Alves F, Martinez V, Smith RC (2017) Physiological and morphological responses to permanent and intermittent waterlogging in seedlings of four evergreen trees of temperate swamp forests. Tree Physiol 37:779–789

Chapter 5
Saline Stress Management

Anand Dave

Abstract Salinity is a significant abiotic stressor that restricts agricultural yield and has an impact on plant development. An obtainable way to assure crop productivity in such unfavorable circumstances is to implement treatments that reduce the effects of salt stress, along with identifying and using salt-tolerant varieties or species in salty environments. It is effectively possible to treat seeds, seedlings, or plants chemically, biologically, or physically prior to their exposure to salt stress. In saline environments, salt-tolerant plant growth-promoting rhizobacteria are thought to provide an alternative means of strengthening crop development. The adaptation of phytohormone status, gene expression, protein function, and metabolite production in plants is how PGPR regulates osmotic balance and ion homeostasis. Following exposure to salt, these treatments stimulate both biological and molecular approaches that allow a plant or seed to react further rapidly and/or extensively. In addition, methods of agricultural management have helped mitigate the negative impacts of too much salt building up in the soil. The improvement in current and efficient methods for salinity alleviation has been made possible by the fundamental understanding of how plants respond to elevated salt concentrations. Here, we give an overview of the main approaches used to mitigate the impact of salt on plants and offer a critical analysis of the primary achievements discussed in this area.

Keywords PGPR · Salinity · Agriculture management · Phytohormone · Crop

---

A. Dave (✉)
Department of Microbiology, Seventh Day Adventist Arts and Science College, Ahmedabad, Gujrat, India
e-mail: ananddave16@gmail.com

P. Singh Chauhan et al. (eds.), *Plant-Microbe Interaction and Stress Management*, Rhizosphere Biology,
https://doi.org/10.1007/978-981-97-4239-4_5

5.1 Introduction

In semiarid locations, where there is a lack of water and insufficient irrigation land drainage significantly impairs crop output, soil salinization poses a danger to agriculture (Hanin et al. 2016). The accumulation of salt affects almost 8% of the land area of Earth, according to data accessible at the Food and Agriculture Organization (FAO) Soils Portal (http://www.fao.org/soils-portal/soil-management). According to the same source, there are an estimated 399 million hectares of saline soils overall compared to 438 million hectares of sodic soils. The current rising climatic changes that are being witnessed in many parts of the world could make this already dangerous situation worse.

According to Bockheim and Gennadiyev (2000), salinization is the general term for the buildup of various salts, such as potassium, magnesium, calcium, and sodium carbonates, bicarbonates, chlorides, and sulfates. This varied ionic composition will therefore produce a broad spectrum of physiochemical properties. Soil salinization and soil sodicity are two terms that are frequently used interchangeably in the literature on salt-affected soils. Based on salt content, salt kind, sodium content, and soil alkalinity, salt-affected soils can be classified as saline, saline–sodic, or sodic, according to the FAO Soils Portal.

An increasing number of studies has examined how plant responses are affected by soil salinity (recent reviews may be found in Hanin et al. 2016, and Negrão et al. 2017). The plant responds quickly to an early stress (a few minutes or a day after stress application) when it senses the osmotic action of the salt in the soil. The osmotic phase is another term for the early responses.

Reduced possibility for soil water of soil and water causes the plant to absorb less amount of water, which closes its stomata, lessens photosynthesis, and prevents the expansion of its leaves. All these effects lower the rate at which the plant grows shoots (Das et al. 2015; Nongpiur et al. 2016; Hanin et al. 2016). The phase of ions occurs when the concentration of ions causes the plant to perceive their harmful effects. The second stage is a more gradual reaction that happens a few days to weeks following harmful Na^+ buildup in photosynthetic tissues. Over time, salt buildup causes a slower restriction of growth, which is particularly noticeable in older leaves and leads to their senescence. Older leaves die because, unlike young leaves, they are unable to dilute the salt that enters their cells. Reduced photosynthetic capacity results in a lower growth rate if the rate of leaf death exceeds the rate of leaf production (Nongpiur et al. 2016). Negrão et al. (2017) divided mechanisms that tolerate salinity in plants into three distinct categories. Tissue tolerance, which describes the compartmentalization of hazardous ions, and ion exclusion, which describes the net exclusion of poisonous ions from the shoot. Review out Munns and Tester for a thorough explanation of the primary responses that are biological, cellular, and metabolic triggered in every scenario. Nevertheless, it is important to emphasize that the primary element influencing salinity tolerance is ion transport through cell membranes (Ismail and Horie 2017). According to El-Mashad and Mohamed (2012), salinity also causes plants under oxidative stress by generating reactive oxygen species (ROS) that can damage DNA and cells, such as superoxide radicals, hydrogen peroxide, and hydroxyl radicals. This is addressed by enzymatic antioxidant mechanisms such as peroxidase (POX), catalase (CAT), and superoxide dismutase (SOD) that initiate antioxidative stress defenses (Gharsallah et al. 2016).

Munns and Tester found that under conditions of salinity the development of specific solutes happens simultaneously with an increase inside the cytosol of solutes known as suitable solutes, low-weight solutes such as proline, polyols, and betaine that are involved in osmotic adjustment as well as osmoprotection when exposed to high salt concentrations (El-Mashad and Mohamed 2012; Gharsallah et al. 2016).

In fact, plants that overproduce proline can withstand stress by preserving osmotic equilibrium or cell turgor, stabilizing membranes to stop electrolytic leaks, as well as delivering ROS amounts that are within optimum limits to avoid phases of oxidation in plants (Hayat et al. 2012). The various levels of salinity tolerance in plants are also significantly shaped by their morphology, biochemistry, and physiology (Nongpiur et al. 2016). The gradient of plant tolerance to NaCl to the environment ranges from exceedingly sensitive to exceedingly tolerant. According to Flowers and Colmer (2015), species belonging to the genera Tecticornia and Atriplex are classified as salt-tolerant and are commonly referred to as halophytes. One of the species that is most vulnerable to salinity is rice (*Oryza sativa*) (Das et al. 2015).

Various treatments to seeds before they are exposed to salt stress are one way to lessen the effects under salt stress on plants. These therapies, which are primarily administered prior to seed growing, will likely occur and trigger biochemical and physiological processes that will supply the seed time to react to an abiotic stressor more swiftly and/or forcefully. This chapter attempts to give a broad summary of the primary methods used in plants to mitigate the impacts of salt, along with a critical analysis of the primary advancements reported in this area. Pigeon pea cultivation, an important biological factor that causes Fusarium wilt, is Pigeon pea wilt, which harms crop growth and yield (Dave and Ingle 2021) (Fig. 5.1).

5.2 Chemical Methods for Reducing Salinity Stress

One technique for improving a plant's resistance to a variety of abiotic stressors, such as salinity, is seed and plant priming. According to Paparella et al. (2015), the priming of seeds is commonly understood to be a presowing treatment given prior to germination with the goal of enhancing germination's speed and uniformity (Nawaz et al. 2013; Jisha and Puthur 2016). However, seedlings and entire plants can also be included in the priming concept. According to Aranega-Bou et al. (2014), priming of plants is a process that results in a physiological condition that permits plants to react to biotic or abiotic challenges more quickly and/or robustly. According to Savvides et al., plant priming can occur spontaneously in reaction, which serves as a signpost indicating a higher likelihood of encountering that stress

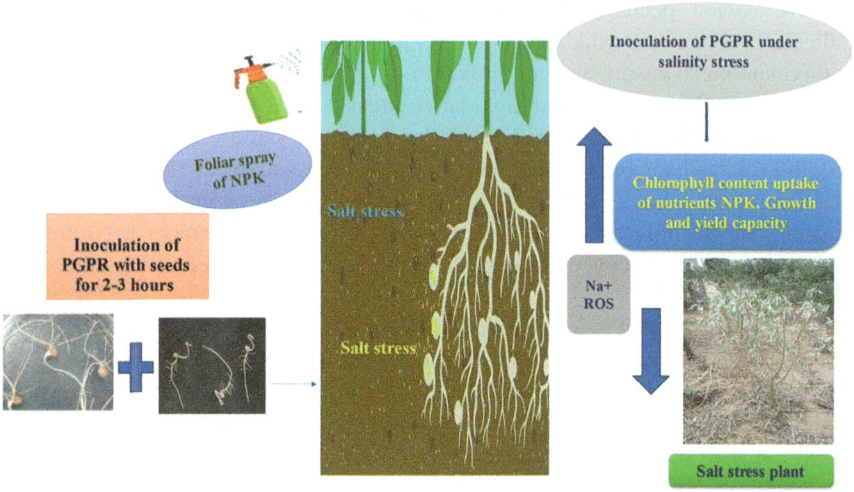

Fig. 5.1 Diagrammatic representation of the effects under salt stress on plants

element in the future. Numerous chemical compounds are effective in reducing salinity stress. Certain plant metabolites are naturally produced in modest quantities, but, when administered exogenously, they can modify gene expression and metabolism regulation, resulting in increased tolerance to salt. These substances include plant growth regulators (like jasmonate (JA)) when responding to an environmental stress, and salicylic acid, polyamines (like spermine, spermidine), vitamins, and sugars (like trehalose), amino acids, and their derivatives (Vaishnav et al. 2016). According to Savvides et al. (2016), using chemical treatment of plants prior to stress incidence boosts the tolerance of unfavorable environmental conditions because it helps the plants respond more quickly and forcefully to unfavorable conditions. Table 5.1 provides a nonexhaustive list of chemical priming treatments aimed at enhancing seed, seedling, and plant salt stress endurance. An essential stage in the growth of plants is seed germination. In many crops, salty environments can hinder seedling growth and cause germination to be delayed. Thus, it is not unexpected that a substantial amount of study is devoted to comprehending seed priming and creating novel priming techniques. Because chemical agents have been demonstrated to be effective at low concentrations, applying them for plant and seed pretreatment appears to be a low-cost means of protecting crops from saline stress (Savvides et al. 2016). Another advantage is that, even though chemical agents are typically only given to the seeds, roots, or leaves, they generate a systemic tolerance to salinity stress. Most of the research, meanwhile, has found that plants can withstand the stress shortly after chemical priming. As a result, it is yet unclear how long the chemical compounds will continue to protect crops.

Table 5.1 Physiological treatment developing salinity stress tolerance in plant

Species	Plant growth stage	Priming agent	Method of application	References
Eggplant (*Solanum melongena*)	Plant	Proline	Foliar spray	Shahbaz et al. (2013)
Soybean (*Glycine max*)	Seed	Proline, glycine betaine	Soaked in solution	Vaishnav et al. (2016)
Wheat (*Triticum aestivum*)	Seed	Cysteine	Soaked in solution	Nasibi et al. (2016)
Mung bean (*Vigna radiata*)	Seed	Chitosan	Soaked in solution	Sen and Mandal (2016)
Common bean (*Phaseolus vulgaris*)	Seed	Vitamin B12	Soaked in solution	Keshavarz and Modares Sanavy (2015)
Maize (*Zea mays*)	Seed	NaCl	Soaked in solution	Gebreegziabher and Qufa (2017)
Cotton (*Gossypium* spp.)	Seed	KNO_3	Soaked in solution	Nazir et al. (2014)

5.3 Enhancing Crop Salinity Tolerance Via Biological Treatments

The discovery of the physiological reactions and biochemical networks underlying the responses of halophytes to salinity has led to the identification of novel genes related to salt stress that are accessible for plant enhancement. This discovery has aided in the creation of fresh strategies to reduce plant salinity and perhaps even encourage crop growth in soils that accumulate more salt (Shrivastava and Kumar 2015). However, field evaluation is necessary to assess the impact of added genes on yield improvement and salinity tolerance. Sadly, these studies have not considered the ecological interactions of the microorganisms that live in the phyllosphere (the leaves) and rhizosphere (the roots) of halophytes, but it is vital to remember that these interactions may play a crucial role in the halophytes' ability to withstand salinity (Ruppel et al. 2013). Plants can attract and establish mutualistic relationships with various soil microbes, which can positively impact their productivity and ability to withstand challenging circumstances (De-La-Peña and Loyola-Vargas 2014; Kasim et al. 2016). Although the role that a particular microbe plays in helping plants endure salt may sometimes be disregarded, it is able to serve as a substitute for genetically modifying crops to produce salt-tolerant ones (Yuan et al. 2016). Growing research indicates that plants can develop some tolerance to salinity stress thanks to rhizospheric fungus and plant growth-promoting rhizobacteria (PGPR) that are associated with these (Campanelli et al. 2012; Habib et al. 2016; Kang 2014; Bal et al. 2013; Gururani et al. 2013; Hajiboland 2013).

5.4 Physical Treatments for Crops Stressed by Salinity

Physical methods of increasing plant output can be applied on a high-throughput scale and have eco-friendly benefits. Often called "physical priming," physical energizing methods offer an alternative to the existing chemical or biologically based approaches. It has been determined that using physical invigoration techniques is a viable method of approaching developing new biotechnology-based alternatives for the increasing demand for seeds (Araújo et al. 2016). Various studies have reported the effective application of several agents as potential presowing seed treatments, including temperature, magnetic fields (MFs), microwaves, ultraviolet (UV), and ionizing radiation (Paparella et al. 2015; Araújo et al. 2016). Exposing people to nonlethal amounts of presowing exposure to nonlethal concentrations of the physical substances generally improves the plants' physiological function and overall quality, including growth and yield, in addition to promoting germination. Breaking dormancy in seeds and improving the total seed quality batches, presowing temperature treatments are a common method. Nevertheless, as thermopriming is typically used in conjunction with other priming techniques (Paparella et al. 2015), we will not be focusing on it in this section. As far as we are aware, there are not many publications that discuss using physically invigorating techniques to promote salt tolerance. However, we have gathered a few instances from academic databases and talked about how they may be used as remedies to lessen plant salinity stress.

5.5 Conclusion

With an enormous ability to provide food security and sustainable agriculture, halotolerant PGPR play an important role in the microbial community using these naturally occurring, beneficial, and environmentally friendly soil microbes to improve plant production and development in salinized soil. Moreover, several signaling components linked to defense are activated by PGPR. Consequently, research is now focused on using salty PGPR instead of fertilizers to increase crop output. Many environmental variables have an impact on saline PGPR. Biological methods that are less sustainable than conventional techniques are being replaced by microbiological approaches as one of the most significant tools for improving the productivity of salty soils.

Acknowledgments The author acknowledges the support of financial support in the form of DST-FIST project (DST/2022/1012) from the Govt. of India.

References

Aranega-Bou P, de la Leyva OM, Finiti I (2014) Priming of plant resistance by natural compounds: hexanoic acid as a model. Front Plant Sci 5:488. https://doi.org/10.3389/fpls.2014.00488

Araújo SS, Paparella S, Dondi D (2016) Physical methods for seed invigoration: advantages and challenges in seed technology. Front Plant Sci 7:646. https://doi.org/10.3389/fpls.2016.00646

Bal HB, Nayak L, Das S, Adhya TK (2013) Isolation of ACC deaminase producing PGPR from rice rhizosphere and evaluating their plant growth promoting activity under salt stress. Plant Soil 366:93–105. https://doi.org/10.1007/s11104-012-140

Bockheim JG, Gennadiyev AN (2000) The role of soil-forming processes in the definition of taxa in soil taxonomy and the world soil reference base. Geoderma 95:53–72. https://doi.org/10.1016/S0016-7061(99)00083-X

Campanelli A, Ruta C, De MG, Morone-Fortunato I (2012) The role of arbuscular mycorrhizal fungi in alleviating salt stress in *Medicago sativa* L. var. icon. Symbiosis 59:65–76. https://doi.org/10.1007/s13199-012-0191-1

Das P, Nutan KK, Singla-Pareek SL, Pareek A (2015) Understanding salinity responses and adopting "omics-based" approaches to generate salinity tolerant cultivars of rice. Front Plant Sci 6:712. https://doi.org/10.3389/fpls.2015.00712

Dave A, Ingle S (2021) *Streptomyces* sp. S-9 promotes plant growth and confers resistance in pigeon pea (*Cajanus cajan*) against *Fusarium* wilt. 3 Biotech 11:459. https://doi.org/10.1007/s13205-021-02989-0

De-la-Peña C, Loyola-Vargas VM (2014) Biotic interactions in the rhizosphere: a diverse cooperative enterprise for plant productivity. Plant Physiol 166:701–719. https://doi.org/10.1104/pp.114.241810

El-Mashad AAA, Mohamed HI (2012) Brassinolide alleviates salt stress and increases antioxidant activity of cowpea plants (*Vigna sinensis*). Protoplasma 249:625–635. https://doi.org/10.1007/s00709-011-0300-7

Flowers TJ, Colmer TD (2015) Plant salt tolerance: adaptations in halophytes. Ann Bot 115:327–331. https://doi.org/10.1093/aob/mcu267

Gebreegziabher BG, Qufa CA (2017) Plant physiological stimulation by seeds salt priming in maize (*Zea mays*): prospect for salt tolerance. Afr J Biotechnol 16:209–223. https://doi.org/10.5897/AJB2016.15819

Gharsallah C, Fakhfakh H, Grubb D, Gorsane F (2016) Effect of salt stress on ion concentration, proline content, antioxidant enzyme activities and gene expression in tomato cultivars. AoB Plants. https://doi.org/10.1093/aobpla/plw055

Gururani MA, Upadhyaya CP, Baskar V (2013) Plant growth–promoting rhizobacteria enhance abiotic stress tolerance in *Solanum tuberosum* through inducing changes in the expression of ROS-scavenging enzymes and improved photosynthetic performance. J Plant Growth Regul 32:245–258. https://doi.org/10.1007/s00344-012-9292-6

Habib SH, Kausar H, Saud HM (2016) Plant growth–promoting rhizobacteria enhance salinity stress tolerance in okra through ROS-scavenging enzymes. Biomed Res Int. https://doi.org/10.1155/2016/6284547

Hajiboland R (2013) Role of arbuscular mycorrhiza in amelioration of salinity. In: Ahmad P, Azooz MA, Prasad MNV (eds) Salt stress in plants: signalling, omics and adaptations. Springer Science Business Media, New York, pp 301–354. https://doi.org/10.1007/978-1-4614-6108-1_13

Hanin M, Ebel C, Ngom M (2016) New insights on plant salt tolerance mechanisms and their potential use for breeding. Front Plant Sci 7:1787. https://doi.org/10.3389/fpls.2016.01787

Hayat S, Hayat Q, Alyemeni MN (2012) Role of proline under changing environments: a review. Plant Signal Behav 7:1456–1466. https://doi.org/10.4161/psb.21949

Ismail AM, Horie T (2017) Genomics, physiology, and molecular breeding approaches for improving salt tolerance. Annu Rev Plant Biol 68.:annurev-arplant-042916-040936. https://doi.org/10.1146/annurev-arplant-042916-040936

Jisha KC, Puthur JT (2016) Seed priming with BABA (beta-amino butyric acid): a cost-effective method of abiotic stress tolerance in *Vigna radiata* (L.) Wilczek. Protoplasma 253:277–289. https://doi.org/10.1007/s00709-015-0804-7

Kang SM (2014) Plant growth promoting rhizobacteria reduce adverse effects of salinity and osmotic stress by regulating phytohormones and antioxidants in *Cucumis sativus*. J Plant Interact 9:673–682. https://doi.org/10.1080/17429145.2014.894587

Kasim WA, Gaafar RM, Abou-Ali RM (2016) Effect of biofilm forming plant growth promoting rhizobacteria on salinity tolerance in barley. Ann Agric Sci 61:217–227. https://doi.org/10.1016/j.aoas.2016.07.003

Keshavarz H, Modares Sanavy SAM (2015) Biochemical and morphological response of common bean (*Phaseolus vulgaris* L.) to salinity stress and vitamin B12. Intl J Farm Alli Sci 4:585–593. https://doi.org/10.1016/j.aoas.2018.10.002

Nasibi F, Kalantari KM, Zanganeh R, Mohammadinejad G, Oloumi H (2016) Seed priming with cysteine modulates the growth and metabolic activity of wheat plants under salinity and osmotic stresses at early stages of growth. Indian J Plant Physiol 21:279–286. https://doi.org/10.1007/s40502-016-0233-4

Nawaz J, Hussain M, Jabbar A, Nadeem GA, Sajid M, Subtain M, Shabbir I (2013) Seed priming: a technique. Int J Agric Crop Sci 6:1373–1381. www.ijagcs.com.IJACS/2013/6-20/1373-1381

Nazir MS, Saad A, Anjum Y, Ahmad W (2014) Possibility of seed priming for good germination of cotton seed under salinity stress. J Biol Agric Health 4:66–68

Negrão S, Schmöckel SM, Tester M (2017) Evaluating physiological responses of plants to salinity stress. Ann Bot 119:1–11. https://doi.org/10.1093/aob/mcw191

Nongpiur RC, Singla-Pareek SL, Pareek A (2016) Genomics approaches for improving salinity stress tolerance in crop plants. Curr Genomics 17:343–357. https://doi.org/10.2174/1389202917666160331202517

Paparella S, Araujo SS, Rossi G, Wijayasinghe M, Carbonera D, Balestrazzi A (2015) Seed priming: state of the art and new perspectives. Plant Cell Rep 34:1281–1293. https://doi.org/10.1007/s00299-015-1784-y

Ruppel S, Franken P, Witzel K (2013) Properties of the halophyte microbiome and their implications for plant salt tolerance. Funct Plant Biol 40:940–951. https://doi.org/10.1071/FP12355

Sen S, Mandal P (2016) Solid matrix priming with chitosan enhances seed germination and seedling invigoration in mung bean under salinity stress. JCEA 17:749–762. https://doi.org/10.5513/jcea01/17.3.1773

Shahbaz M, Mushtaq Z, Andaz F, Masood A (2013) Does proline application ameliorate adverse effects of salt stress on growth, ions and photosynthetic ability of eggplant (*Solanum melongena* L.)? Scientia Hortic 164:507–511. https://doi.org/10.1016/j.scienta.2013.10.001

Shrivastava P, Kumar R (2015) Soil salinity: a serious environmental issue and plant growth promoting bacteria as one of the tools for its alleviation. Saudi J Biol Sci 22:123–131. https://doi.org/10.1016/j.sjbs.2014.12.001

Vaishnav A, Kumari S, Jain S, Choudhary DK, Sharma KP (2016) Exogenous chemical mediated induction of salt tolerance in soybean plants. IJALS 2:43–47. https://doi.org/10.9379/sf.ijals.122064-007-0081-x

Yuan Z, Druzhinina IS, Labbé J (2016) Specialized microbiome of a halophyte and its role in helping non-host plants to withstand salinity. Sci Rep 6:1–13. https://doi.org/10.1038/srep32467

Chapter 6
Plant–Microbe Interaction for Heavy Metal Tolerance and Detoxification

Govind Gupta, Riddhi Shrivastava, and Naveen Kumar Singh

Abstract Metal contamination in soil and water is increasing at an alarming rate due to geogenic and anthropogenic activities. Increasing human population leads to extensive developmental activities and industrialization, which results in the generation and discharge of wastewater- containing metal polluting the aquatic and soil. Continuous release and discharge of metal-containing waste lead to metal enrichment in soil and water. Plants absorb metals from soil and water with nutrients and concentrate in the tissues and show phytotoxic symptoms, which hampers the plant growth and development and overall productivity. The microbes associated with plant's rhizosphere play a very important role in metal mobilization, uptake, and detoxification. Understanding plant–microbe interactions under metal stress would help in developing better agronomic and environmental practices for detoxifying metal toxicity to promote better plant growth for more productivity and environmental sustainability.

Keywords Metal · Uptake · Tolerance · Phytotoxicity · Waste treatment · Soil contamination

---

G. Gupta · N. K. Singh (✉)
Department of Chemistry, Environmental Science Discipline, Manipal University Jaipur, Dehmi Kalan, Jaipur, Rajasthan, India

R. Shrivastava
Department of Chemistry, Environmental Science Discipline, Manipal University Jaipur, Dehmi Kalan, Jaipur, Rajasthan, India

Poornima College of Engineering, Jaipur, Rajasthan, India

P. Singh Chauhan et al. (eds.), *Plant-Microbe Interaction and Stress Management*, Rhizosphere Biology,
https://doi.org/10.1007/978-981-97-4239-4_6

6.1 Introduction

Increasing levels of heavy metal contamination are a serious global problem (Briffa et al. 2020). Rapid urbanization and the industrial revolution led to the accumulation of heavy metals in the soil. As a result, increasing industrialization has negative effects on the productivity of crops (Tiwari and Lata 2018). Plant metabolism is inhibited and plant growth is decreased as a result of the prolonged high levels of heavy metal in the soil being excessively absorbed by plant roots and transported to aboveground portions (Mishra et al. 2017). Soil is contaminated with various heavy metals, such as lead, arsenic, chromium, cadmium, and aluminum (Rashid et al. 2023). These harmful substances can affect the plant's metabolic and morphological characteristics, leading to a reduction in plant production (Amari et al. 2017). Long-term soil environmental contamination is viewed as posing a risk to the ecosystem functions provided by soil. Various factors, such as the pH level, moisture content, and the kind of pollutants present in the soil, can affect the bioavailability of contaminants (Dua et al. 2002). As a result, a lot of focus has been given to the plant–microbe–heavy metal relationship in the context of contaminated site remediation. The ability to remove toxins utilizing bacteria and plants is a promising technique for treating ecosystems that have been damaged. The most common method for remediating contaminated sites uses bioremediation techniques since they are more practical from an economic, environmental, and aesthetic standpoint. The initial stages of microbial root colonization frequently involve the discovery of unusual compounds in the root exudates produced by microorganisms. These substances probably also control the interactions between underground communities (Compant et al. 2010). Ecology defines microbial interactions into three categories: positive, neutral, and negative (Acosta-Martínez et al. 2010). These relationships are present in every ecosystem. Most plant-associated microbes do not interact with their host plants in a harmful manner (Beattie 2006). The purpose of this chapter is to discuss the role of the PGPR in mitigating environmental stressors, such as salinity, drought, nutritional imbalance, and heavy metal toxicity.

6.2 Heavy Metal Stress

Heavy metals (HM) have high-density and in biological sense, the term "HM" is usually used to refer to such metals and metalloids with high potential toxicity rates (Tchounwon et al. 2012). Although soil is the most common and natural source of heavy metals, at present some geological and anthropogenic activities continuously increase the HM to the next level, which is harmful to plants, animals, as well as human beings (Gautam et al. 2016). These heavy metals have a high tendency to accumulate in the soil, water, and biological systems due to their high transportation ratio either in gas or in particulate form (Adriano et al. 2004). Although less amount of HMs is essential for the optimal growth of plants, excessive amount adversely affects the plants and hence organisms by entering into the food chain. Some common anthropogenic actions like municipal sewage waste, manufacturing waste treatment, incineration, and long-term use of excessive amounts of pesticides are the common sources of high HMs in agricultural soils. These heavy metals are directly transferred through the food chain to humans due to the accumulation of HMs by plants (Zhuang et al. 2014). HM can be absorbed by the surface and subsurface areas of plants, which can have an indirect or direct effect on their health. Inhibition of enzymes and structural harm toward cells brought on by oxidative stress are examples of direct impacts (Jadia and Fulekar 2009). Stress from oxidation with reactive oxygen species production (ROS) and cytotoxic substances such as methyl glyoxal (MG) disturbs ionic homeostasis symmetry in plant cells (Hossain) (Sytar et al. 2013) Several HM unwittingly enforce numerous methods, including glutathione, to combat oxidative stress depletion and attachment to protein sulfhydryl groups, blocking antioxidant enzymes, or triggering the production of ROS such as of NADPH oxidases. Plants have been exposed, directly or indirectly, to large amounts of HM.

6.2.1 Metal Stress in Soil and Water

One of the biggest concerns of industrialized nations is the threat of heavy metal poisoning on the soil (Hinojosa et al. 2004). Besides harming the plants' quality and yield, heavy metal pollution can also alter the composition, size, and activity of the microbial community (Yao et al. 2003). Thus, HM is one of the key factors contributing to soil pollution. Ni, Cu, Cd, Zn, Cr, and Pb are a few of the elements that contribute to soil heavy metal pollution (Hinojosa et al. 2004). These metals alter the biological and chemical properties of soil detrimentally. The common features of the soil like organic matter, clay concentration, and pH have a substantial impact on how much of an impact metals have on biological and biochemical traits (Speir et al. 1999). Heavy metals influence the microbial community that creates enzymes, which indirectly affects the enzymatic activity in soil (Huang et al. 2009). An increased level of HM increases actinomycetes in the soil, whereas it decreases the biomass of bacteria in soils that had been adulterated. The xylem network carries water taken up by the roots from the soil to the leaves. Water travels passively through the roots during transpiration in response to a water potential gradient that is created by transpiration rather than being actively absorbed by them (Vernay et al. 2007). Lowered transpiration rates and/or altered stomal resistance in leaves are two ways that an excess of heavy metals affects the effectiveness of water flow. Additionally, it was found that endoplasmic breakage, such as the rupturing of intercellular connections, may result in a reduction in water transport throughout the intercellular endoplasm system (plasmodesmata) (Chen et al. 2004). Different metals have variable effects on enzyme activity because the soil system's enzymes have an extensive range of chemical affinities. Cd is known to be hazardous to enzymes due to its weaker affinity for colloids compared to Pd. Compared to cellulose activity, copper (Cu) reduces β-glucosidase activity (Chen et al. 2010). Pb considerably reduces the activity of acid phosphatase, urease, catalase, and invertase. Cd contamination can affect the activity of some enzymes, such as phosphatase and sulfatase. In spite of its negative effects on the activity of protease, arylsulfatase invertase, and urease, it does not seriously damage these enzymes (Wang et al. 2007). The sensitivity of each soil enzyme to heavy metals varies. In general, the order of decreasing urease activity inhibition was Cr > Cd > Zn > Mn > Pb. Both forms of chromium, Cr (III) and Cr (VI), which are widely found in soils and have different chemical properties and toxicities, are present (Garrido et al. 2005).

6.2.2 Plant Growth Responses and Toxicity

The productivity of soils and crops is negatively impacted by large-scale industrialization because HM deposition damages soil texture by affecting its pH level. The presence of different elements in soil has an impact on the molecular and physiological functions of plants, either directly or indirectly (Hassan et al. 2017; Panuccio et al. 2009). When their concentration surpasses the unfavorable level, metals like arsenic (As), lead (Pb), cadmium (Cd), mercury (Hg), aluminum, Cr, and beryllium (Be) can drastically lower the plant efficiency (Pierart et al. 2015; Xiong et al. 2014). Metabolic issues and morphological defects are brought on by these hazardous substances that reduced plant yields (Amari et al. 2017). The abnormality can lead to the development of reactive substances such as hydroxyl radicals, superoxide anions, and H_2O_2 (Pourrut et al. 2011; Gill and Tuteja 2010). These substances can cause cell homeostasis and destruction. HM toxicity is one of the main factors that can affect the functioning of plants. In the case that the concentration of heavy metals exceeds the recommended levels, it can severely reduce the crop yield (Xiong et al. 2014). This issue is linked to the imbalance of the state of redox, which is known to be a contributing factor to the toxicity of HM in plants (Neumann et al. 2010; Nie et al. 2002).

6.2.3 Biochemical and Physiological Changes

The earliest response of a plant to biotic or abiotic stress is the generation of reactive oxygen species (ROS), which is brought on by the auto-oxidation of redox-active heavy metals, cellular damage, or perhaps ion transposition. When metabolic processes are carried out normally, reactive oxygen species (ROS) are continuously formed at low levels (Arora et al. 2002). Though one of the early reactions to various stressors in biological systems is an increase in ROS generation (Singh and Sinha 2005). Through the peroxidation of membrane lipids, modification of protein activities, DNA mutation, oxidation of chlorophyll, and disruption of some metabolic processes (ATP synthesis and electron transport chain), ROS causes damage to biomolecules (Semane et al. 2010). Intricate ROS scavenging mechanisms exist in plants that inhibit the oxidation of biomolecules and oxidative chain reactions (Michalak 2006). Physiological mechanisms like photosynthesis, ion absorption, and transport account for the majority of a plant's reaction to environmental stress (Orcutt and Nilsen 2000; Cseh 2002). Plants under Cd stress do not exhibit signs like leaf rolling, chlorosis, or stunting. It caused structural problems in several species, which prevented growth. Additionally, it has been shown to have an impact on metabolism, as seen by decreased food intake (Djebali et al. 2002, 2005) modifications to nitrogen metabolism (Ghnaya et al. 2007; Wang et al. 2008), changes to water balance, and suppression of stomatal opening (Sandalio et al. 2001). Moreover, Cd^{2+} ions may alter the permeability of membranes by influencing the lipid content and specific membrane-associated enzymes like H^+-ATPase (Fodor et al. 1995). Cd harms chlorophyll production (Li et al. 2008) and the healthy growth of chloroplasts (Jin et al. 2008; Djebali et al. 2002) and is also sensitive to net photosynthesis. It caused a decrease in shoot dry weight, which was accompanied by a considerable loss in leaf area and thickness, as well as in the length and width of the internodes, but it depends on the age of the plant. The presence of cadmium in the leaves caused additional physiological and structural harm. The reduced expansion of mesophyll cubicles hindered leaf growth and lowered blade thickness, which increased the tissue dry weight/fresh weight ratio, particularly in young (third) plants (Perfus-Barbeoch et al. 2002). Changes in actin filament organization and the ratio of polymeric and monomeric forms of actin are frequently seen in conjunction with changes in cellular metabolism. In the green alga *Spirogyra*, cadmium disrupted and destroyed microfilaments and microtubules, and aluminum led to the meristematic cells of *Triticum turgidum* roots to form tight bundles of actin filaments.

6.3 Plant–Microbe Interaction under Metal Stress

The interactions between soil microbes and plant roots are the most critical factors that contribute to developing the phytoremediation process. In a plant–bacterial relationship, the plant gives the bacteria a carbon source that can reduce the harmful effects associated with heavy metal exposure. This relationship is referred to as plant–bacterial symbiosis. The metabolic processes of plants may stimulate the development of bacteria in the soil through the root exudates. This process allows the microbes to eliminate the contaminants. In general, plants can select the microbial community within their own environment. Each plant species has its own group of linked microbes (Hartmann et al. 2009). The interactions between plants and microbes in the soil can help develop a beneficial relationship. The information about the development of these mechanisms suggests that the plants and microbes have evolved ways to respond to each other. In addition, the plants excrete various chemicals, which are collectively known as exudates, through the roots. These include amino acids, organic acids, and sugars (Penrose and Glick 2001). The nutrients found in the root exudates provide beneficial nutrients to soil microbes, which could help increase the number of microbial populations within the rhizosphere. The presence of these nutrients can affect the metal availability in the soil through its mobility, solubility, and accessibility (Luo et al. 2014; Chiang et al. 2011). Microbes can also help improve the growth of plants by producing a wide range of phytohormones, which include vitamins, amino acids, enzymes, antibiotics, and pathogens (Ahmad et al. 2008; Ma et al. 2016). Studies have been conducted on the interactions between microbes and plants (Tak et al. 2013). Microbes and plants are being studied under metal stress due to the characteristics of these organisms that allow them to utilize metal from the soil. Since these organisms are the oldest living creatures on Earth, they can be ideal materials for studying how metal resistance mechanisms work. The knowledge about the interaction between microbes and plants, which was previously limited, was finally gained through the last two decades. Recent studies on the application of microbial agents in the reclamation of metal-contaminated soil have provided valuable information about the interactions between plants, microbes, and metal (Nies 1999). For instance, by altering the soil's pH, the level of reactive oxygen, and the release of certain chelators, the microbes can improve the metal bioavailability (Zarei et al. 2010; Miransari 2011). Table 6.1 summarizes the various metabolites generated by microbes that

Table 6.1 Potential of microbial metabolites' reactions to mobilize/immobilize metals by plants

Microorganisms	Metabolites or reactions	Microbial potential	References
Azotobacter vinelandii	Azoto chelin and azotobactin	Helps in Mo and V acquisition	Wichard et al. (2009)
Pseudomonas aeruginosa	Pyochelin	Chelates many metals like Cd^{2+}, Cr^{2+}, Al^{3+}, Mn^{2+}, Zn^{2+}	Braud et al. (2009)
Streptomyces tendae	Desferrioxamine and coelichelin	Enhanced uptake of Cd and Fe by plants	Dimkpa et al. (2009)
A. niger, *Burkholderia cepacia*, *Beauveria caledonica*, *Oidiodendron maius*, *Pseudomonas fluorescens*, *Penicillium bilaiae*	Oxalic acid, tartaric acid, formic acid, acetic acid, malic acid	*A. niger*, *Burkholderia cepacia*, *Beauveria caledonica*, *Oidiodendron maius*, *Pseudomonas fluorescens*, *Penicillium bilaiae*	Arwidsson et al. (2010), and Hoberg et al. (2005)
Pseudomonas aeruginosa	Rhamnolipids, dirhamnolipid	Mobilized Cu, Cd, and Pb	Venkatesh and Vedaraman (2012), and Juwarkar et al. (2007)
Azotobacter spp.	Polymeric substances (extracellular)	Immobilized Cd and Cr	Joshi and Juwarkar (2009)
Glomus mosseae	Glomalin	Immobilized Cu, Pb, and Cd	Gonzalez-Chavez et al. (2004)

interact with plants. Some microbes that are associated with plants can also produce siderophores with iron chelators. However, these siderophores can also form complex structures with other heavy metals, such as Cu, Zn, and Al (Schalk et al. 2011). These compounds can improve the effectiveness of phytoextraction by causing the solubilization of the unavailable heavy metal (Rajkumar et al. 2010). The study conducted by the researchers revealed that the production of these compounds by the microbe can improve the levels of various heavy metals in the maize's rhizosphere. Braud et al. (2009) reported that the addition of various metals, such as Ni, Cu, Al, and Ga, boosted the bioavailability of heavy metals like Cr and Pb. Their study demonstrates how microbes and plants' low molecular weight can enhance the absorption of trace elements and HM in soil. As ligands, they can form complex structures with heavy metals. However, the stability of complex structures made of organic acids and heavy metals depends on various factors, such as the number of carboxyl groups in the compounds, their form, and the pH level of the solution (Ryan et al. 2001). Various types of acids, such as formic acid, oxalic, tartaric, and succinic acid, are produced by the rhizospheric bacteria in *Sedum alfredii*'s rhizosphere and can solubilize zinc and Cd (Li et al. 2010). The plant root can also absorb organic acids, which help accelerate the absorption of various heavy metals such as Cu (Chen et al. 2005) and Pb (Sheng et al. 2008). For instance, the ability of fungi, such as the ericoid mycorrhizal fungi and the *Beauveria caledonica*, to release malic and citric acids, can increase the Zn solubility of insoluble materials. These organic acids can be used either by acidolysis or by chelating to improve the Zn availability from insoluble sources, such as zinc oxide, pyromorphite, and $Zn_3(PO_4)_2$ (Fomina et al. 2005). The process of microbial self-surfactant creation leads to enhanced dissolution and bioavailability of metals by introducing them into the subsurface, leading to metal sorption from the soil's organic matrix. Studies undertaken by the *P. aeruginosa* revealed the remarkable ability to transform various heavy metals, including zinc, copper, and Pb, into biosurfactants (Juwarkar et al. 2007; Venkatesh and Vedaraman 2012). In addition, studies on the effects of biosurfactants produced by the plant–microorganism association on the uptake of Cd by different plants such as maize, tomato, and rape have shown promising results. In 2008, a biosurfactant made by the *Bacillus* sp. J119 species was able to increase the uptake of Cd by soil subjected to varying levels of this heavy metal (Sheng et al. 2008). Redox transformations can be initiated by plant-associated microbes, which can increase the mobility of heavy metals. From a phytoextraction perspective, the effects of rhizospheric microbes on metal oxidation are particularly interesting. For instance, the presence of a sulfur-oxidizing bacterium in the rhizosphere enhanced the uptake of Cu by the plant tissue and in contaminated soil (Shi et al. 2011). The reduction in the soil pH level caused by the reduction in sulfur to sulfates resulted in the copper uptake by the sulfur-oxidizing microbes. Chen and Lin (2001) discovered that the presence of an oxidizing bacteria could enhance the bioavailability of metals by causing an acidification response. The reduction in the heavy metal mobility caused by microbial reduction can also immobilize the heavy metals in the microenvironment. For instance, Chatterjee et al. (2009) reported that the green chili plants grown in contaminated soil had a significant decrease in their metal uptake. Abou-Shanab et al. (2007) reported that the reduction in the Cr levels in the water hyacinth's root to the shoots was an indication that the rhizosphere could be affected by the microbial reduction. Di Gregorio et al. (2005) also reported that the *Stenotrophicoma maltophilia* strain from the microenvironment of the *Astragalus bisulcatus* could also reduce the heavy metals in the plant. Ma et al. (2016) defined biosorption as the process of adsorbing a soluble or insoluble organic or inorganic contaminant by either a passive or metabolism-driven process. The biosorption process is carried out through a solid phase, which usually involves a biosorbent or biological material, and a liquid, usually water. The higher the affinity of a sorbent for a particular metal, the more it can bind to that metal. The bioremediation process continues until the equilibrium between the sorbate's portion remaining in the mixture and the amount of solid-bound species is reached (Das et al. 2008). The biosorption process' efficiency is dependent on various factors such as the concentration of the initial metal, the pH level, and the biomass weight of the solution. Researchers have noted that biosorption can reduce the plant's metal uptake due to its restricted entry. Madhaiyan et al. (2007) reported the development of a bioremediation strategy that involved the use of bacteria and fungi that can bind to metal. Vivas et al. (2006) reported that the application of a repens containing *Brevibacillus* sp. B-I significantly decreased the accumulation of zinc in the shoots and roots of tomato. The extent to which plant-associated microbes can alter the bioavailability and metabolism of heavy metals by altering the metabolites or processes that they produce is different. Their survival and colonization affect the amount of metal that plants acquire in contaminated soil. The soil's biological properties, which include the pH level, mineral composition, toxicity, and nutrient abundance, are responsible for its accumulation of heavy metals. Figure 6.1 shows the biotic and abiotic stress responses in plants.

6.3.1 Plant–Microbe Association

The presence of plant-associated microbes in the rhizospheric or rhizoplane soil or in the endophytes of the plant can be seen as a component of the complex and dynamic interactions between microbes and plants. Endophytes are the plant tissues that are responsible for plant development and protection. The endophytes may escape the plant's defense system and establish an inside presence without causing any harm. These microbes are known to produce various bioactive molecules that help protect plants from different types of pathogens (Datta et al. 2019). The interactions between plants and microbes are said to be as old as Earth itself. Over the years, the interactions between plants and various microbes have resulted in the establishment of an ecological unit called Holobiont, which is a discrete cluster of nonhost and host species. In natural and agricultural settings, plants are often attacked by pathogenic and beneficial microbes, such as fungi and bacteria (Wille et al. 2019). The interactions between plants and various microbes can be defined as indirect or direct mechanisms that help plants obtain mineral nutrients and improve

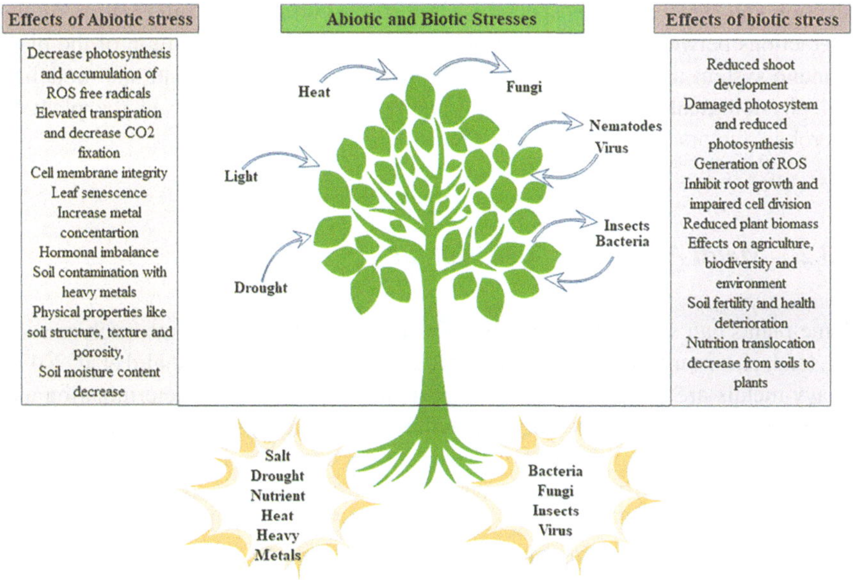

Fig. 6.1 Biotic and abiotic stress response in plants

their growth. These include the actions of the plant-associated microbes, such as the transfer of nutrients, the avoidance of pathogens, and the mitigation of stresses. The harmful effects of the interactions between microbes and plants are usually more severe as invading organisms may cause the plant to develop necrotic conditions. Understanding the interactions between microbes and plants is therefore very important to identify the beneficial and harmful effects of these organisms. In order to survive, microbes need to produce protein and carbohydrate-based signals known as MAMPs (Boller and Felix 2009). These signals are essential for the survival of the microorganisms. Due to their conservation, plants have evolved various types of receptor proteins that can bind to these signals. In response to the PAMPs, the plants produce a defense mechanism known as the PAMP-Triggered Immunity. It is the first stage of defense that helps limit the presence of pathogens in a plant species (Jones and Dangl 2006). During the course of an invasion, the microbes secrete the effector molecules, which play a vital role in the development of the diseases (Oliva et al. 2019). Plant resistance genes have evolved to encode R proteins, which enable plants to recognize avirulence proteins. When a plant recognizes avirulence, it triggers a series of immune responses. Comprehending the intricate molecular structures of the interaction between microbes and plants may help develop new strategies for mutualism, disease resistance, and plant–microbe cooperation through gene editing, silencing, and other means. This special issue looks into both hostile and friendly encounters between plants and microbes. The findings of this special issue provide new insights into the complex interactions between plants and microbes. They also help us understand how these interactions affect the plant community's function and structure. This special issue features cutting-edge information on the interactions between microbes and plants. It also covers the evolution of the plant immune system and how it functions through the use of computational modeling and physiological principles. It additionally explores variations in the plant or microbes' factors and defense mechanisms.

6.3.2 Metal Absorption and Uptake

Some plants can successfully remove various types of contaminants, such as arsenic, lead, chromium, and other radioactive substances from soil (Khalid et al. 2017). Heavy metals are a broad category of elements. Although their determination and defirmation are subject to debate, one of the main consensus statements about them states that an excessive amount of these substances can be harmful (Aliyev et al. 2018). One of the most effective ways to remove heavy metals is through phytoremediation, which involves taking advantage of the plant's ability to uptake certain heavy metals. However, this method can also be used to remove other nonmetal elements, such as Cu, Zn, and Mo (Cho-Ruk et al. 2006). The ability of certain plant species to concentrate heavy metals significantly exceeds that of nonaccumulator plants. Usually, microorganisms living in the rhizosphere, which are closely related to plants, help mobilize the metal ions in the soil. In terms of how they eliminate organic contaminants, plants are much more effective than inorganic compounds when it comes to absorbing heavy metals. Their ability to accomplish this through various mechanisms, such as phytoextraction, rhizofiltration, stabilization, and phytovolatilizing, is shown in Fig. 6.2.

The various plant species' abilities are utilized in the field of plant phytomediation, which is a constantly evolving area of phytotechnologies. It allows them to accumulate, stabilize, and degrade various substances while maintaining a

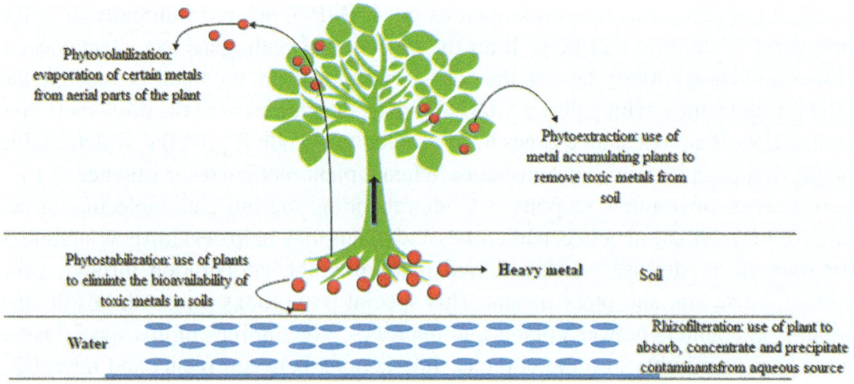

Fig. 6.2 Mechanism of phytoremediation of heavy metals in plants

long-term green cover (Parihar et al. 2021). Unlike conventional techniques, this method does not disturb the environment around the affected area (Ali et al. 2013). One of the main components of this process is the process of plant accumulation of heavy metals. This process can be carried out through phytoextraction, in which plants are extracted from tissues. It is based on how some plants can accumulate heavy metals through their roots. They then store them underground (Castañeda-Espinoza et al. 2023; Antoniadis et al. 2021). Recent studies on the plant species belonging to the Salicaceae family revealed that they can accumulate large amounts of heavy metals. One of the well-mapped genera in this area is the Populus (Marmiroli et al. 2011). The various characteristics of the Populus that are considered when it comes to developing effective and efficient plant strategies for dealing with environmental issues are its high biomass production rate and continuous metal accumulation. These characteristics allow them to perform their duties effectively and efficiently (Villette et al. 2019). Unfortunately, inconsistencies can be observed in the published findings regarding the accumulation of metal in different parts of the plant. Adams et al. (2011) reported that the plant's accumulation of heavy metals was affected by the levels of zinc contamination in the environment. Another study conducted by Algreen et al. (2014) revealed that the plant's exposure to different levels of heavy metals was not consistent with the plant's overall composition. Numerous studies have demonstrated that the acidity level can affect the intensity of heavy metal buildup in woody plants.

6.3.3 *Metal Tolerance*

The ability of a plant to safely and effectively eliminate or prevent HMs from entering the shoot determines its tolerance and response to this issue. This is why it is important that plant species are categorized into two categories: the excluders and the hyperaccumulators. The former refers to those that prevent HM uptake and transport into the shoot, while the latter refers to those that can endure extremely high metal concentrations (Shrivastava et al. 2019). Depending on the metal's chemical reactivity, the hyperaccumulators can accumulate large amounts of HMs. For instance, they can accumulate up to 100 g of Tl or Se, 300 g of Cu or Cr, 1000 g of Ni, and 10,000 g of Zn or molybdenum per gram of the leaves. When grown in their natural habitat, these plants can accumulate these heavy metals at a rate of around 10,000 g per gram of foliage (van der Ent et al. 2013; Reeves et al. 2018). Depending on the plant's survival in contaminated soil, the quantity of metal remediation can also be affected by the kinetic characteristics of the uptake process and the presence of HM transporters with low affinity or high affinity. In *Centella asiatica* and *G. issima* fragrant, the accumulation capacity of hyperaccumulators varied by around 20 days to a year. The presence of high root uptake and rapid transport to the shoot of HMs can also be affected by the different factors that affect the plant's response to this issue. For instance, the higher concentration of chelators and ligands can help speed up the process of depositing and transporting HMs to the shoot (Manara et al. 2020). A closer look at the uptake rates of different HMs shows that important elements, such as Cu, Fe, and Mn, require less time to build up in hyperaccumulators. On the other hand, nonessential elements, such as Hg, Al, and Cd, require a lot of time to get absorbed from contaminated soil. Besides the competitive inhibition, other factors such as the lack of strong chelators or ligands that can facilitate in-plant translocation and the low affinity of uptake transporters are also being studied to determine why nonmetals are being taken less.

6.3.4 Metal Detoxification

After translocating, the hyperaccumulators can remove heavy metals from the plant and allow them to live in areas contaminated with toxic substances. This process, which is known as detoxification, occurs in the plant's vacuole (Kanoun-Boulé et al. 2009). Some of the key players in this process include the CDF, NRAMP, and ABC transporter families. The two main ABC families are involved in the transport of heavy metals into the vacuole. The first known example of an ABC transporter in this process is the HMT1, which is located in the tonoplast of *Schizosaccharomyces pombe*. It helps in the transport of PC-Cd, which is composed of phytochelatins and cadmium, in the cytosol (Kuriakose and Prasad 2008). Studies in *A. thaliana* revealed that the AtMRP1 and AtMRP2 are two transport systems that help in the transport of PC-Cd complexes into the vacuole. These two transport systems confer metal tolerance. The CDF family, which is also involved in this process, is known to move metal ions such as Zn^{2+}, Co^{2+}, Ni^{2+}, and Co^{2+} from the cytosol to the vacuole (Montanini et al. 2007). The four groups of transport systems are known to be important in the process. They are grouped into two main groups: group I and group III. In studies on the different types of hyperaccumulators and non-hyperaccumulators, the expression of the MTP1 and the MTP8 groups was higher than that of the non-hyperaccumulators (van de Mortel et al. 2006). The AhMTP1 protein exhibited a higher level of expression in *A. halleri's* leaves when exposed to an exogenous supply of zinc (Dräger et al. 2004). The close homologs of the MTP11 and the MTP8 were also confirmed in *A. thaliana* (Delhaize et al. 2003), which implies that these proteins play a part in the plant's ability to tolerate metal ions. Other transport proteins, such as NRAMP and HMA, are also involved in the process of transporting heavy metals into the vacuole from the cytosol. However, the overexpression of HMAs may be responsible for its role in the plant's detoxification mechanisms. In addition to the transport system's role, organic acids play a vital role in the process of entrapping and removing the metal ions. Kramer et al. (2000) reported that the citrate–Ni interaction in the leaves of *Thlasi goesingense* allows the formation of an organic acid complex that is capable of removing metal ions. The malate–Zn interaction in the leaves of *A. halleri* and *T. caerulescens* can also contribute to the plant's ability to remove heavy metals. The mechanism by which hyperaccumulators can achieve this is dependent on the overexpression of certain genes that are involved in antioxidant activity (Anjum et al. 2014). The studies also showed that elevated concentrations of salicylic acid can boost serine transferase activity and GSH level. NgSAT gene's overexpression in *Nocaea goingsense* increased the levels of various nutrients (Freeman and Salt 2007).

6.4 Metal Removal and Treatment Technologies

The cleanup of heavy metals is a vital part of the environment's preservation (Glick 2010). Various biological and physicochemical techniques have been utilized to remove heavy metals. Although their speed is generally faster, physical–chemical methods are considered to be difficult due to their expense. Physical–chemical methods can have negative effects on soil's physical, chemical, and biological characteristics. On the other hand, biological remediation is believed to be the more efficient method of removing heavy metals since it involves the use of natural and environmentally benign processes (Doble and Kumar 2005). One of the most effective strategies for bioremediation of soil contaminated with heavy metals is by utilizing microbes that promote the growth of plants. This technique can be used in response to the excessive use of fertilizers in agricultural areas and climate change (Tiwari et al. 2016; Nautiyal et al. 2013). Because they can consume waste and degrade complex waste into straightforward, nontoxic by-products, microbes are known to benefit plant development and survival when subjected to heavy metal stress. This is possible because microbes have created a variety of defense mechanisms that enable them to survive the occurrence of poisonous heavy metals.

For instance, bacteria and plant roots assist in chelation, acidification, and precipitation to sequester metal ions and lower soil pH, which enhances the bioavailability of metals from the soil (Mishra et al. 2017).

6.5 Restoration of Metal Contaminated Land and Water Ecosystem

The main source of metal toxicity in terrestrial ecosystems is soil (Gadd 2010). The average soil can contain anywhere between 1 and 100,000 mg/kg of metal, of which a sizable portion is actively altered by soil bacteria. In contaminated soils, soil microorganisms, particularly those in the rhizosphere, are crucial for HM detoxification. This contribution of the rhizomicrobial community is known as rhizoremediation. This is a result of higher metabolic activity among prokaryotes, eukaryotes, and other microbes near plant roots. Firmicutes, Proteobacteria, and Actinobacteria make up the majority of the bacteria in HM-contaminated areas, and the most common genera are *Bacillus* (Checcucci et al. 2017), *Pseudomonas*, and *Arthrobacter*. Additionally, significant PGP microorganisms, rhizobia originate in the rhizosphere. HM stress can be extremely harmful to nodulation and nitrogenase activities, even though HMT rhizobial strains have been found in polluted areas and successfully carried out symbiotic nitrogen fixation. It is well identified that the symbiotic relationship between legumes and rhizobia recovers the quality of contaminated soils and detoxifies HM. Arbuscular mycorrhizal (AM) fungi have also been seen to predominantly colonize poor in nutrient but contaminated HM soils (Khan et al. 2000). The movement of metal ions in the surrounding environment or their movement within the cell is what drives the growth of AM fungus and other rhizosphere microorganisms. These bacteria can convert a wide variety of metals. The interaction of metal microorganisms in the rhizosphere is extremely complex and is influenced by the physicochemical makeup of the soil, metabolic activity, and the variety of the bacteria. By chelation, acidification, and precipitation, for instance, bacteria and plant roots help sequester metal ions and lower the pH of the soil, which increases the bioavailability of metals from the soil (Mishra et al. 2017).

6.6 Conclusion

In this chapter, more emphasis is given to the various mechanisms in the microbes and plants to cope with heavy metal stress and detoxification in metal-contaminated sites. Understanding plant–microbes' interactions could enhance defense mechanisms and tolerance of plants against elevated concentration of heavy metals. Contaminated sites could be restored by revegetating degraded land through metal-tolerant microbes that provide essential nutrients to help plants grow and survive. Biotic and abiotic stresses affecting plants and crops are altering their productivities, attributes, and survival. The plants and crops endure stress conditions that result in the development of drought, salinity, heat, and cold-tolerant proteins. Various factors that affect the development and growth of plants are becoming increasingly detrimental to global food security, crop yield, and food quality. The multiple stressors that are currently affecting plants are causing them to develop an imbalance of nutrients and hormonal levels. The use of mycorrhizal fungi and PGPM strains combined with stress-tolerant microbes can help improve the growth of plants. These types of organisms can help regulate the development and growth of plants by producing siderophores, improving the nutrition of plants, and enhancing their antioxidant capacity. One of the mechanisms that plants and microbes evolved to cope with the multiple stresses is the induced stress response (ISR). This process increases the water and nutrient supply during the stress condition. The use of microbes can help solve the food security and soil health issues that are currently affecting the global population. In this chapter, we identified the potential of microbes to serve as ecological engineers to address environmental stress issues. In the current scenario, more studies are required to identify PGPM strains that could endure multiple stress. Further testing of microbial consortia is necessary to develop effective solutions to global environmental problems.

Acknowledgments The authors thank the Department of Chemistry, Manipal University Jaipur, India, for providing facilities and encouragement.

References

Abou-Shanab RA, Angle JS, van Berkum P (2007) Chromate-tolerant bacteria for enhanced metal uptake by *Eichhornia crassipes* (Mart.). Int J Phytoremediation 9:91–105. https://doi.org/10.1080/15226510701232708

Acosta-Martínez V, Bell CW, Morris BEL, Zak J, Allen VG (2010) Long-term soil microbial community and enzyme activity responses to an integrated cropping-livestock system in a semi-arid region. Agric Ecosyst Environ 137:231–240. https://doi.org/10.1016/j.agee.2010.02.008

Adams JP, Adeli A, Hsu CY, Harkess RL, Page GP, dePamphilis CW, Schultz EB, Yuceer C (2011) Poplar maintains zinc homeostasis with heavy metal genes HMA4 and PCS1. J Exp Bot 62:3737–3752. https://doi.org/10.1093/jxb/err025

Adriano DC, Bolan NS, Vangronsveld J, Wenzel WW (2004) Heavy metals. In Encyclopedia of soils in the environment. Elsevier-Hanley and Belfus Inc 175–182. doi:https://doi.org/10.3389/fmicb.2017.01706

Ahmad F, Ahmad I, Khan MS (2008) Screening of free-living rhizospheric bacteria for their multiple plant growth promoting activities. Microbiol Res 163:173–181. https://doi.org/10.1016/j.micres.2006.04.001

Algreen M, Trapp S, Rein A (2014) Phytoscreening and phytoextraction of heavy metals at Danish polluted sites using willow and poplar trees. Environ Sci Pollut R 21:8992–9001. https://doi.org/10.1007/s11356-013-2085-z

Ali H, Khan E, Sajad MA (2013) Phytoremediation of heavy metals-concepts and applications. Chemosphere 91:869–881. https://doi.org/10.1016/j.chemosphere.2013.01.075

Aliyev F, Khalilova H, Aliyev F (2018) Heavy metal pollution of ecosystem in an industrialized and urbanized region of the Republic of Azerbaijan. INTECH. https://doi.org/10.5772/intechopen.74600

Amari T, Ghnaya T, Abdelly C (2017) Nickel, cadmium and lead phytotoxicity and potential of halophytic plants in heavy metal extraction. S Afr J Bot 111:99–110. https://doi.org/10.1016/j.sajb.2017.03.011

Anjum NA, Umar S, Iqbal M (2014) Assessment of cadmium accumulation, toxicity, and tolerance in Brassicaceae and Fabaceae plants—implications for phytoremediation. Environ Sci Pollut Res 21:10286–10293. https://doi.org/10.1007/s11356-014-2889-5

Antoniadis V, Shaheen SM, Stärk HJ, Wennrich R, Levizou E, Merbach I, Rinklebe J (2021) Phytoremediation potential of twelve wild plant species for toxic elements in a contaminated soil. Environ Int 146:106233. https://doi.org/10.1016/j.envint.2020.106233

Arora A, Sairam RK, Srivastava GC (2002) Oxidative stress and antioxidative system in plants. Curr Sci:1227–1238. https://doi.org/10.4236/abb.2013.411A2003

Arwidsson Z, Johansson E, von Kronhelm T, Allard B, van Hees P (2010) Remediation of metal contaminated soil by organic metabolites from fungi—production of organic acids. Water Air Soil Pollut 205:215–226. https://doi.org/10.1007/s11270-009-0067-z

Beattie GA (2006) Plant-associated bacteria: survey, molecular phylogeny, genomics and recent advances. In: Gnanamanickam SS (ed) Plant-associated bacteria. Springer, Dordrecht, pp 1–56. https://doi.org/10.1007/978-1-4020-4538-7

Boller T, Felix G (2009) A renaissance of elicitors: perception of microbe-associated molecular patterns and danger signals by pattern-recognition receptors. Annu Rev Plant Biol 60:379–406. https://doi.org/10.1146/annurev.arplant.57.032905.105346

Braud A, Hannauer M, Milsin GLA, Schalk IJ (2009) The *Pseudomonas aeruginosa* pyocheliniron uptake pathway and its metal specificity. J Bacteriol 191:5317–5325. https://doi.org/10.1128/jb.00010-09
Briffa J, Sinagra E, Blundell R (2020) Heavy metal pollution in the environment and their toxicological effects on humans. Heliyon 6:e04691. https://doi.org/10.1016/j.heliyon.2020.e04691
Castañeda-Espinoza J, David Osvaldo SS, Patricia MG, María Luisa CG, Alexis R, Manasés GC, Alejandro ZÁ, Efraín TS (2023) D odonaea viscosa (Sapindaceae) as a phytoremediator for soils contaminated by heavy metals in abandoned mines. Environ Sci Pollut Res 30:2509–2529. https://doi.org/10.1007/s11356-022-22374-5
Chatterjee S, Sau GB, Mukherjee SK (2009) Plant growth promotion by a hexavalent chromium reducing bacterial strain, *Cellulosimicrobium cellulans* KUCr3. World J Microbiol Biotechnol 25:1829–1836. https://doi.org/10.1007/s11274-009-0084-5
Checcucci A, Bazzicalupo M, Mengoni A (2017) Exploiting nitrogen-fixing rhizobial symbionts genetic resources for improving phytoremediation of contaminated soils. In Enhancing cleanup of environmental pollutants 275–288. https://doi.org/10.1007/978-3-319-55426-6_13
Chen SY, Lin JG (2001) Effect of substrate concentration on bioleaching of metal-contaminated sediment. J Hazard Mater 82:77–89. https://doi.org/10.1016/S0304-3894(00)00357-5
Chen CT, Chen TH, Lo KF, Chiu CY (2004) Effects of proline on copper transport in rice seedlings under excess copper stress. Plant Sci 166:103–111. https://doi.org/10.1016/j.plantsci.2003.08.015
Chen YX, Wang YP, Lin Q, Luo YM (2005) Effect of copper-tolerant rhizosphere bacteria on mobility of copper in soil and copper accumulation by *Elsholtzia splendens*. Environ Int 31:861–866. https://doi.org/10.1016/j.envint.2005.05.044
Chen GQ, Chen Y, Zeng GM, Zhang JC, Chen YN, Wang L, Zhang WJ (2010) Speciation of cadmium and changes in bacterial communities in red soil following application of cadmium-polluted compost. Environ Eng Sci 27:1019–1026. https://doi.org/10.1089/ees.2010.0015
Chiang PN, Chiu CY, Wang MK, Chen BT (2011) Low-molecular-weight organic acids exuded by Millet (*Setaria italica* (L.) Beauv.) roots and their effect on the remediation of cadmium contaminated soil. Soil Sci 176:33–38. https://doi.org/10.1097/SS.0b013e318202fdc9
Cho-Ruk K, Kurukote J, Supprung P (2006) Perennial plants in the phytoremediation of lead-contaminated soils. Biotechnol 5:1–4. https://doi.org/10.3923/biotech.2006.1.4
Compant S, Clément C, Sessitsch A (2010) Plant growth-promoting bacteria in the rhizo-and endosphere of plants: their role, colonization, mechanisms involved and prospects for utilization. Soil Biol Biochem 42:669–678. https://doi.org/10.1016/j.soilbio.2009.11.024
Cseh E (2002) Metal permeability, transport and efflux in plants. In Physiology and biochemistry of metal toxicity and tolerance in plants 1–36. https://doi.org/10.1007/978-94-017-2660-3_1
Das N, Vimala R, Karthika P (2008) Biosorption of heavy metals–an overview. Indian J Biotechnol 7:159–169
Datta R, Meena RS, Pathan SI, Ceccherini MT (eds) (2019) Carbon and nitrogen cycling in soil. Springer Nature
Delhaize E, Kataoka T, Hebb DM, White RG, Ryan PR (2003) Genes encoding proteins of the cation diffusion facilitator family that confer manganese tolerance. Plant Cell 15:1131–1142. https://doi.org/10.1105/tpc.009134
Di Gregorio S, Lampis S, Vallini G (2005) Selenite precipitation by a rhizospheric strain of *Stenotrophomonas* sp isolated from the root system of *Astragalus bisulcatus*: a biotechnological perspective. Environ Int 31:233–241. https://doi.org/10.1016/j.envint.2004.09.021
Dimkpa CO, Merten D, Svatoš A, Büchel G, Kothe E (2009) Siderophores mediate reduced and increased uptake of cadmium by Streptomyces tendae F4 and sunflower (*Helianthus annuus*), respectively. J Appl Microbiol 107:1687–1696. https://doi.org/10.1111/j.1365-2672.2009.04355
Djebali W, Chaïbi W, Ghorbel MH (2002) Croissance, activité peroxydasique et modifications ultrastructurales induites par le cadmium dans la racine de tomate. Canad J Bot 80:942–953. https://doi.org/10.1139/b02-062

Djebali W, Zarrouk M, Brouquisse R, El Kahoui S, Limam F, Ghorbel MH, Chaïbi W (2005) Ultrastructure and lipid alterations induced by cadmium in tomato (*Lycopersicon esculentum*) chloroplast membranes. Plant Biol 7:358–368. https://doi.org/10.1055/s-2005-837696

Doble M, Kumar A (2005) Biotreatment of industrial effluents. Elsevier. https://doi.org/10.1016/j.resconrec.2007.01.012

Dräger DB, Desbrosses-Fonrouge AG, Krach C, Chardonnens AN, Meyer RC, Saumitou-Laprade P, Krämer U (2004) Two genes encoding *Arabidopsis halleri* MTP1 metal transport proteins co-segregate with zinc tolerance and account for high MTP1 transcript levels. Plant J 39:425–439. https://doi.org/10.1111/j.1365-313X.2004.02143.x

Dua M, Sethunathan N, Johri AK (2002) Biotechnology and biormediation: success and limitations. Appl Microbiol Biotechnol 59:143–152. https://doi.org/10.1007/s00253-002-1024-6

Fodor E, Szabó-Nagy A, Erdei L (1995) The effects of cadmium on the fluidity and H+-ATPase activity of plasma membrane from sunflower and wheat roots. J Plant Physiol 147:87–92. https://doi.org/10.1016/S0176-1617(11)81418-5

Fomina MA, Alexander IJ, Colpaert JV, Gadd GM (2005) Solubilization of toxic metal minerals and metal tolerance of mycorrhizal fungi. Soil Biol Biochem 37:851–866. https://doi.org/10.1016/j.soilbio.2004.10.013

Freeman JL, Salt DE (2007) The metal tolerance profile of *Thlaspi goesingense* is mimicked in *Arabidopsis thaliana* heterologously expressing serine acetyl-transferase. BMC Plant Biol 7:1–10. https://doi.org/10.1186/1471-2229-7-63

Gadd GM (2010) Metals, minerals and microbes: geomicrobiology and bioremediation. Microbiology 156:609–643. https://doi.org/10.1186/1471-2229-7-63

Garrido S, Campo G, Esteller MV, Vaca R, Lugo J (2005) Heavy metals in soil treated with sewage sludge composting, their effect on yield and uptake of broad bean seeds (*Vicia faba* L.). Water Air Soil Pollut 166:303–319. https://doi.org/10.1007/s11270-005-5269-4

Gautam PK, Gautam RK, Banerjee S, Chattopadhyaya MC, Pandey JD (2016) Heavy metals in the environment: fate, transport, toxicity and remediation technologies. Nova Science Publishers 60:101–130

Ghnaya T, Slama I, Messedi D, Grignon C, Ghorbel MH, Abdelly C (2007) Effects of Cd^{2+} on K^+, Ca^{2+} and N uptake in two halophytes *Sesuvium portulacastrum* and *Mesembryanthemum crystallinum*: consequences on growth. Chemosphere 67:72–79. https://doi.org/10.1016/j.chemosphere.2006.09.064

Gill SS, Tuteja N (2010) Reactive oxygen species and antioxidant machinery in abiotic stress tolerance in crop plants. Plant Physiol Biochem 48:909–930. https://doi.org/10.1016/j.plaphy.2010.08.016

Glick BR (2010) Using soil bacteria to facilitate phytoremediation. Biotechnol Adv 28:367–374. https://doi.org/10.1016/j.biotechadv.2010.02.001

Gonzalez-Chavez MC, Carrillo-Gonzalez R, Wright SF, Nichols KA (2004) The role of glomalin, a protein produced by arbuscular mycorrhizal fungi, in sequestering potentially toxic elements. Environ Pollut 130:317–323. https://doi.org/10.1016/j.envpol.2004.01.004

Hartmann A, Schmid M, Van Tuinen D, Berg G (2009) Plant-driven selection of microbes. Plant Soil 321:235–257. https://doi.org/10.1007/s11104-008-9814-y

Hassan TU, Bano A, Naz I (2017) Alleviation of heavy metals toxicity by the application of plant growth promoting rhizobacteria and effects on wheat grown in saline sodic field. Int J Phytoremediation 19:522–529. https://doi.org/10.1080/15226514.2016.1267696

Hinojosa MB, Carreira JA, García-Ruíz R, Dick RP (2004) Soil moisture pre-treatment effects on enzyme activities as indicators of heavy metal-contaminated and reclaimed soils. Soil Biol Biochem 36:1559–1568. https://doi.org/10.1080/15226514.2016.1267696

Hoberg E, Marschner P, Lieberei R (2005) Organic acid exudation and pH changes by *Gordonia* sp. and *Pseudomonas* fluorescens grown with P adsorbed to goethite. Microbiol Res 160:177–187. https://doi.org/10.1016/j.micres.2005.01.003

Huang SH, Bing PENG, Yang ZH, Chai LY, Zhou LC (2009) Chromium accumulation, microorganism population and enzyme activities in soils around chromium-containing slag

heap of steel alloy factory. Trans Nonferrous Met Soc 19:41–248. https://doi.org/10.1016/S1003-6326(08)60259-9
Jadia CD, Fulekar MH (2009) Phytoremediation of heavy metals: recent techniques. Afr J Biotechnol 8:921–928
Jin X, Yang X, Islam E, Liu D, Mahmood Q (2008) Effects of cadmium on ultrastructure and antioxidative defense system in hyperaccumulator and non-hyperaccumulator ecotypes of *Sedum alfredii* Hance. J Hazard Mater 156:387–397. https://doi.org/10.1016/j.jhazmat.2007.12.064
Jones JD, Dangl JL (2006) The plant immune system. Nature 444:323–329. https://doi.org/10.1038/nature05286
Joshi PM, Juwarkar AA (2009) In vivo studies to elucidate the role of extracellular polymeric substances from Azotobacter in immobilization of heavy metals. Environ Sci Technol 43:5884–5889. https://doi.org/10.1021/es900063b
Juwarkar AA, Nair A, Dubey KV, Singh SK, Devotta S (2007) Biosurfactant technology for remediation of cadmium and lead contaminated soils. Chemosphere 10:1996–2002. https://doi.org/10.1016/j.chemosphere.2007.02.027
Kanoun-Boulé M, Vicente JA, Nabais C, Prasad MNV, Freitas H (2009) Ecophysiological tolerance of duckweeds exposed to copper. Aquat Toxicol 91:1–9. https://doi.org/10.1016/j.aquatox.2008.09.009
Khalid M, Bilal M, Hassani D, Iqbal HMN, Wang H, Huang D (2017) Mitigation of salt stress in white clover (*Trifolium repens*) by *Azospirillum brasilense* and its inoculation effect. Bot Stud 58:5. https://doi.org/10.1186/s40529-016-0160-8
Khan AG, Kuek C, Chaudhry TM, Khoo CS, Hayes WJ (2000) Role of plants, mycorrhizae and phytochelators in heavy metal contaminated land remediation. Chemosphere 41:197–207. https://doi.org/10.1016/S0045-6535(99)00412-9
Kramer U, Pickering IJ, Prince RC, Raskin I, Salt DE (2000) Subcellular localization and speciation of nickel in hyperaccumulator and non-accumulator *Thlaspi* species. Plant Physiol 122:1343–1354. https://doi.org/10.1104/pp.122.4.1343
Kuriakose SV, Prasad MNV (2008) 16 cadmium as an environmental contaminant: consequences to plant and human health. Trace elements as contaminants and nutrients: Consequences in ecosystems and human Health 373. doi:https://doi.org/10.1002/9780470370124.ch16
Li M, Zhang LJ, Tao L, Li W (2008) Ecophysiological responses of *Jussiaea rapens* to cadmium exposure. Aquat Bot 88:347–352. https://doi.org/10.1016/j.aquabot.2008.01.002
Li WC, Ye ZH, Wong MH (2010) Metal mobilization and production of short-chain organic acids by rhizosphere bacteria associated with a cd/Zn hyperaccumulating plant *sedum alfredii*. Plant Soil 326:453–467. https://doi.org/10.1007/s11104-009-0025-y
Luo ZB, Wu C, Zhang C, Li H, Lipka U, Polle A (2014) The role of ectomycorrhizas in heavy metal stress tolerance of host plants. Environ Exp Bot 108:47–62. https://doi.org/10.1016/j.envexpbot.2013.10.018
Ma Y, Oliveira RS, Freitas H, Zhang C (2016) Biochemical and molecular mechanisms of plantmicrobe-metal interactions: relevance for phytoremediation. Front Plant Sci 7:1–19. https://doi.org/10.3389/fpls.2016.00918
Madhaiyan M, Poonguzhali S, Sa T (2007) Metal tolerating methylotrophic bacteria reduces nickel and cadmium toxicity and promotes plant growth of tomato (*Lycopersicon esculentum* L.). Chemosphere 69:220–228. https://doi.org/10.1016/j.chemosphere.2007.04.017
Manara A, Fasani E, Furini A, DalCorso G (2020) Evolution of the metal hyperaccumulation and hypertolerance traits. Plant Cell Environ 43:2969–2986. https://doi.org/10.1111/pce.13821
Marmiroli M, Pietrini F, Maestri E, Zacchini M, Marmiroli N, Massacci A (2011) Growth, physiological and molecular traits in Salicaceae trees investigated for phytoremediation of heavy metals and organics. Tree Physiol 31:1319–1334. https://doi.org/10.1093/treephys/tpr090
Michalak A (2006) Phenolic compounds and their antioxidant activity in plants growing under heavy metal stress. Polish J Environ Stud 15
Miransari M (2011) Hyperaccumulators, arbuscular mycorrhizal fungi and stress of heavy metals. Biotechnol Adv 29:645–653. https://doi.org/10.1016/j.biotechadv.2011.04.006

Mishra J, Singh R, Arora NK (2017) Alleviation of heavy metal stress in plants and remediation of soil by rhizosphere microorganisms. Front Microbiol 8:1706. https://doi.org/10.3389/fmicb.2017.01706

Montanini B, Blaudez D, Jeandroz S, Sanders D, Chalot M (2007) Phylogenetic and functional analysis of the cation diffusion facilitator (CDF) family: improved signature and prediction of substrate specificity. BMC Genomics 8:1–16. https://doi.org/10.1186/1471-2164-8-107

Nautiyal CS, Srivastava S, Chauhan PS, Seem K, Mishra A, Sopory SK (2013) Plant growth-promoting bacteria *bacillus amyloliquefaciens* NBRISN13 modulates gene expression profile of leaf and rhizosphere community in rice during salt stress. Plant Physiol Biochem 66:1–9. https://doi.org/10.1016/j.plaphy.2013.01.020

Neumann RB, Ashfaque KN, Badruzzaman ABM, Ashraf Ali M, Shoemaker JK, Harvey CF (2010) Anthropogenic influences on groundwater arsenic concentrations in Bangladesh. Nat Geosci 3:46–52. https://doi.org/10.1038/ngeo685

Nie L, Shah S, Rashid A, Burd GI, Dixon DG, Glick BR (2002) Phytoremediation of arsenate contaminated soil by transgenic canola and the plant growth-promoting bacterium *Enterobacter cloacae* CAL2. Plant Physiol Biochem 40:355–361. https://doi.org/10.1016/S0981-9428(02)01375-X

Nies DH (1999) Microbial heavy metal resistance. Appl Microbiol Biotechnol 51:730–750. https://doi.org/10.1007/s002530051457

Oliva R, Ji C, Atienza-Grande G, Huguet-Tapia JC, PerezQuintero A, Li T, Eom JS, Li C, Nguyen H, Liu B, Auguy F, Sciallano C, Luu VT, Dossa GS, Cunnac S, Schmidt SM, Slamet-Loedin IH, Vera Cruz C, Szurek B, Frommer WB, White FF, Yang B (2019) Broad-spectrum resistance to bacterial blight in rice using genome editing. Nat Biotechnol 37:1344–1350. https://doi.org/10.1038/s41587-019-0267-z

Orcutt DM, Nilsen ET (2000) Physiology of plants under stress: soil and biotic factors, vol 2. John Wiley & Sons

Panuccio MR, Sorgonà A, Rizzo M, Cacco G (2009) Cadmium adsorption on vermiculite, zeolite and pumice: batch experimental studies. J Environ Manag 90:364–374. https://doi.org/10.1016/j.jenvman.2007.10.005

Parihar JK, Parihar PK, Pakade YB, Katnoria JK (2021) Bioaccumulation potential of indigenous plants for heavy metal phytoremediation in rural areas of Shaheed Bhagat Singh Nagar, Punjab (India). Environ Sci Pollut Res 28:2426–2442. https://doi.org/10.1007/s11356-020-10454-3

Penrose DM, Glick BR (2001) Levels of 1-aminocyclopropane-1-carboxylic acid (ACC) in exudates and extracts of canola seeds treated with plant growth-promoting bacteria. Can J Microbiol 47:368–372. https://doi.org/10.1139/w01-014

Perfus-Barbeoch L, Leonhardt N, Vavasseur A, Forestier C (2002) Heavy metal toxicity: cadmium permeates through calcium channels and disturbs the plant water status. Plant J 32:539–548. https://doi.org/10.1046/j.1365-313X.2002.01442.x

Pierart A, Shahid M, Sejalon-Delmas N, Dumat C (2015) Antimony bioavailability: knowledge and research perspectives for sustainable agricultures. J Hazard Mater 289:219–234. https://doi.org/10.1016/j.jhazmat.2015.02.011

Pourrut B, Jean S, Silvestre J, Pinelli E (2011) Lead-induced DNA damage in *Vicia faba* root cells: potential involvement of oxidative stress. Mutat Res Genet Toxicol Environ Mutagen 726:123–128. https://doi.org/10.1016/j.mrgentox.2011.09.001

Rajkumar M, Ae N, Prasad MNV, Freitas H (2010) Potential of siderophore-producing bacteria for improving heavy metal phytoextraction. Trends Biotechnol 28:142–149. https://doi.org/10.1016/j.tibtech.2009.12.002

Rashid A, Schutte BJ, Ulery A, Deyholos MK, Sanogo S, Lehnhoff EA, Beck L (2023) Heavy metal contamination in agricultural soil: environmental pollutants affecting crop health. Agronomy 13:1521. https://doi.org/10.3390/agronomy13061521

Reeves RD, Baker AJ, Jaffré T, Erskine PD, Echevarria G, van Der Ent A (2018) A global database for plants that hyperaccumulate metal and metalloid trace elements. New Phytol 218:407–411. https://doi.org/10.1111/nph.14907

Ryan PR, Delhaize E, Jones DL (2001) Function and mechanism of organic anion exudation from plant roots. Annu Rev Plant Physiol Plant Mol Biol 52:527–560. https://doi.org/10.1146/annurev.arplant.52.1.527

Sandalio LM, Dalurzo HC, Gomez M, Romero-Puertas MC, Del Rio LA (2001) Cadmium-induced changes in the growth and oxidative metabolism of pea plants. J Exp Bot 52:2115–2126. https://doi.org/10.1093/jexbot/52.364.2115

Schalk IJ, Hannauer M, Braud A (2011) New roles for bacterial siderophores in metal transport and tolerance. Environ Microbiol 13:2844–2854. https://doi.org/10.1111/j.1462-2920.2011.02556.x

Semane B, Dupae J, Cuypers A, Noben JP, Tuomainen M, Tervahauta A, Vangronsveld J (2010) Leaf proteome responses of *Arabidopsis thaliana* exposed to mild cadmium stress. J Plant Physiol 167:247–254. https://doi.org/10.1016/j.jplph.2009.09.015

Sheng XF, Xia JJ, Jiang CY, He LY, Qian M (2008) Characterization of heavy metal-resistant endophytic bacteria from rape (*Brassica napus*) roots and their potential in promoting the growth and lead accumulation of rape. Environ Pollut 156:1164–1170. https://doi.org/10.1016/j.envpol.2008.04.007

Shi JY, Lin HR, Yuan XF, Chen XC, Shen CF, Chen YX (2011) Enhancement of copper availability and microbial community changes in rice rhizospheres affected by sulfur. Molecules 16:1409–1417. https://doi.org/10.3390/molecules16021409

Shrivastava M, Khandelwal A, Srivastava S (2019) Heavy metal hyperaccumulator plants: the resource to understand the extreme adaptations of plants towards heavy metals. Plant-metal interactions 79–97. doi:https://doi.org/10.1007/978-3-030-20732-8_5

Singh S, Sinha S (2005) Accumulation of metals and its effects in *Brassica juncea* (L.) *Czern.*(cv. Rohini) grown on various amendments of tannery waste. Ecotoxicol Environ Saf 62:118–127. https://doi.org/10.1016/j.ecoenv.2004.12.026

Speir TW, Kettles HA, Percival HJ, Parshotam A (1999) Is soil acidification the cause of biochemical responses when soils are amended with heavy metal salts? Soil Biol Biochem 31:1953–1961. https://doi.org/10.1016/S0038-0717(99)00115-7

Sytar O, Kumar A, Latowski D, Kuczynska P, Strzałka K, Prasad MNV (2013) Heavy metal-induced oxidative damage, defense reactions, and detoxification mechanisms in plants. Acta Physiol Plant 35:985–999. https://doi.org/10.1007/s11738-012-1169-6

Tak HI, Ahmad F, Babalola OO (2013) Advances in the application of plant growth-promoting rhizobacteria in phytoremediation of heavy metals. Rev Environ Contam Toxicol 223:33–52. https://doi.org/10.1007/978-1-4614-5577-6_2

Tchounwon PB, Yedjou CG, Patlolla AK, Sutton DJ (2012) Heavy metal toxicity and the environment. Molecular, clinical and environmental toxicology, experentia supplementum 101:133–164. doi:https://doi.org/10.1007/978-3-7643-8340-4_6

Tiwari S, Lata C (2018) Heavy metal stress, signaling, and tolerance due to plant-associated microbes: an overview. Front Plant Sci 9. https://doi.org/10.3389/fpls.2018.00452

Tiwari S, Lata C, Chauhan PS, Nautiyal CS (2016) *Pseudomonas putida* attunes morphophysiological, biochemical and molecular responses in *Cicer arietinum* L. during drought stress and recovery. Plant Physiol Biochem 99:108–117. https://doi.org/10.1016/j.plaphy.2015.11.001

Van De Mortel, Judith E, Laia Almar V, Henk S, Jeroen K, Sean C, Perry D, Ver Loren ME, van Themaat MK, Mark GMA (2006) Large expression differences in genes for iron and zinc homeostasis, stress response, and lignin biosynthesis distinguish roots of *Arabidopsis thaliana* and the related metal hyperaccumulator *Thlaspi caerulescens*. Plant Physiol 142:1127–1147. https://doi.org/10.1104/pp.106.082073

Van der Ent A, Baker AJ, Reeves RD, Pollard AJ, Schat H (2013) Hyperaccumulators of metal and metalloid trace elements: facts and fiction. Plant Soil 362:319–334. https://doi.org/10.1007/s11104-012-1287-3

Venkatesh NM, Vedaraman N (2012) Remediation of soil contaminated with copper using rhamnolipids produced from *Pseudomonas aeruginosa* MTCC 2297 using waste frying rice bran oil. Ann Microbiol 62:85–91. https://doi.org/10.1007/s13213-011-0230-9

Vernay P, Gauthier-Moussard C, Hitmi A (2007) Interaction of bioaccumulation of heavy metal chromium with water relation, mineral nutrition and photosynthesis in developed leaves of Lolium perenne L. Chemosphere 68:1563–1575. https://doi.org/10.1016/j.chemosphere.2007.02.052

Villette C, Maurer L, Delecolle J, Zumsteg J, Erhardt M, Heintz D (2019) In situ localization of micropollutants and associated stress response in *Populus nigra* leaves. Environ Int 126:523–532. https://doi.org/10.1016/j.envint.2019.02.066

Vivas A, Biro B, Ruíz-Lozanoa JM, Azcon R (2006) Two bacterial strains isolated from a Zn-polluted soil enhance plant growth and mycorrhizal efficiency under Zn toxicity. Chemosphere 52:1523–1533. https://doi.org/10.1016/j.chemosphere.2005.06.053

Wang Y, Shi J, Wang H, Lin Q, Chen X, Chen Y (2007) The influence of soil heavy metals pollution on soil microbial biomass, enzyme activity, and community composition near a copper smelter. Ecotoxicol Environ Saf 67:75–81. https://doi.org/10.1016/j.ecoenv.2006.03.007

Wang L, Zhou Q, Ding L, Sun Y (2008) Effect of cadmium toxicity on nitrogen metabolism in leaves of *Solanum nigrum* L. as a newly found cadmium hyperaccumulator. J Hazard Mater 154:818–825. https://doi.org/10.1016/j.jhazmat.2007.10.097

Wichard T, Bellenger JP, Morel FM, Kraepiel AM (2009) Role of the siderophore azotobactin in the bacterial acquisition of nitrogenase metal cofactors. Environ Sci Technol 43:7218–7224. https://doi.org/10.1021/es8037214

Wille L, Messmer MM, Studer B, Hohmann P (2019) Insights to plant–microbe interactions provide opportunities to improve resistance breeding against root diseases in grain legumes. Plant Cell Environ 42:20–40. https://doi.org/10.1111/pce.13214

Xiong T, Leveque T, Shahid M, Foucault Y, Mombo S, Dumat C (2014) Lead and cadmium phytoavailability and human bioaccessibility for vegetables exposed to soil or atmospheric pollution by process ultrafine particles. J Environ Qual 43:1593–1600. https://doi.org/10.2134/jeq2013.11.0469

Yao H, Xu J, Huang C (2003) Substrate utilization pattern, biomass and activity of microbial communities in a sequence of heavy metal-polluted paddy soils. Geoderma 115:139–148. https://doi.org/10.1016/S0016-7061(03)00083-1

Zarei M, Hempel S, Wubet T, Schäfer T, Savaghebi G, Jouzani GS, Nekouei MK, Buscot F (2010) Molecular diversity of arbuscular mycorrhizal fungi in relation to soil chemical properties and heavy metal contamination. Environ Pollut 158:2757–2765. https://doi.org/10.1016/j.envpol.2010.04.017

Zhuang P, Lu H, Li Z, Zou B, McBride MB (2014) Multiple exposure and effects assessment of heavy metals in the population near mining area in South China. PLoS One 9:94484. https://doi.org/10.1371/journal.pone.0094484

Chapter 7
Nature's Protectors: A Biofilm Perspective on Bacterial Disease Control in Plants

Daniel Gómez-Pérez, Leonie M. Zott, Monja Schmid, and Vasvi Chaudhry

Abstract Bacterial diseases significantly threaten agricultural productivity, necessitating innovative and sustainable approaches to disease management. Above- and belowground biofilms are essential in mitigating plant bacterial diseases. We examine aboveground biofilms, primarily located on leaf surfaces, for their ability to form protective barriers and produce antimicrobial compounds, thereby impeding the establishment of pathogenic bacteria. Concurrently, we explore belowground biofilms in the rhizosphere for their contributions to nutrient cycling, enhanced nutrient uptake, and the deployment of biocontrol agents against soilborne bacterial pathogens. Here, we employ a multidisciplinary approach by integrating molecular, microbiological, and ecological analyses to unravel the mechanisms underlying the formation and function of these beneficial biofilms. We investigate quorum sensing, microbial communication, and the intricate interplay between plant hosts and beneficial microbes to elucidate the complex networks orchestrating disease resistance. We consider environmental factors such as surface structure, temperature, and nutrient availability. We acknowledge their impact on biofilm-mediated disease management strategies relevant for controlling plant disease. Furthermore, we emphasize the importance of microbial diversity within biofilms for its role in enhancing plant resilience to bacterial diseases. Overall, this chapter provides a foundation for developing targeted and sustainable strategies in bacterial disease management, underpinning the critical role of above- and belowground beneficial biofilms.

D. Gómez-Pérez
Center for Organismic Adaptation, Faculty of Biology, LMU Munich, Planegg-Martinsried, Germany

Earlham Institute, Norwich, UK

L. M. Zott · M. Schmid · V. Chaudhry (✉)
Microbial Interactions in Plant Ecosystems, Center for Plant Molecular Biology, University of Tübingen, Tübingen, Germany

P. Singh Chauhan et al. (eds.), *Plant-Microbe Interaction and Stress Management*, Rhizosphere Biology,
https://doi.org/10.1007/978-981-97-4239-4_7

Insights gained from beneficial biofilms contribute to the fundamental understanding of synergistic relationships in the plant microbiome environment and their implications for sustainable agriculture

Keywords Biofilm · Quorum Sensing · Amyloids · Biocontrol · Phyllosphere · Rhizosphere

7.1 Introduction

Feeding the fast-growing human population, which will reach approximately 9.5 billion by the year 2050 (OECD 2012), will pose challenges for food and agricultural scientists in the coming decades. Global crop production faces alarming threats from both abiotic and biotic stresses, which are increasingly exacerbated by the effects of climate change (Teshome et al. 2020). Regarding biotic factors, plants are under constant attack from diverse organisms, including fungi, bacteria, oomycetes, nematodes, and herbivores. It is essential to mitigate plant disease to produce nutritionally improved food in sufficient quantities for the rapidly growing global population. Several disease control approaches are helpful in the management of plant diseases, including crop rotation, resistance breeding, and the use of biological control substances. Regarding the latter, commonly used are chemical fertilizers and pesticides to improve crop quality and quantity. However, these methods carry disadvantages, including disruption of ecosystems and downstream contamination (Aktar et al. 2009).

Plant disease occurs due to the complex interaction of a pathogen and its host plant. Within the plant-associated microbial diversity, only a fraction of the microbes are considered pathogenic, while the majority of them are beneficial and promote host plant growth and its fitness (van Elsas et al. 2012; Mendes et al. 2013; Philippot et al. 2013). Detrimental microbes have adapted to infect plants as specialized pathogens, causing numerous diseases and severely damaging crop plants. These are a serious concern to agriculture and crop productivity as they are directly responsible for crop losses (Mansfield et al. 2012). Therefore, the understanding of plant–pathogen interactions paves the way for disease management.

The percentage of bacterial phytopathogens causing plant diseases can vary significantly based on numerous factors, including geographical region, crop type, agricultural practices, and environmental conditions. Phytopathogens of bacterial origin are responsible for a notable portion of plant diseases; generally, it is estimated that bacterial phytopathogens contribute to roughly 10–20% of plant diseases worldwide. Factors such as the pervasiveness of specific pathogens, crop susceptibility, and climate conditions can influence the prevalence of bacterial diseases in plants. Bacterial phytopathogens pose a significant threat to agricultural production as plant diseases that trace their etiology to bacteria amount to a large percentage of all crop losses (Savary et al. 2019). Amid these challenges, innovative approaches such as the harnessing of biofilms have emerged as sustainable means to protect crops from bacterial pathogens.

7.2 Plant–Microbe Interactions: Mechanisms of Pathogen Success and Host Defense

Plants frequently encounter pathogenic microbes that challenge their growth and development throughout their life cycle. The intricate innate immune system in plants safeguards against the proliferation of harmful microbes, employing two interconnected defense mechanisms rooted in pathogen detection. These defense strategies encompass microbe-associated molecular pattern (MAMP)-triggered immunity and microbial effector-triggered immunity (ETI). Both MAMP and ETI responses activate multiple defense mechanisms to restrict pathogen invasion, safeguard the plant against pathogens, and curtail their growth (Thrall et al. 2007; Pieterse et al. 2014; Rodriguez et al. 2019). After exposure to pathogens, plants often develop immunological memory, leading to systemic acquired resistance. In contrast to their interactions with harmful microbes, plants establish and promote symbiotic relationships with beneficial microbes to fortify their immune system. The success of bacterial phytopathogens can be attributed to several factors, including their ability to colonize the plant, evade its defense mechanisms, and obtain essential nutrients for growth and reproduction. Bacterial pathogens also exhibit less specific physiological adaptations to ensure survival outside the host plant, which include metabolic responses to abiotic stress. Some of these are based on the production of exopolysaccharides, which can protect against desiccation and cold, and pigments, which protect against ultraviolet (UV) radiation.

Bacterial pathogens cause disease symptoms, including leaf spots, blights, cankers, and wilts. Notable bacterial pathogens, including species of *Xanthomonas*, *Pseudomonas*, *Agrobacterium*, *Ralstonia*, *Erwinia*, and others, are known to cause diseases in a wide range of crops, affecting agriculture globally (Mansfield et al. 2012). Bacterial diseases impact crops significantly, leading to yield losses and affecting food security. Bacterial phytopathogens have adapted in response to developments in plant resistance, resulting in an evolutionary arms race (Wang et al. 2021). This has led to several strategies for successful infection, including the expansion of effectors released through the type III secretion system in Gram-negative bacteria (Tang et al. 2006). The latter encompass some of the most important groups of bacterial phytopathogens, the *Pseudomonas syringae*'s pathovars (pv.) (Chen et al. 2022).

The entry of bacterial phytopathogens into plants determines their fitness and disease development. Pathogens typically penetrate plants through natural openings, such as stomata and hydathodes, wounds, or by using specialized mechanisms such as the secretion of enzymes or toxins that facilitate their entry into plant tissues (van der Wolf and De Boer 2014). Some pathogens can swim or move along moisture gradients to reach these openings. Wounds on plant tissues, caused by mechanical damage from herbivore feeding, also act as entry points for bacteria to infect plants. Some bacterial phytopathogens secrete enzymes such as cellulases and pectinases that break down the structural barriers and degrade plant cell walls, allowing them to penetrate plant tissues. Some bacterial phytopathogens have evolved mechanisms to bypass plant defenses inside plant tissues (Abramovitch and Martin 2004).

Advancements in understanding these pathogens, improved disease management practices, and the development of resistant crop varieties are essential for minimizing the impact of bacterial phytopathogens on plant health and agricultural productivity. In addition, the plant microbiome plays a key role in controlling and regulating the virulence of these phytopathogens (Gao et al. 2021). Plants have been shown to actively recruit bacteria from their environment to overcome and prevent disease (Liu et al. 2021). Similarly, dysbiosis in the phyllosphere, which refers to the lack of stability in the microbiota composition, has been associated with a higher likelihood of infection (Gong and Xin 2021). The mechanisms by which microbes can control pathogen-prone bacteria in the plant include direct inhibition of these through, for example, the release of antimicrobials or indirectly boosting plant defenses (Legein et al. 2020).

7.3 Microbial Adaptation: From Planktonic Cells to Biofilm Communities

Microbes can remarkably transform and adapt to their environments by converting key nutrients and energy into accessible forms for survival. Well-adapted microbes participate in many microbe–microbe and microbe–environment interactions. In nature, microbial cells exist as solitary planktonic cells or are assembled into structured codependent communities known as biofilms. This transition from planktonic to biofilm is a widespread phenomenon observed in the environment and in bacteria associated with hosts such as humans, animals, and plants (Jefferson 2004). Biofilms consist of either single- or multispecies communities that join forces to promote their survival. Biofilms are a dominant form of microbial lifestyle that can persist for many generations wherein microbial communities are shaped by common evolutionary pressures (Steenackers et al. 2016). The community performs different tasks through stable co-occurrence, intercellular interactions, and communication, which can hardly be attained by free-living microbes (Moons et al. 2009). Biofilms impact different fields of study, from clinical and plant infection to beneficial colonization in humans and plants. Compared to monospecies biofilms, multispecies biofilms with a complex mix of prokaryotes and eukaryotes are predominant and, in general, exhibit higher stability (Costerton et al. 1987; León-Romero et al. 2015; Flemming and Wuertz 2019).

Biofilm formation is a complex and dynamic process where microbial communities are embedded in a self-produced matrix with cell densities ranging between 10^8–10^{11} cells/gram wet weight. Matrix components contribute > 90% of dry weight, whereas microbial dry weight accounts for < 10% (Flemming et al. 2016). This matrix forms during the attachment stage, stabilizing the interacting microbes within the biofilm (Jamal et al. 2018). The properties of biofilm microbial communities are distinct from those of free-living microbes due to their presence in a matrix (Berlanga and Guerrero 2016). Biofilms in natural settings develop viable but nonculturable subpopulation phenotypes that are successful in situ but sometimes fail in culture-based isolations. Predicting the conditions under which microbial biofilm communities are successful or benefit the host remains challenging in microbiome research.

7.4 Key Factors Governing Biofilm Formation

The formation of beneficial biofilms is a complex process mainly divided into four stages: (1) initial reversible attachment: adhesion of planktonic cells to the surface; (2) irreversible attachment: formation of microcolonies; (3) maturation stage: macrocolonies formation and development of matured biofilm; and (4) dispersal stage: detachment of microbial cells. Different factors influence these steps, and their understanding is crucial for both applied and fundamental aspects of microbial research (Fig. 7.1).

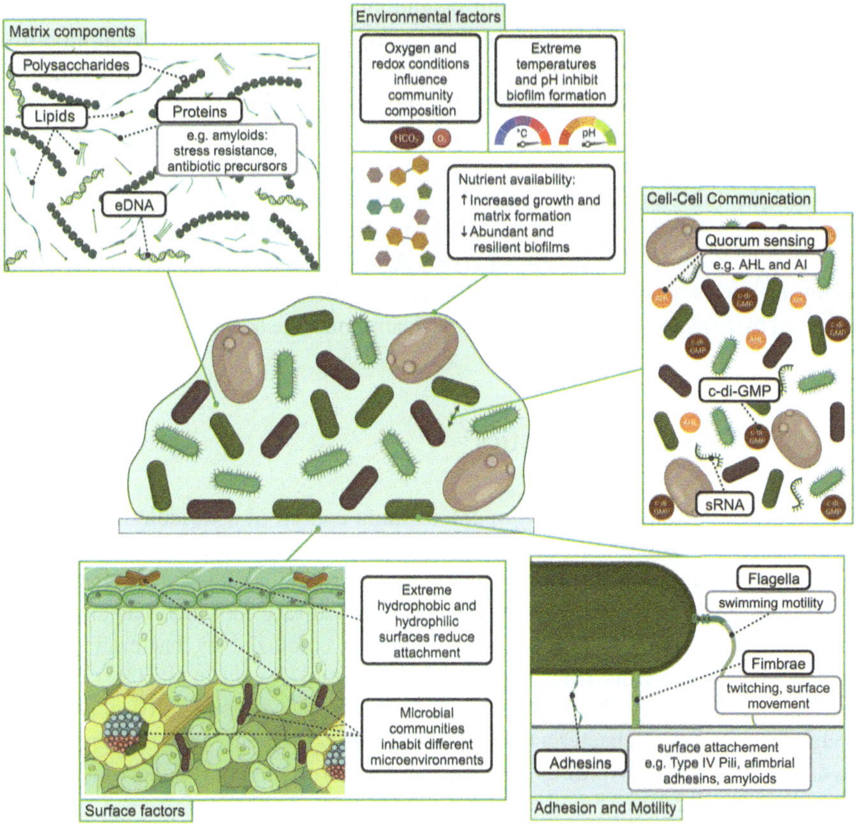

Fig. 7.1 Overview of the key factors influencing biofilm formation. Biofilm formation, whether in plants or on other surfaces, is a complex process, influenced by many biotic and abiotic factors. Biofilm community composition, attachment, and persistence are shaped by environmental conditions and surface structure. Biofilm structure is, furthermore, determined by the producing microbes, their associated matrix, communication and attachment genes. *eDNA* extracellular DNA, *AHL* acyl-homoserine lactone, *AI* autoinducer (e.g., autoinducing Peptides AIP, AI-2), *c-di-GMP* cyclic diguanylate, *sRNA* small RNA. (Graphic created with BioRender.com)

7.5 Bacterial Mobility and Adhesins

Bacterial mobility is considered an important factor for biofilm formation though its mechanism differs between strains and it is mostly required for the earlier stages of biofilm formation such as the initial attachment (Ahmad et al. 2017a). Bacterial mobility is mostly facilitated by two types of proteins: flagella and fimbriae. Flagella are long, spiral structures on bacterial cell surfaces that facilitate movement in liquid media. They act like propellers, allowing bacteria to swim and navigate the liquid environment (Nakamura and Minamino 2019). On the other hand, fimbriae are short, straight projections on bacterial cytoplasmic membranes that enable limited, twitching motility of bacteria on surfaces. They also facilitate bacterial attachment and can be involved in the crawling or gliding motion on solid surfaces (Jin and Marshall 2020). Both flagella and fimbriae play crucial roles in bacterial mobility, allowing bacteria to adapt and respond to different environments. Bacterial cells first establish contact with a substrate surface through adhesins, which are specific cell membrane proteins that play a crucial role in the initial attachment. Once the initial interaction forms, adhesins help maintain a stable connection between the bacterial cell and the substrate surface. Adhesins or adhesin-like structures can be of a fimbrial nature, like type IV pili, which aid in the adhesion of several Gram-positive bacteria (Piepenbrink and Sundberg 2016). Afimbrial adhesins are often secreted by Gram-negative bacteria via the type V secretion system. Type VIII secretion systems are used by *Escherichia coli* to secrete curli, amyloid-like adhesins (Leo et al. 2012).

7.6 Surface Characteristics

Plant surfaces available for biofilm colonization consist of a variable range of different microenvironments. Each type of plant tissue has a different surface structure, and bacteria must adapt their biofilm-forming strategies to the unique conditions and characteristics. The nature of the surface to which bacteria attach shapes biofilm development. In plants, surfaces colonized by bacterial biofilms include the leaves, stems, roots, seeds, and vasculature (Ramey et al. 2004). Bacteria may prefer specific surface characteristics based on hydrophobicity, charge, and roughness. In general, bacteria tend to form biofilms on moderately hydrophobic or hydrophilic surfaces, avoiding extremes, and prefer relatively rougher surfaces for increased attachment (Ahmad et al. 2017b; Zheng et al. 2021).

7.7 Cell–Cell Communication and Regulation

Formation of biofilms requires many individual microbial cells to work together toward the same goal. Microbes interact with each other through specialized communication systems, known broadly as quorum sensing (QS). Bacteria use QS molecules to communicate and coordinate with each other in monospecies and multispecies communities. QS allows bacteria to sense their population density and regulate gene expression accordingly (Miller and Bassler 2001). One of the best characterized QS systems consists in the response against *N*-acyl-homoserine lactone molecules (AHLs). It is used, among others, by symbiotic rhizobia and during biofilm formation in plant-associated pseudomonads (Whitehead et al. 2001; Loh et al. 2002). Shrestha et al. (2020) showed that long-chain AHLs induce expression of several plant defense genes and also resistance to the bacterial pathogen *P. syringae* (Shrestha et al. 2020). In addition to AHLs, Gram-negative and Gram-positive bacteria engage in interspecific communication through autoinducer-2 (AI-2) molecules (Li and Nair 2012).

Apart from QS, other regulatory factors that govern biofilm formation include cyclic diguanosine-5′-monophosphate (c-di-GMP) signaling and small non-coding RNA (sRNA) release (Chambers and Sauer 2013; Fazli et al. 2014). These both bind and interact with a large number of effector molecules and gene targets, modifying the expression of many genes related to motility and biofilm formation. Ultimately, sRNAs and c-di-GMP, together with AI-2- or AHL-based QS, determine the balance between planktonic and biofilm stages in bacteria.

7.8 Extracellular Polymeric Substance Production

Bacteria in biofilms produce extracellular polymeric substances (EPS), which consist of an intricate mix of lipids, polysaccharides, extracellular DNA (eDNA), and proteins (Yin et al. 2019). The composition of EPS can vary between species and influences the organization of the biofilm and its protective effect on plants (Fong and Yildiz 2015; Heredia-Ponce et al. 2021a, b). The EPS enables the adhesion and aggregation of cells, as well as retention of nutrients and water within the biofilm matrix (Flemming and Wingender 2010).

A major part of the proteinaceous component of the EPS is composed of proteins of amyloidogenic nature, meaning that they spontaneously form molecular amyloid structures. Amyloids play key roles in the integrity and functionality of biofilms, including those that are found in plants (Garcia et al. 2011; Gómez-Pérez et al. 2021). The main function of amyloids is the establishment of a scaffold whereupon the rest of the EPS components and cells can be assembled. This is due to secreted amyloidogenic monomers aggregating and forming robust amyloid fibrils. Fibrils provide stability and resilience against environmental pressures since the mature filaments are highly resistant to chemical and thermal denaturation, as well as mechanical stress (Taglialegna et al. 2016). Furthermore, the monomeric state of some amyloids exhibits toxic properties. The amyloid fibrils in the biofilm can, therefore, also act as reservoirs for antimicrobial peptides, which may be liberated if needed (Syed and Boles 2014). This can be triggered by changes in the surrounding areas such as pH shifts, potentially caused by the arrival of bacterial pathogens. Although most components of the EPS are variable, amyloids are a ubiquitous central part of biofilm formation for many microbes, most recently described for cyanobacterial biofilms (Erskine et al. 2018; Frenkel et al. 2023). The synthesis of amyloids often entails the development of dedicated systems for their production since they carry the risk of cytoplasmic polymerization, leading to cell toxicity (Taglialegna et al. 2016).

Apart from playing several roles as part of the biofilm structure, such as nucleation of amyloids and binding to hydrophobic surfaces, eDNA also contributes to the exchange of genetic information (Heijstra et al. 2009; Peterson et al. 2013; Gallo et al. 2015). The release of eDNA, similar to other aspects of the formation of biofilms, is promoted by QS molecule signaling (Spoering and Gilmore 2006). The ability of cells within a biofilm to integrate eDNA enables genetic material exchange related to, for example, resistance to a particular antibiotic, allowing certain phenotypes to quickly spread throughout the biofilm population (Bender et al. 2022). Finally, other components of the EPS such as polysaccharides and lipids are also essential for ensuring the structural integrity and cohesion of the matrix (Karygianni et al. 2020).

7.9 Environmental Conditions

Various abiotic factors influence biofilm formation, including nutrient availability and environmental conditions. Nutrient-rich conditions may promote biofilm development as bacteria utilize resources for growth and matrix production. However, in general, biofilms formed under nutrient-scarce conditions, such as in soil, are more prevalent and resilient than those formed under abundance conditions, likely as a way to promote survival in adverse conditions (Petrova and Sauer 2012).

Temperature and pH are both critical environmental factors affecting biofilm formation. The optimal temperature for the production of biofilms varies per strain, but it is typically within the range of 25–37 °C. A lower temperature can decrease bacterial motility, reducing the first attachment phase and biofilm dispersal. The pH of the environment can affect the charge of surfaces, potentially reducing attachment, particularly in the initial phases of biofilm formation (Palmer et al. 2007). Generally, bacterial biofilm production is optimal under near-neutral pH. Extreme conditions relating to temperature or pH can both negatively affect biofilm production.

Relating to the microenvironmental conditions within the biofilm, oxygen and redox states can vary, influencing the diversity and abundance of bacterial species. Different bacteria may thrive in specific microenvironments within the biofilm structure, which is particularly relevant concerning multispecies biofilms (Mitri et al. 2011). In turn, biofilms have been shown to also directly impact their own and their host's environmental conditions. For example, in a grass leaf, bacteria within biofilms have been shown to increase O_2 levels and pH, decreasing the photosynthetic potential of the plant (Zhang et al. 2022).

7.10 Beneficial Biofilms

The significance of beneficial biofilms associated with a plant host lies in their protective role for the contained bacteria against environmental stresses, including UV radiation, pH fluctuation, osmotic stress, and dehydration. The adaptive capability of biofilms enhances the resilience of bacterial communities, allowing them to thrive in diverse ecological niches. Though not just the microbes but also the host plant profit from the biofilm formation. The role of aboveground and belowground plant-associated biofilms in bacterial disease management is complex and has been the subject of increased research interest in recent years (Fig. 7.2). To study plant biofilms, it is often convenient to separately look at the microbial communities associated with the aboveground (phyllosphere) and belowground (rhizosphere) parts of the plant (Fig. 7.3). In both habitats, a number of beneficial biofilm-forming bacteria have been identified (Table 7.1).

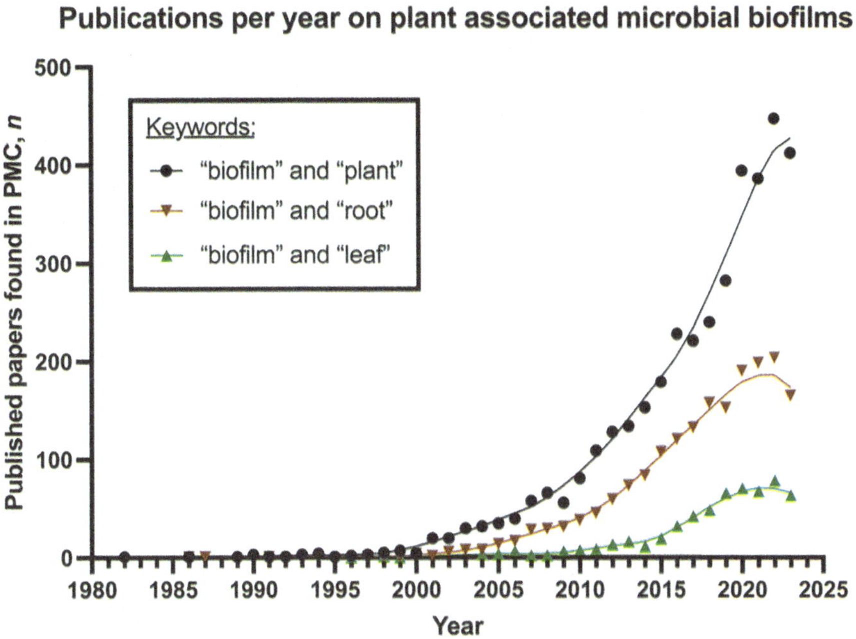

Fig. 7.2 Publications per year relating to plant-associated biofilms. The plots depict the number of publications found in the PMC database with the keywords "biofilm" together with either "plant," "root," or "leaf" in the title and abstract per year (as of December 4th, 2023). The number of publications (*n*) is represented by each dot of the scatter plot and trend lines were drawn with local regression based on the LOWESS method

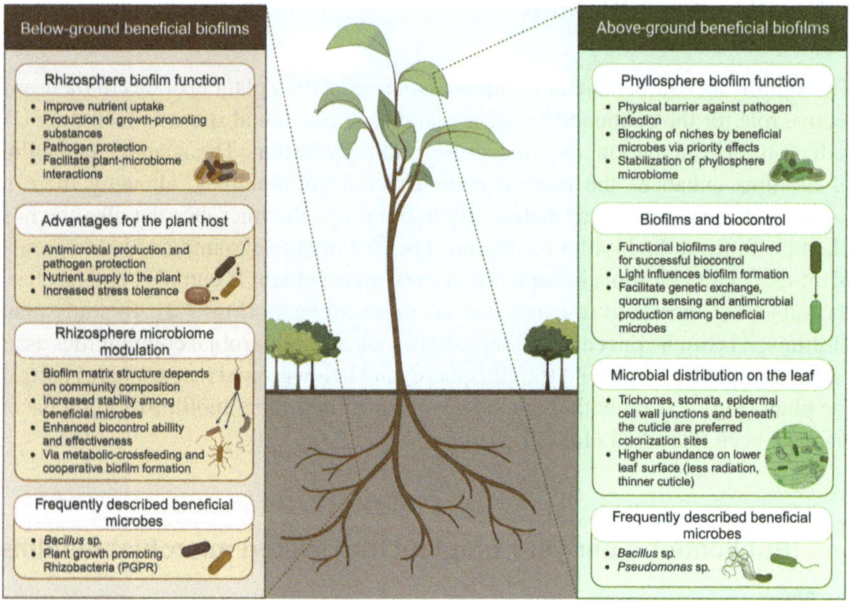

Fig. 7.3 Comparison of belowground and aboveground beneficial biofilms in plants. In both environments, beneficial biofilms are important for plant health and plant growth. Aboveground and belowground biofilms both offer pathogen protection and stabilize their respective microbial communities. While belowground biofilms have primarily been reported to enhance nutrient uptake, aboveground biofilms are known to adapt their benefits to a varying environment. (Graphic created with BioRender.com)

Table 7.1 Summary of important biofilm-forming plant-beneficial microbes

Microbe	Host plant	Colonization site	Reference
Azospirillum sp.	Cereals	Root	Burdman et al. (2000)
Bacillus amyloliquefaciens	Cucumber, maize, tomato, barley	Rhizosphere, root, seed, soil	Liu et al. (2014), Kasim et al. (2016), Wang et al. (2019), Hazarika et al. (2021)
Bacillus aryabhattai	Tomato	Root	Haque et al. (2020)
Bacillus subtilis	Tomato, *Arabidopsis*, lettuce, rice	Root, rhizosphere, seed	Luo et al. (2015), Arnaouteli et al. (2021), Sarti et al. (2023)
Bacillus pumilus	Masson pine	Roots	Zhu et al. (2020)
Bacillus velezensis	Tomato	Root, leaf	Stoll et al. (2021)
Bacillus sp.	Maize	Leaf	Fessia et al. (2022b)
Paenibacillus polymyxa	Peanut	Seed	Haggag and Timmusk (2008)
Pseudomonas aeruginosa	Tomato	Root	Ghadamgahi et al. (2022)
Pseudomonas chlororaphis	Avocado	Root	Heredia-Ponce et al. (2021a, b)

(continued)

Table 7.1 (continued)

Microbe	Host plant	Colonization site	Reference
Pseudomonas putida	Maize, chickpea, tomato	Rhizosphere, root	Espinosa-Urgel et al. (2002), Ramos-González et al. (2005), Srivastava et al. (2008), Sun et al. (2017)
Pseudomonas synxantha, *Pseudomonas brassicacearum*	*Arabidopsis*	Root	Harting et al. (2021)
Pseudomonas sp.	Tea	Root	Hazarika et al. (2021)
Rhizobium leguminosarum	Clover	Root nodulation	Janczarek et al. (2010)

7.11 Belowground Beneficial Biofilms

The rhizosphere is the region of soil influenced by root secretions and comprises a vast diversity of microorganisms. Being a microbially dense and diverse habitat, constant competition for niches and nutrients among root- and rhizosphere-associated beneficial microbes and pathogens is essential for the mechanisms behind protection (Lugtenberg and Kamilova 2009). It is a critical zone for beneficial biofilms that can improve nutrient uptake, produce growth-promoting substances, and compete with pathogenic organisms for resources. Root biofilms are a complex and coexisting community of microorganisms that can form a protective layer on the root surface that interacts directly with the plant (Pandit et al. 2020). The biofilm-structured microbial communities attach to the plant root surface forming a three-dimensional matrix (Castiblanco and Sundin 2016). Microbe–microbe interactions are essential for the stability and functional potential of root biofilms. These interactions can help maintain the structural integrity of the biofilm and regulate the diversity or evenness of the community of microbes that can carry out various functions (Ajijah et al. 2023). For instance, some microbes focus on producing EPS that help bind the biofilm together. Besides, others produce antimicrobials that inhibit the growth of competing or "cheater" microorganisms. In addition, some microorganisms form symbiotic relationships with plant roots, providing the plant with essential nutrients and helping to promote plant growth. This specialization of bacterial cells within biofilms is known as division of labor and is a staple of biofilms produced by many different bacteria (van Gestel et al. 2015). Root biofilms contribute to plant health by, for example, providing a protective barrier against pathogens, improving water and nutrient uptake, and facilitating the exchange of signaling molecules between the plant and the microbiome (Bogino et al. 2013). Finally, the root biofilm can also influence the plant's tolerance to stress (Timmusk and Nevo 2011).

The successful root colonization and community assemblage by rhizosphere and root-associated microbial communities depend on critical steps, including uptake of plant metabolites, chemotaxis, cell motility, and biofilm formation (Trivedi et al.

2020). The colonized beneficial microbes on the root surface are crucial to ward off soilborne pathogens from invading the root (Soni and Keharia 2021). These microbes in the rhizosphere can directly inhibit pathogenic organisms by producing antibiotics or competing for space and nutrients. *Bacillus* is the most studied beneficial microbial genus and usually colonizes plant roots as cell aggregates enclosed in a self-produced extracellular matrix, a biofilm (Arnaouteli et al. 2021). Multiple studies using fluorescence reporter and microscopy techniques have shown that biofilm formation on roots by *Bacillus* strains involved in competition with plant pathogens leads to disease suppression (Chen et al. 2013; Pandin et al. 2017; Molina-Santiago et al. 2019; Xu et al. 2019; Berlanga-Clavero et al. 2022).

Competition for utilizing root and rhizosphere nutrients also significantly confers pathogen suppression (Xia et al. 2022). One of the essential micronutrients is iron, which is highly abundant in soils but has low solubility and is unsuitable for microbial uptake, particularly in acidic environments (Lugtenberg and Kamilova 2009). Therefore, iron bioavailability is a limiting factor and promotes nutrient competition among living microbes. Some microorganisms synthesize specialized iron-chelating compounds, siderophores, that restrict the available iron to pathogens and, hence, their growth (Dimopoulou et al. 2021; Ghazy and El-Nahrawy 2021; Lahlali et al. 2022). Members of *Bacillus* species synthesize iron-chelating compounds, such as the bacillibactin siderophore (Yu et al. 2011).

Some *Bacillus* strains interact with plant roots, elicit induced systemic resistance (ISR), activate host defense responses, and induce the resistance of the entire plant against pathogenic bacteria and other pathogens (Kloepper et al. 2004). Examples of molecules from *Bacillus* that function as "elicitors" belong to lipopeptides, polyketides, exopolysaccharides, and volatile organic compounds. These activate salicylic acid, jasmonic acid (JA), or ethylene (ET) signaling pathways involved in priming the JA/ET-dependent defense-related genes, callose deposition, and stomatal closure, and help the plant's resistance against pathogen infection (Niu et al. 2012; Gowtham et al. 2018; Vanthana et al. 2019). Root inoculation of *B. velezensis* SQR9 activated plant systemic resistance in *Arabidopsis* and reduced the disease incidence of the foliar pathogens *P. syringae* pv. tomato (Pst DC3000) and *B. cinerea* (Wu et al. 2018a, b).

Plant growth-promoting rhizobacteria (PGPR) associations in the rhizosphere and phyllosphere protect plants from pathogens and promote plant growth by facilitating robust biofilm formation and surfactin production (Bais et al. 2004; Hashem et al. 2019). By colonizing the root surface, some bacteria trigger ISR in aerial portions of the plant (Kloepper et al. 2004). Research showed that pathogens like Pst DC3000 can induce root secretion in *Arabidopsis thaliana*, recruiting beneficial biofilm-formers such as *Bacillus subtilis*, which can then aid the plant in controlling the pathogen (Rudrappa et al. 2008). *P. syringae* infection in *A. thaliana* leaves imparts chemical signals belowground through malic acid as root exudate secretions. Further, *P. syringae* stimulates the root secretions of the tricarboxylic acid cycle pathway intermediate, L-malic acid, which selectively attracts beneficial rhizobacterium *B. subtilis* FB17 by promoting its root colonization and biofilm formation, which is dependent on the sensor histidine kinase KinD, specifically on the extracellular CACHE domain (Chen et al. 2012).

Another well-studied biofilm-forming genus that has been associated with a beneficial effect for the plant is *Pseudomonas*. Strain KT2440 of *P. putida* forms biofilms on maize roots, resulting in plant growth promotion and pathogen protection (Espinosa-Urgel et al. 2002). Another strain of *P. putida*, A1, has been shown to decrease bacterial wilt disease in tomato seedlings and to produce biofilms in vitro (Sun et al. 2017). *P. aeruginosa* strain FG106, isolated from the tomato rhizosphere, forms biofilms and exhibits broad-spectrum pathogen control, including antagonism against fungal diseases caused by *Alternaria alternata* and *B. cinerea* (Ghadamgahi et al. 2022).

Biofilm formation as monoassociation and in cooperative interaction of specific taxa can enhance the stability and effectiveness of these beneficial microbes. For example, a biofilm-deficient mutant strain of *Bacillus pumilus* was shown to have decreased biocontrol capacity against damping-off disease in pine when compared to the same strain with the capacity to form biofilms (Zhu et al. 2020). Some studies have shown that *Bacillus* species can modulate the native rhizosphere microbiome, which involves the stimulation of specific taxa exhibiting direct biocontrol activity and cooperative interactions with other beneficial microbes. Collectively, these modulations improve plant disease suppression (Tao et al. 2020; Moreira et al. 2023). One such example is the cooperation by biofilm formation of beneficial *Bacillus* and *Pseudomonas* strains through metabolic cross-feeding where *Bacillus* assists *Pseudomonas* by providing valeric acid and levulinic acid and *Pseudomonas* in return provides branched-chain amino acids (Sun et al. 2021). In another study, cocultivation of *B. velezensis* with helper bacteria *Flavobacterium* showed enhanced biofilm formation under in vitro assay. Further, the study revealed that *Flavobacterium* improved *B. velezensis* biocontrol efficiency against *Ralstonia solanacearum*, causing tomato bacterial wilt to decrease, reducing the colonization of *R. solanacearum* in the rhizosphere, and increasing transcription of plant defense gene PR1α in tomato (Wang et al. 2023).

7.12 Aboveground Beneficial Biofilms

Compared to the rhizosphere, microbes exhibit a nonuniform distribution in the phyllosphere. In particular, on leaf surfaces, bacterial cells preferentially attach to specific locations, such as the base of trichomes, stomata, epidermal cell wall junctions, grooves along veins, depressions, and beneath the cuticle (Beattie and Lindow 1999; Lindow and Brandl 2003; Chaudhry et al. 2020). Bacteria are more abundant on the lower leaf surface, potentially due to reduced radiation exposure, a higher density of stomata or trichomes, and a thinner cuticular layer, creating favorable conditions for their survival and growth, increasing their likelihood of persistence compared to upper leaf parts (Karamanoli et al. 2012). There is a general lack of experimental evidence regarding resistance against biotic stresses, particularly related to in vivo studies as outlined in (Fessia et al. 2022a). However, phyllosphere biofilms, analogously to rhizosphere biofilms, could act as a protective mechanism for plant growth and increased productivity. The main mechanism by which this could take place is by acting as a physical barrier that blocks biological niches or seals plant surface injuries potentially sought after by a pathogen to start an infection. This is related to the priority effect in ecology, which postulates that early-arriving organisms have a fitness advantage that, after the community has stabilized, is hard for newcomers to disrupt (Debray et al. 2022). Consistent with the latter, the higher resistance to infection of stable microbial communities (as opposed to a dysbiosis state) is known for many host organisms, including crop plants (Shade et al. 2012; Fu et al. 2019).

Biofilms originating from bacteria in the phyllosphere have the potential to act as a biocontrol agent, as shown for some *Bacillus* strains in maize (Fessia et al. 2022b). For a *Bacillus subtilis* strain, it was shown that the biocontrol activity depends on the formation of a functional biofilm to effectively antagonize bacterial pathogens in the phyllosphere (Zeriouh et al. 2014). Similar observations were made for *Pseudomonas piscium* in the wheat head microbiome, where biofilm formation was necessary to induce the production of antimicrobial substances (Chen et al. 2018). An effective establishment on the plant is key for the formation of viable biofilms composed of bacteria with biocontrol activity. For this, the influence of abiotic factors should not be underestimated. One essential difference between the phyllosphere and soil or rhizosphere is the exposure to light. It has been shown that non-phototrophic, biocontrol-associated bacteria in the phyllosphere respond to access and quality of light. Certain wavelengths, particularly in the UV range, can promote the biofilm formation of these bacteria in the phyllosphere (Karlsson et al. 2023). Factors like light are relatively easy to control during certain kinds of agricultural production as light conditions in greenhouses could be adjusted to promote biofilm formation of desired microbes and thereby enhance biocontrol activity.

Additionally, biofilms promote genetic exchange among their members, enabling the faster spread of resistance genes among the population through horizontal gene transfer, increasing their chance of fending off invaders (Madsen et al. 2012). The latter is enhanced by QS within the members of the biofilm, which promotes a distribution of roles, allowing a subset of specialized cells to focus on the production of antimicrobials, ultimately increasing their biocontrol potential (van Gestel et al. 2015). However, drawbacks related to the lack of dispersal of biofilm-prone biocontrol agents could limit their effectiveness in practice (Wei et al. 2016). Nevertheless, looking forward, the elucidation of the dynamics of these biocontrol mechanisms remains a promising avenue for future research, offering insights into their practical applications in agriculture and plant protection.

7.13 Challenges and Considerations: Harnessing Biofilms for Plant Health

Despite progress in recent years, harnessing beneficial biofilms for effective disease management and general plant health promotion presents a multifaceted landscape, marked by challenges and exciting prospects (Dewi et al. 2023). A critical challenge is the need for a comprehensive understanding of the mechanisms underlying biofilm-mediated biocontrol. The intricate interplay between the biofilm-forming microbes and pathogenic bacteria within the plant microbiome demands in-depth studies to unravel the specific biochemical and molecular processes involved in pathogen inhibition (Patil et al. 2022). Furthermore, application-oriented studies are needed to ensure biofilm treatments are successful in the field with minimal disruption of the native ecosystem.

7.14 Biofilm Stability and Specificity

As described earlier, environmental conditions such as soil type, temperature, pH, and nutrient availability influence biofilm stability (Ansari et al. 2017). Understanding the impact of these factors on biofilm formation and persistence is necessary for successful disease management. For example, pathogenic bacteria may also form biofilms on plant surfaces and compete with beneficial biofilms (Carezzano et al. 2023). For the persistence of beneficial biofilms and the prevalence of their producing microbes, these environmental conditions should be kept within optimal ranges to prevent disruption and displacement by pathogens.

Identifying and selecting microbial strains with beneficial traits and pathogen-suppressing properties is another crucial point. Ensuring that the biofilm-forming microbes have a higher affinity and dominance in the habitat than target pathogens can enhance the efficacy of disease suppression (Afridi et al. 2022). For example, this can be achieved through a combination of compatible biocontrol agents and certain chemical pesticides (Ons et al. 2020; Djaenuddin et al. 2021). In this case, the pesticides carve out a niche for the biocontrol agents to colonize and prevent loss of beneficial bacteria.

A complex multispecies biofilm, including beneficial bacteria and fungi, could contribute to a more robust defense against pathogens. Consideration must be given to the host plant's compatibility with the biofilm-forming microbes to avoid negative effects. Most bacteria have a narrow host range and only colonize certain areas within their host (Tovi et al. 2019). To ensure successful colonization of beneficial microbes, treatment strategies have to be adjusted to the individual plant species.

7.15 Biofilm Application

Determining the optimal timing for biofilm application is essential for effective disease management. Factors such as plant growth and disease development stages should be taken into consideration. This challenge overlaps with the optimization of environmental conditions as these also determine application success (Lopes et al. 2021). Developing effective formulations for delivering biofilm-forming microbes to plant surfaces is challenging. Inoculation success depends strongly on the individual microbe, as well as on the plant host (Lopes et al. 2021). Antagonistic interactions within the inoculation members must be taken into account to ensure effectiveness (Xu et al. 2011). Additionally, formulations should ideally also have a long microbial shelf life and ensure host adherence and biofilm formation in natural settings.

Developing reliable methods for monitoring the establishment and effectiveness of beneficial biofilms in the field is essential for the success of the treatment. This includes nondestructive techniques for assessing biofilm coverage and activity (Romano et al. 2020). Most methods in use have their downsides, hence the appropriate method has to be chosen based on individual needs. Current methods include culture isolation (Morris et al. 1998) microscopy (Velmourougane et al. 2017), quantitative PCR (Stets et al. 2015), and next-generation or whole-genome sequencing (Guzmán et al. 2020). Continuous monitoring of plant health and incidence (Roper et al. 2021) is necessary to evaluate the long-term impact of biofilm applications on disease suppression and plant productivity. Modern techniques suggest using "phytosensor" plants, which can immediately alter their phenotype upon pathogen recognition (Mazarei et al. 2008). Other simple and noninvasive methods include imaging (Mahlein 2016) and spectroscopy (Altangerel et al. 2017). Additionally, approaches have been described based on measuring the electrical conductivity of nutrient solutions (Jeon et al. 2017).

Biofilm-forming microbes should be cautiously introduced to avoid disrupting the natural microbial balance in the ecosystem (Cook 1996). Risks include gene transfer and uncontrolled spread of, for example, antibiotic resistances among other microbes native to an ecosystem. Other concerns may relate to the production of unintended secondary metabolites, causing unwanted antagonism among the microbial community and leading to elicitation of plant defenses (Woo and Lorito 2007). Assessing the potential environmental impact and long-term consequences of biofilm applications is essential as well as challenging in itself, and is an area of ongoing research and legal regulation (Liu et al. 2022; Vassileva et al. 2022).

7.16 Conclusions and Future Perspective

Biofilm research has grown substantially over the past few decades. Initially centered on preventive and destructive strategies against the detrimental effects, mainly in the medical domain, the focus has shifted toward harnessing biofilms for biotic and abiotic stress protection. This presents a promising avenue for sustainable agriculture. The diverse outcomes of these biofilms, ranging from disease suppression to enhanced nutrient uptake, underscore their potential to revolutionize plant health management. As research advances, it is evident that understanding the intricate microbial interactions within biofilms and their influence on plant hosts is pivotal. Notably, research has expanded to explore monospecies biofilms and artificial synthetic communities that combine diverse metabolic activities. Future directions in this field should emphasize the exploration of microbial diversity within biofilms and the integration of biofilm engineering to tailor these microbial communities for optimal plant–microbe interactions. Additionally, they should not neglect the development of monitoring techniques and investigation into climate-resilient bacterial biofilms. Global collaborative initiatives will further accelerate progress, facilitating the commercialization and widespread adoption of beneficial biofilm technologies in agriculture. The convergence of these future directions holds the key to unlocking the full potential of beneficial biofilms, offering sustainable solutions for plant protection, and contributing to the resilience of agricultural systems in the face of a growing global demand for food.

Acknowledgments The authors warmly acknowledge Prof. Eric Kemen. VC acknowledges Alexander von Humboldt-Stiftung, Cluster of Excellence Controlling Microbes to Fight Infections (CMFI) University of Tübingen for support. DG-P and MS would like to acknowledge support from the European Research Council (ERC) under the DeCoCt research program (grant agreement: ERC-2018-COG 820124).

References

Abramovitch RB, Martin GB (2004) Strategies used by bacterial pathogens to suppress plant defenses. Curr Opin Plant Biol 7:356–364. https://doi.org/10.1016/j.pbi.2004.05.002

Afridi MS, Ali S, Salam A, Terra WC, Hafeez A, Sumaira AB, AlTami MS, Ameen F, Ercisli S, Marc RA, Medeiros FHV, Karunakaran R (2022) Plant microbiome engineering: hopes or hypes. Biology 11:1782. https://doi.org/10.3390/biology11121782

Ahmad I, Khan MS, Altaf MM, Qais FA, Ansari FA, Rumbaugh KP (2017a) Biofilms: an overview of their significance in plant and soil health. Biofilms in plant and soil health pp.1–25. doi:https://doi.org/10.1002/9781119246329.ch1

Ahmad M, Liu S, Mahmood N, Mahmood A, Ali M, Zheng M, Ni J (2017b) Effects of porous carrier size on biofilm development, microbial distribution and nitrogen removal in microaerobic bioreactors. Bioresour Technol 234:360–369. https://doi.org/10.1016/j.biortech.2017.03.076

Ajijah N, Fiodor A, Pandey AK, Rana A, Pranaw K (2023) Plant growth-promoting bacteria (PGPB) with biofilm-forming ability: a multifaceted agent for sustainable agriculture. Diversity 15:112. https://doi.org/10.3390/d15010112

Aktar W, Sengupta D, Chowdhury A (2009) Impact of pesticides use in agriculture: their benefits and hazards. Interdiscip Toxicol 2:1–12. https://doi.org/10.2478/v10102-009-0001-7

Altangerel N, Ariunbold GO, Gorman C, Alkahtani MH, Borrego EJ, Bohlmeyer D, Hemmer P, Kolomiets MV, Yuan JS, Scully MO (2017) In vivo diagnostics of early abiotic plant stress response via Raman spectroscopy. Proc Natl Acad Sci 114:3393–3396. https://doi.org/10.1073/pnas.1701328114

Ansari FA, Jafri H, Ahmad I, Abulreesh HH, (2017) Factors affecting biofilm formation in in vitro and in the rhizosphere. Biofilms in plant and soil health, pp.275–290. https://doi.org/10.22573/spg.023.978-93-90357-07-9/2

Arnaouteli S, Bamford NC, Stanley-Wall NR, Kovács ÁT (2021) Bacillus subtilis biofilm formation and social interactions. Nat Rev Microbiol 19:600–614. https://doi.org/10.1038/s41579-021-00540-9

Bais HP, Fall R, Vivanco JM (2004) Biocontrol of *Bacillus subtilis* against infection of arabidopsis roots by pseudomonas syringae is facilitated by biofilm formation and surfactin production. Plant Physiol 134:307–319. https://doi.org/10.1104/pp.103.028712

Beattie GA, Lindow SE (1999) Bacterial colonization of leaves: a Spectrum of strategies. Phytopathology 89:353–359. https://doi.org/10.1094/phyto.1999.89.5.353

Bender N, Hennes M, Maier B (2022) Mobility of extracellular DNA within gonococcal colonies. Biofilms 4:100078. https://doi.org/10.1016/j.bioflm.2022.100078

Berlanga M, Guerrero R (2016) Living together in biofilms: the microbial cell factory and its biotechnological implications. Microb Cell Factories 15:165. https://doi.org/10.1186/s12934-016-0569-5

Berlanga-Clavero MV, Molina-Santiago C, Caraballo-Rodríguez AM, Petras D, Díaz-Martínez L, Pérez-García A, Vicente A de, Carrión VJ, Dorrestein PC, Romero D (2022) *Bacillus subtilis* biofilm matrix components target seed oil bodies to promote growth and anti-fungal resistance in melon. Nat Microbiol 1–15. doi:https://doi.org/10.1038/s41564-022-01134-8

Bogino PC, Oliva MDLM, Sorroche FG, Giordano W (2013) The role of bacterial biofilms and surface components in plant-bacterial associations. Int J Mol Sci 14:15838–15859. https://doi.org/10.3390/ijms140815838

Burdman S, Okon Y, Jurkevitch E (2000) Surface characteristics of *Azospirillum brasilense* in relation to cell aggregation and attachment to plant roots. Crit Rev Microbiol 26:91–110. https://doi.org/10.1080/10408410091154200

Carezzano ME, Rovey MFP, Cappellari LDR, Gallarato LA, Bogino P, Oliva MDLM, Giordano W (2023) Biofilm-forming ability of Phytopathogenic bacteria: a review of its involvement in plant stress. Plan Theory 12:2207. https://doi.org/10.3390/plants12112207

Castiblanco LF, Sundin GW (2016) New insights on molecular regulation of biofilm formation in plant-associated bacteria. J Integr Plant Biol 58:362–372. https://doi.org/10.1111/jipb.12428

Chambers JR, Sauer K (2013) Small RNAs and their role in biofilm formation. Trends Microbiol 21:39–49. https://doi.org/10.1016/j.tim.2012.10.008

Chaudhry V, Runge P, Sengupta P, Doehlemann G, Parker JE, Kemen E (2020) Shaping the leaf microbiota: plant–microbe–microbe interactions. J Exp Bot 72:36–56. https://doi.org/10.1093/jxb/eraa417

Chen Y, Cao S, Chai Y, Clardy J, Kolter R, Guo J, Losick R (2012) A *Bacillus subtilis* sensor kinase involved in triggering biofilm formation on the roots of tomato plants. Mol Microbiol 85:418–430. https://doi.org/10.1111/j.1365-2958.2012.08109.x

Chen F, Gao Y, Chen X, Yu Z, Li X (2013) Quorum quenching enzymes and their application in degrading signal molecules to block quorum sensing-dependent infection. Int J Mol Sci 14:17477–17500. https://doi.org/10.3390/ijms140917477

Chen Y, Wang J, Yang N, Wen Z, Sun X, Chai Y, Ma Z (2018) Wheat microbiome bacteria can reduce virulence of a plant pathogenic fungus by altering histone acetylation. Nat Commun 9:3429. https://doi.org/10.1038/s41467-018-05683-7

Chen H, Chen J, Zhao Y, Liu F, Fu ZQ (2022) *Pseudomonas syringae* pathovars. Trends Microbiol 30:912–913. https://doi.org/10.1016/j.tim.2022.03.002

Cook RJ (1996) Assuring the safe use of microbial biocontrol agents: a need for policy based on real rather than perceived risks. Can J Plant Pathol 18:439–445. https://doi.org/10.1080/07060669609500602

Costerton JW, Cheng KJ, Geesey GG, Ladd TI, Nickel JC, Dasgupta M, Marrie TJ (1987) Bacterial biofilms in nature and disease. Ann Rev Microbiol 41:435–464. https://doi.org/10.1146/annurev.mi.41.100187.002251

Debray R, Herbert RA, Jaffe AL, Crits-Christoph A, Power ME, Koskella B (2022) Priority effects in microbiome assembly. Nat Rev Microbiol 20:109–121. https://doi.org/10.1038/s41579-021-00604-w

Dewi WS, Amalina DD, Romadhon MR (2023) Microbial biofilm for soil health, plant growth, and productivity under multi stress. A review. IOP Conf Ser Earth Environ Sci 1162:012008. https://doi.org/10.1088/1755-1315/1162/1/012008

Dimopoulou A, Theologidis I, Benaki D, Koukounia M, Zervakou A, Tzima A, Diallinas G, Hatzinikolaou DG, Skandalis N (2021) Direct antibiotic activity of Bacillibactin broadens the biocontrol range of *Bacillus amyloliquefaciens* MBI600. mSphere 6:00376–21. https://doi.org/10.1128/msphere.00376-21

Djaenuddin N, Sebayang A, Nonci N, Muis A (2021) Compatibility of biocontrol agent formulas and synthetic fungicides in controlling maydis leaf blight on corn caused by Bipolaris maydis. IOP Conf Ser Earth Environ Sci 911:012062. https://doi.org/10.1088/1755-1315/911/1/012062

Erskine E, MacPhee CE, Stanley-Wall NR (2018) Functional amyloid and other protein fibers in the biofilm matrix. J Mol Biol 430:3642–3656. https://doi.org/10.1016/j.jmb.2018.07.026

Espinosa-Urgel M, Kolter R, Ramos J-L (2002) Root colonization by *Pseudomonas putida*: love at first sight. Microbiology 148:341–343. https://doi.org/10.1099/00221287-148-2-341

Fazli M, Almblad H, Rybtke ML, Givskov M, Eberl L, Tolker-Nielsen T (2014) Regulation of biofilm formation. Environ Microbiol 16:1961–1981. https://doi.org/10.1111/1462-2920.12448

Fessia A, Barra P, Barros G, Nesci A (2022a) Could *Bacillus* biofilms enhance the effectivity of biocontrol strategies in the phyllosphere. J Appl Microbiol 133:2148–2166. https://doi.org/10.1111/jam.15596

Fessia A, Sartori M, García D, Fernández L, Ponzio R, Barros G, Nesci A (2022b) In vitro studies of biofilm-forming *Bacillus* strains, biocontrol agents isolated from the maize phyllosphere. Biofilms 4:100097. https://doi.org/10.1016/j.bioflm.2022.100097

Flemming HC, Wingender J (2010) The biofilm matrix. Nat Rev Microbiol 8:623–633. https://doi.org/10.1038/nrmicro2415

Flemming HC, Wuertz S (2019) Bacteria and archaea on earth and their abundance in biofilms. Nat Rev Microbiol 17:247–260. https://doi.org/10.1038/s41579-019-0158-9

Flemming HC, Wingender J, Szewzyk U, Steinberg P, Rice SA, Kjelleberg S (2016) Biofilms: an emergent form of bacterial life. Nat Rev Microbiol 14:563–575. https://doi.org/10.1038/nrmicro.2016.94

Fong JNC, Yildiz FH (2015) Biofilm matrix proteins. Microbial Biofilm:201–222. https://doi.org/10.1128/microbiolspec.mb-0004-2014

Frenkel A, Zecharia E, Gómez-Pérez D, Sendersky E, Yegorov Y, Jacob A, Benichou JIC, Stierhof YD, Parnasa R, Golden SS, Kemen E, Schwarz R (2023) Cell specialization in cyanobacterial biofilm development revealed by expression of a cell-surface and extracellular matrix protein. NPJ Biofilms Microbiomes 9:10. https://doi.org/10.1038/s41522-023-00376-6

Fu L, Ou Y, Shen Z, Wang B, Li R, Shen Q (2019) Stable microbial community and specific beneficial taxa associated. J Microbiol Biotechnol 29:1624–1628. https://doi.org/10.4014/jmb.1904.04061

Gallo PM, Rapsinski GJ, Wilson RP, Oppong GO, Sriram U, Goulian M, Buttaro B, Caricchio R, Gallucci S, Tükel Ç (2015) Amyloid-DNA composites of bacterial biofilms stimulate autoimmunity. Immunity 42:1171–1184. https://doi.org/10.1016/j.immuni.2015.06.002

Gao M, Xiong C, Gao C, Tsui CKM, Wang M-M, Zhou X, Zhang A-M, Cai L (2021) Disease-induced changes in plant microbiome assembly and functional adaptation. Microbiome 9:187. https://doi.org/10.1186/s40168-021-01138-2

Garcia MC, Lee JT, Ramsook CB, Alsteens D, Dufrêne YF, Lipke PN (2011) A role for amyloid in cell aggregation and biofilm formation. PLoS One 6:17632. https://doi.org/10.1371/journal.pone.0017632

Ghadamgahi F, Tarighi S, Taheri P, Saripella GV, Anzalone A, Kalyandurg PB, Catara V, Ortiz R, Vetukuri RR (2022) Plant growth-promoting activity of *Pseudomonas aeruginosa* FG106 and its ability to act as a biocontrol agent against potato, tomato and taro pathogens. Biology 11:140. https://doi.org/10.3390/biology11010140

Ghazy N, El-Nahrawy S (2021) Siderophore production by *Bacillus subtilis* MF497446 and *Pseudomonas koreensis* MG209738 and their efficacy in controlling Cephalosporium maydis in maize plant. Arch Microbiol 203:1195–1209. https://doi.org/10.1007/s00203-020-02113-5

Gómez-Pérez D, Chaudhry V, Kemen A, Kemen E (2021) Amyloid proteins in plant-associated microbial communities. Microb Physiol 31:88–98. https://doi.org/10.1159/000516014

Gong T, Xin X (2021) Phyllosphere microbiota: community dynamics and its interaction with plant hosts. J Integr Plant Biol 63:297–304. https://doi.org/10.1111/jipb.13060

Gowtham HG, Murali M, Singh SB, Lakshmeesha TR, Murthy KN, Amruthesh KN, Niranjana SR (2018) Plant growth promoting rhizobacteria-*Bacillus amyloliquefaciens* improves plant growth and induces resistance in chilli against anthracnose disease. Biol Control 126:209–217. https://doi.org/10.1016/j.biocontrol.2018.05.022

Guzmán AC, Hurtado MIG, Cuesta-Astroz Y, Torres G (2020) Metagenomic characterization of bacterial biofilm in four food processing plants in Colombia. Braz J Microbiol 51:1259–1267. https://doi.org/10.1007/s42770-020-00260-x

Haggag WM, Timmusk S (2008) Colonization of peanut roots by biofilm-forming *Paenibacillus polymyxa* initiates biocontrol against crown rot disease. J Appl Microbiol 104:961–969. https://doi.org/10.1111/j.1365-2672.2007.03611.x

Haque MM, Mosharaf MK, Khatun M, Haque MA, Biswas MS, Islam MS, Islam MM, Shozib HB, Miah MMU, Molla AH, Siddiquee MA (2020) Biofilm producing rhizobacteria with multiple plant growth-promoting traits promote growth of tomato under water-deficit stress. Front Microbiol 11:542053. https://doi.org/10.3389/fmicb.2020.542053

Harting R, Nagel A, Nesemann K, Höfer AM, Bastakis E, Kusch H, Stanley CE, Stöckli M, Kaever A, Hoff KJ, Stanke M, deMello AJ, Künzler M, Haney CH, Braus-Stromeyer SA, Braus GH (2021) *Pseudomonas* strains induce transcriptional and morphological changes and reduce root colonization of *Verticillium* spp. Front Microbiol 12:652468. https://doi.org/10.3389/fmicb.2021.652468

Hashem A, Tabassum B, Abd-Allah EF (2019) *Bacillus subtilis*: a plant-growth promoting rhizobacterium that also impacts biotic stress. Saudi J Biol Sci 26:1291–1297. https://doi.org/10.1016/j.sjbs.2019.05.004

Hazarika SN, Saikia K, Borah A, Thakur D (2021) Prospecting endophytic bacteria endowed with plant growth promoting potential isolated from *Camellia sinensis*. Front Microbiol 12:738058. https://doi.org/10.3389/fmicb.2021.738058

Heijstra BD, Pichler FB, Liang Q, Blaza RG, Turner SJ (2009) Extracellular DNA and Type IV pili mediate surface attachment by *Acidovorax temperans*. Antonie Van Leeuwenhoek 95:343–349. https://doi.org/10.1007/s10482-009-9320-0

Heredia-Ponce Z, Gutiérrez-Barranquero JA, Purtschert-Montenegro G, Eberl L, Vicente A, Cazorla FM (2021a) Role of extracellular matrix components in the formation of biofilms and their contribution to the biocontrol activity of *Pseudomonas chlororaphis* PCL1606. Environ Microbiol 23:2086–2101. https://doi.org/10.1111/1462-2920.15355

Heredia-Ponce Z, de Vicente A, Cazorla FM, Gutiérrez-Barranquero JA (2021b) Beyond the wall: exopolysaccharides in the biofilm lifestyle of pathogenic and beneficial plant-associated *Pseudomonas*. Microorganisms 9:445. https://doi.org/10.3390/microorganisms9020445

Jamal M, Ahmad W, Andleeb S, Jalil F, Imran M, Nawaz MA, Hussain T, Ali M, Rafiq M, Kamil MA (2018) Bacterial biofilm and associated infections. J Chin Méd Assoc 81:7–11. https://doi.org/10.1016/j.jcma.2017.07.012

Janczarek M, Kutkowska J, Piersiak T, Skorupska A (2010) *Rhizobium leguminosarum* bv. *trifolii rosR* is required for interaction with clover, biofilm formation and adaptation to the environment. BMC Microbiol 10:284–284. https://doi.org/10.1186/1471-2180-10-284

Jefferson KK (2004) What drives bacteria to produce a biofilm? FEMS Microbiol Lett 236:163–173. https://doi.org/10.1111/j.1574-6968.2004.tb09643.x

Jeon E, Choi S, Yeo K-H, Park KS, Rathod ML, Lee J (2017) Development of electrical conductivity measurement technology for key plant physiological information using microneedle sensor. J Micromech Microeng 27:085009. https://doi.org/10.1088/1361-6439/aa7362

Jin X, Marshall JS (2020) Mechanics of biofilms formed of bacteria with fimbriae appendages. PLoS One 15:e0243280. https://doi.org/10.1371/journal.pone.0243280

Karamanoli K, Thalassinos G, Karpouzas D, Bosabalidis AM, Vokou D, Constantinidou HI (2012) Are leaf glandular trichomes of oregano hospitable habitats for bacterial growth? J Chem Ecol 38:476–485. https://doi.org/10.1007/s10886-012-0117-7

Karlsson ME, Hellström M, Flöhr A, Bergstrand K-J, Alsanius BW (2023) The power of light: impact on the performance of biocontrol agents under minimal nutrient conditions. Front Microbiol 14:1087639. https://doi.org/10.3389/fmicb.2023.1087639

Karygianni L, Ren Z, Koo H, Thurnheer T (2020) Biofilm matrixome: extracellular components in structured microbial communities. Trends Microbiol 28:668–681. https://doi.org/10.1016/j.tim.2020.03.016

Kasim WA, Gaafar RM, Abou-Ali RM, Omar MN, Hewait HM (2016) Effect of biofilm forming plant growth promoting rhizobacteria on salinity tolerance in barley. Ann Agric Sci 61:217–227. https://doi.org/10.1016/j.aoas.2016.07.003

Kloepper JW, Ryu CM, Zhang S (2004) Induced systemic resistance and promotion of plant growth by *Bacillus* spp. Phytopathology 94:1259–1266. https://doi.org/10.1094/phyto.2004.94.11.1259

Lahlali R, Ezrari S, Radouane N, Kenfaoui J, Esmaeel Q, Hamss HE, Belabess Z, Barka EA (2022) Biological control of plant pathogens: a global perspective. Microorganisms 10:596. https://doi.org/10.3390/microorganisms10030596

Legein M, Smets W, Vandenheuvel D, Eilers T, Muyshondt B, Prinsen E, Samson R, Lebeer S (2020) Modes of action of microbial biocontrol in the phyllosphere. Front Microbiol 11:1619. https://doi.org/10.3389/fmicb.2020.01619

Leo JC, Grin I, Linke D (2012) Type V secretion: mechanism(s) of autotransport through the bacterial outer membrane. Philos Trans R Soc 367:1088–1101. https://doi.org/10.1098/rstb.2011.0208

León-Romero Á, Domínguez-Manzano J, Garrido-Fernández A, Arroyo-López FN, Jiménez-Díaz R (2015) Formation of in vitro mixed-species biofilms by *Lactobacillus pentosus* and yeasts isolated from Spanish-style green table olive fermentations. Appl Environ Microbiol 82:689–695. https://doi.org/10.1128/aem.02727-15

Li Z, Nair SK (2012) Quorum sensing: how bacteria can coordinate activity and synchronize their response to external signals? Protein Sci 21:1403–1417. https://doi.org/10.1002/pro.2132

Lindow SE, Brandl MT (2003) Microbiology of the phyllosphere. Appl Environ Microbiol 69:1875–1883. https://doi.org/10.1128/aem.69.4.1875-1883.2003

Liu Y, Zhang N, Qiu M, Feng H, Vivanco JM, Shen Q, Zhang R (2014) Enhanced rhizosphere colonization of beneficial *Bacillus amyloliquefaciens* SQR9 by pathogen infection. FEMS Microbiol Lett 353:49–56. https://doi.org/10.1111/1574-6968.12406

Liu H, Li J, Carvalhais LC, Percy CD, Verma JP, Schenk PM, Singh BK (2021) Evidence for the plant recruitment of beneficial microbes to suppress soil-borne pathogens. New Phytol 229:2873–2885. https://doi.org/10.1111/nph.17057

Liu X, Roux XL, Salles JF (2022) The legacy of microbial inoculants in agroecosystems and potential for tackling climate change challenges. iScience 25:103821. https://doi.org/10.1016/j.isci.2022.103821

Loh J, Pierson EA, Pierson LS, Stacey G, Chatterjee A (2002) Quorum sensing in plant-associated bacteria. Curr Opin Plant Biol 5:285–290. https://doi.org/10.1016/s1369-5266(02)00274-1

Lopes MJDS, Dias-Filho MB, Gurgel ESC (2021) Successful plant growth-promoting microbes: inoculation methods and abiotic factors. Front Sustain Food Syst 5:606454. https://doi.org/10.3389/fsufs.2021.606454

Lugtenberg B, Kamilova F (2009) Plant-growth-promoting rhizobacteria. Ann Rev Microbiol 63:541–556. https://doi.org/10.1146/annurev.micro.62.081307.162918

Luo C, Zhou H, Zou J, Wang X, Zhang R, Xiang Y, Chen Z (2015) Bacillomycin L and surfactin contribute synergistically to the phenotypic features of *Bacillus subtilis* 916 and the biocontrol of rice sheath blight induced by *Rhizoctonia solani*. Appl Microbiol Biotechnol 99:1897–1910. https://doi.org/10.1007/s00253-014-6195-4

Madsen JS, Burmølle M, Hansen LH, Sørensen SJ (2012) The interconnection between biofilm formation and horizontal gene transfer. FEMS Immunol Méd Microbiol 65:183–195. https://doi.org/10.1111/j.1574-695x.2012.00960.x

Mahlein AK (2016) Plant disease detection by imaging sensors—parallels and specific demands for precision agriculture and plant phenotyping. Plant Dis 100:241–251. https://doi.org/10.1094/pdis-03-15-0340-fe

Mansfield J, Genin S, Magori S, Citovsky V, Sriariyanum M, Ronald P, Dow M, Verdier V, Beer SV, Machado MA, Toth I, Salmond G, Foster GD (2012) Top 10 plant pathogenic bacteria in molecular plant pathology. Mol Plant Pathol 13:614–629. https://doi.org/10.1111/j.1364-3703.2012.00804.x

Mazarei M, Teplova I, Hajimorad MR, Stewart CN (2008) Pathogen phytosensing: plants to report plant pathogens. Sensors 8:2628–2641. https://doi.org/10.3390/s8042628

Mendes R, Garbeva P, Raaijmakers JM (2013) The rhizosphere microbiome: significance of plant beneficial, plant pathogenic, and human pathogenic microorganisms. FEMS Microbiol Rev 37:634–663. https://doi.org/10.1111/1574-6976.12028

Miller MB, Bassler BL (2001) Quorum sensing in bacteria. Ann Rev Microbiol 55:165–199. https://doi.org/10.1146/annurev.micro.55.1.165

Mitri S, Xavier JB, Foster KR (2011) Social evolution in multispecies biofilms. Proc Natl Acad Sci 108:10839–10846. https://doi.org/10.1073/pnas.1100292108

Molina-Santiago C, Pearson JR, Navarro Y, Berlanga-Clavero MV, Caraballo-Rodriguez AM, Petras D, García-Martín ML, Lamon G, Haberstein B, Cazorla FM, de Vicente A, Loquet A, Dorrestein PC, Romero D (2019) The extracellular matrix protects *Bacillus subtilis* colonies from *Pseudomonas* invasion and modulates plant co-colonization. Nat Commun 10:1919. https://doi.org/10.1038/s41467-019-09944-x

Moons P, Michiels CW, Aertsen A (2009) Bacterial interactions in biofilms. Crit Rev Microbiol 35:157–168. https://doi.org/10.1080/10408410902809431

Moreira ZPM, Chen MY, Ortuno DLY, Haney CH (2023) Engineering plant microbiomes by integrating eco-evolutionary principles into current strategies. Curr Opin Plant Biol 71:102316. https://doi.org/10.1016/j.pbi.2022.102316

Morris CE, Monier JM, Jacques MA (1998) A technique to quantify the population size and composition of the biofilm component in communities of bacteria in the phyllosphere. Appl Environ Microbiol 64:4789–4795. https://doi.org/10.1128/aem.64.12.4789-4795.1998

Nakamura S, Minamino T (2019) Flagella-driven motility of bacteria. Biomolecules 9:279. https://doi.org/10.3390/biom9070279

Niu DD, Wang CJ, Guo YH, Jiang CH, Zhang WZ, Wang Y, Guo JH (2012) The plant growth-promoting rhizobacterium *Bacillus cereus* AR156 induces resistance in tomato with induction and priming of defence response. Biocontrol Sci Tech 22:991–1004. https://doi.org/10.1080/09583157.2012.706595

OECD (2012) World population by age class, 1970-2050. https://doi.org/10.1787/env_outlook-2012-graph8-en

Ons L, Bylemans D, Thevissen K, Cammue BPA (2020) Combining biocontrol agents with chemical fungicides for integrated plant fungal disease control. Microorganisms 8:1930. https://doi.org/10.3390/microorganisms8121930

Palmer J, Flint S, Brooks J (2007) Bacterial cell attachment, the beginning of a biofilm. J Ind Microbiol Biotechnol 34:577–588. https://doi.org/10.1007/s10295-007-0234-4

Pandin C, Coq DL, Canette A, Aymerich S, Briandet R (2017) Should the biofilm mode of life be taken into consideration for microbial biocontrol agents? Microb Biotechnol 10:719–734. https://doi.org/10.1111/1751-7915.12693

Pandit A, Adholeya A, Cahill D, Brau L, Kochar M (2020) Microbial biofilms in nature: unlocking their potential for agricultural applications. J Appl Microbiol 129:199–211. https://doi.org/10.1111/jam.14609

Patil A, Gondi R, Rale V, Saroj SD (2022) Microbial biofilms in plant disease management. In Biocontrol Mechanisms of Endophytic Microorganisms pp 239–259. doi:https://doi.org/10.1016/b978-0-323-88478-5.00005-5

Peterson BW, van der Mei HC, Sjollema J, Busscher HJ, Sharma PK (2013) A distinguishable role of eDNA in the viscoelastic relaxation of biofilms. MBio 4:10–1128. https://doi.org/10.1128/mbio.00497-13

Petrova OE, Sauer K (2012) Sticky situations: key components that control bacterial surface attachment. J Bacteriol 194:2413–2425. https://doi.org/10.1128/jb.00003-12

Philippot L, Raaijmakers JM, Lemanceau P, van der Putten WH (2013) Going back to the roots: the microbial ecology of the rhizosphere. Nat Rev Microbiol 11:789–799. https://doi.org/10.1038/nrmicro3109

Piepenbrink KH, Sundberg EJ (2016) Motility and adhesion through type IV pili in gram-positive bacteria. Biochem Soc Trans 44:1659–1666. https://doi.org/10.1042/bst20160221

Pieterse CMJ, Zamioudis C, Berendsen RL, Weller DM, van Wees SCM, Bakker PAHM (2014) Induced systemic resistance by beneficial microbes. Annu Rev Phytopathol 52:1–29. https://doi.org/10.1146/annurev-phyto-082712-102340

Ramey BE, Koutsoudis M, von Bodman SB, Fuqua C (2004) Biofilm formation in plant–microbe associations. Curr Opin Microbiol 7:602–609. https://doi.org/10.1016/j.mib.2004.10.014

Ramos-González MI, Campos MJ, Ramos JL (2005) Analysis of *Pseudomonas putida* KT2440 gene expression in the maize rhizosphere: in vivo expression technology capture and identification of root-activated promoters. J Bacteriol 187:5504–5504. https://doi.org/10.1128/jb.187.15.5504.2005

Rodriguez PA, Rothballer M, Chowdhury SP, Nussbaumer T, Gutjahr C, Falter-Braun P (2019) Systems biology of plant-microbiome interactions. Mol Plant 12:804–821. https://doi.org/10.1016/j.molp.2019.05.006

Romano I, Ventorino V, Pepe O (2020) Effectiveness of plant beneficial microbes: overview of the methodological approaches for the assessment of root colonization and persistence. Front Plant Sci 11:6. https://doi.org/10.3389/fpls.2020.00006

Roper JM, Garcia JF, Tsutsui H (2021) Emerging technologies for monitoring plant health in vivo. ACS Omega 6:5101–5107. https://doi.org/10.1021/acsomega.0c05850

Rudrappa T, Czymmek KJ, Paré PW, Bais HP (2008) Root-secreted malic acid recruits beneficial soil bacteria. Plant Physiol 148:1547–1556. https://doi.org/10.1104/pp.108.127613

Sarti GC, Galelli ME, Arreghini S, Cristóbal-Miguez JAE, Curá JA, Paz-González A (2023) Inoculation with biofilm of *Bacillus subtilis* promotes the growth of *Lactuca sativa*. Sustain For 15:15406. https://doi.org/10.3390/su152115406

Savary S, Willocquet L, Pethybridge SJ, Esker P, McRoberts N, Nelson A (2019) The global burden of pathogens and pests on major food crops. Nat Ecol Evol 3:430–439. https://doi.org/10.1038/s41559-018-0793-y

Shade A, Peter H, Allison SD, Baho DL, Berga M, Bürgmann H, Huber DH, Langenheder S, Lennon JT, Martiny JBH, Matulich KL, Schmidt TM, Handelsman J (2012) Fundamentals of microbial community resistance and resilience. Front Microbiol 3:417. https://doi.org/10.3389/fmicb.2012.00417

Shrestha A, Grimm M, Ojiro I, Krumwiede J, Schikora A (2020) Impact of quorum sensing molecules on plant growth and immune system. Front Microbiol 11:1545. https://doi.org/10.3389/fmicb.2020.01545

Soni R, Keharia H (2021) Phytostimulation and biocontrol potential of gram-positive endospore forming Bacilli. Planta 254:49. https://doi.org/10.1007/s00425-021-03695-0

Spoering AL, Gilmore MS (2006) Quorum sensing and DNA release in bacterial biofilms. Curr Opin Microbiol 9:133–137. https://doi.org/10.1016/j.mib.2006.02.004

Srivastava S, Yadav A, Seem K, Mishra S, Chaudhary V, Nautiyal CS (2008) Effect of high temperature on *pseudomonas putida* NBRI0987 biofilm formation and expression of stress sigma factor RpoS. Curr Microbiol 56:453–457. https://doi.org/10.1007/s00284-008-9105-0

Steenackers HP, Parijs I, Dubey A, Foster KR, Vanderleyden J (2016) Experimental evolution in biofilm populations. FEMS Microbiol Rev 40:373–397. https://doi.org/10.1093/femsre/fuw002

Stets MI, Alqueres SMC, Souza EM, Pedrosa FDO, Schmid M, Hartmann A, Cruz LM (2015) Quantification of *Azospirillum brasilense* FP2 bacteria in wheat roots by strain-specific quantitative PCR. Appl Environ Microbiol 81:6700–6709. https://doi.org/10.1128/aem.01351-15

Stoll A, Salvatierra-Martínez R, González M, Araya M (2021) The role of surfactin production by *bacillus velezensis* on colonization, biofilm formation on tomato root and leaf surfaces and subsequent protection (ISR) against *Botrytis cinerea*. Microorganisms 9:2251. https://doi.org/10.3390/microorganisms9112251

Sun C, Guo Y, Tang K, Wen Z, Li B, Zeng Z, Wang X (2017) MqsR/MqsA toxin/antitoxin system regulates persistence and biofilm formation in *Pseudomonas putida* KT2440. Front Microbiol 8:840. https://doi.org/10.3389/fmicb.2017.00840

Sun X, Xu Z, Xie J, Hesselberg-Thomsen V, Tan T, Zheng D, Strube ML, Dragoš A, Shen Q, Zhang R, Kovács ÁT (2021) *Bacillus velezensis* stimulates resident rhizosphere *Pseudomonas stutzeri* for plant health through metabolic interactions. ISME J 16:774–787. https://doi.org/10.1038/s41396-021-01125-3

Syed AK, Boles BR (2014) Fold modulating function: bacterial toxins to functional amyloids. Front Microbiol 5:401. https://doi.org/10.3389/fmicb.2014.00401

Taglialegna A, Lasa I, Valle J (2016) Amyloid structures as biofilm matrix scaffolds. J Bacteriol 198:2579–2588. https://doi.org/10.1128/jb.00122-16

Tang X, Xiao Y, Zhou J-M (2006) Regulation of the type III secretion system in phytopathogenic bacteria. Mol Plant Microbe Interact 19:1159–1166. https://doi.org/10.1094/mpmi-19-1159

Tao C, Li R, Xiong W, Shen Z, Liu S, Wang B, Ruan Y, Geisen S, Shen Q, Kowalchuk GA (2020) Bio-organic fertilizers stimulate indigenous soil *Pseudomonas* populations to enhance plant disease suppression. Microbiome 8:137. https://doi.org/10.1186/s40168-020-00892-z

Teshome DT, Zharare GE, Naidoo S (2020) The threat of the combined effect of biotic and abiotic stress factors in forestry under a changing climate. Front Plant Sci 11:601009. https://doi.org/10.3389/fpls.2020.601009

Thrall PH, Hochberg ME, Burdon JJ, Bever JD (2007) Coevolution of symbiotic mutualists and parasites in a community context. Trends Ecol Evol 22:120–126. https://doi.org/10.1016/j.tree.2006.11.007

Timmusk S, Nevo E (2011) Bacteria in agrobiology: plant nutrient. Management 285–300. https://doi.org/10.1007/978-3-642-21061-7_12

Tovi N, Frenk S, Hadar Y, Minz D (2019) Host specificity and spatial distribution preference of three pseudomonas isolates. Front Microbiol 9:3263. https://doi.org/10.3389/fmicb.2018.03263

Trivedi P, Leach JE, Tringe SG, Sa T, Singh BK (2020) Plant–microbiome interactions: from community assembly to plant health. Nat Rev Microbiol 18:607–621. https://doi.org/10.1038/s41579-020-0412-1

van der Wolf J, De Boer SH (2014) Phytopathogenic bacteria. In Principles of plant-microbe interactions: Microbes for Sustainable Agriculture 65–77. doi:https://doi.org/10.1007/978-3-319-08575-3_9

van Elsas JD, Chiurazzi M, Mallon CA, Elhottovā D, Krištůfek V, Salles JF (2012) Microbial diversity determines the invasion of soil by a bacterial pathogen. Proc Natl Acad Sci 109:1159–1164. https://doi.org/10.1073/pnas.1109326109

van Gestel J, Vlamakis H, Kolter R (2015) Division of labor in biofilms: the ecology of cell differentiation. Microbiol Spectr 3:MB-0002-2014. https://doi.org/10.1128/microbiolspec.mb-0002-2014

Vanthana M, Nakkeeran S, Malathi VG, Renukadevi P, Vinodkumar S (2019) Induction of in planta resistance by flagellin (Flg) and elongation factor-TU (EF-Tu) of *Bacillus amyloliquefaciens* (VB7) against groundnut bud necrosis virus in tomato. Microb Pathog 137:103757. https://doi.org/10.1016/j.micpath.2019.103757

Vassileva M, Mocali S, Canfora L, Malusá E, Moral LFG, del Martos V, Flor-Peregrin E, Vassilev N (2022) Safety level of microorganism-bearing products applied in soil-plant systems. Front Plant Sci 13:862875. https://doi.org/10.3389/fpls.2022.862875

Velmourougane K, Prasanna R, Singh S, Chawla G, Kumar A, Saxena AK (2017) Modulating rhizosphere colonisation, plant growth, soil nutrient availability and plant defense enzyme activity through *Trichoderma viride-Azotobacter chroococcum* biofilm inoculation in chickpea. Plant Soil 421:157–174. https://doi.org/10.1007/s11104-017-3445-0

Wang D-C, Jiang C-H, Zhang L-N, Chen L, Zhang X-Y, Guo J-H (2019) Biofilms positively contribute to *Bacillus amyloliquefaciens* 54-induced drought tolerance in tomato plants. Int J Mol Sci 20:6271. https://doi.org/10.3390/ijms20246271

Wang Q, Shakoor N, Boyher A, Veley KM, Berry JC, Mockler TC, Bart RS (2021) Escalation in the host-pathogen arms race: a host resistance response corresponds to a heightened bacterial virulence response. PLoS Pathog 17:e1009175. https://doi.org/10.1371/journal.ppat.1009175

Wang N, Ding J, Chen Y, Zhu Y, Zhang L, Wei Y, Li J, Xu T, Ding G (2023) *Bacillus velezensis* BER1 enriched *Flavobacterium daejeonense*-like bacterium in the rhizosphere of tomato against bacterial wilt. FEMS Microbiol Ecol 99:fiad054. https://doi.org/10.1093/femsec/fiad054

Wei F, Hu X, Xu X (2016) Dispersal of *Bacillus subtilis* and its effect on strawberry phyllosphere microbiota under open field and protection conditions. Sci Rep 6:22611. https://doi.org/10.1038/srep22611

Whitehead NA, Barnard AML, Slater H, Simpson NJL, Salmond GPC (2001) Quorum-sensing in Gram-negative bacteria. FEMS Microbiol Rev 25:365–404. https://doi.org/10.1111/j.1574-6976.2001.tb00583.x

Woo SL, Lorito M (2007) Novel biotechnologies for biocontrol agent enhancement and management. NATO Secur Sci Ser:107–130. https://doi.org/10.1007/978-1-4020-5799-1_6

Wu G, Liu Y, Xu Y, Zhang G, Shen Q, Zhang R (2018a) Exploring elicitors of the beneficial Rhizobacterium *Bacillus amyloliquefaciens* SQR9 to induce plant systemic resistance and their interactions with plant signaling pathways. Mol Plant-Microbe Interact 31:560–567. https://doi.org/10.1094/mpmi-11-17-0273-r

Wu L, Huang Z, Li X, Ma L, Gu Q, Wu H, Liu J, Borriss R, Wu Z, Gao X (2018b) Stomatal closure and SA-, JA/ET-signaling pathways are essential for *Bacillus amyloliquefaciens* FZB42 to restrict leaf disease caused by *Phytophthora nicotianae* in *Nicotiana benthamiana*. Front Microbiol 9:847. https://doi.org/10.3389/fmicb.2018.00847

Xia L, Miao Y, Cao A, Liu Y, Liu Z, Sun X, Xue Y, Xu Z, Xun W, Shen Q, Zhang N, Zhang R (2022) Biosynthetic gene cluster profiling predicts the positive association between antagonism and phylogeny in *bacillus*. Nat Commun 13:1023. https://doi.org/10.1038/s41467-022-28668-z

Xu X-M, Jeffries P, Pautasso M, Jeger MJ (2011) Combined use of biocontrol agents to manage plant diseases in theory and practice. Phytopathology 101:1024–1031. https://doi.org/10.1094/phyto-08-10-0216

Xu Z, Mandic-Mulec I, Zhang H, Liu Y, Sun X, Feng H, Xun W, Zhang N, Shen Q, Zhang R (2019) Antibiotic Bacillomycin D affects iron acquisition and biofilm formation in *Bacillus velezensis* through a Btr-mediated FeuABC-dependent pathway. Cell Rep 29:1192–1202.e5. https://doi.org/10.1016/j.celrep.2019.09.061

Yin W, Wang Y, Liu L, He J (2019) Biofilms: the microbial "protective clothing" in extreme environments. Int J Mol Sci 20:3423. https://doi.org/10.3390/ijms20143423

Yu X, Ai C, Xin L, Zhou G (2011) The siderophore-producing bacterium, *Bacillus subtilis* CAS15, has a biocontrol effect on Fusarium wilt and promotes the growth of pepper. Eur J Soil Biol 47:138–145. https://doi.org/10.1016/j.ejsobi.2010.11.001

Zeriouh H, Vicente A, Pérez-García A, Romero D (2014) Surfactin triggers biofilm formation of *Bacillus subtilis* in melon phylloplane and contributes to the biocontrol activity. Environ Microbiol 16:2196–2211. https://doi.org/10.1111/1462-2920.12271

Zhang Q, Kühl M, Brodersen KE (2022) Effects of epiphytic biofilm activity on the photosynthetic activity, pH and inorganic carbon microenvironment of seagrass leaves (*Zostera marina* L.). Front Mar Sci 9:835381. https://doi.org/10.3389/fmars.2022.835381

Zheng S, Bawazir M, Dhall A, Kim HE, He L, Heo J, Hwang G (2021) Implication of surface properties, bacterial motility, and hydrodynamic conditions on bacterial surface sensing and their initial adhesion. Front Bioeng Biotechnol 9:643722. https://doi.org/10.3389/fbioe.2021.643722

Zhu ML, Wu XQ, Wang YH, Dai Y (2020) Role of biofilm formation by *Bacillus pumilus* HR10 in biocontrol against pine seedling damping-off disease caused by *Rhizoctonia solani*. Forests 11:652. https://doi.org/10.3390/f11060652

Chapter 8
Rhizosphere Microbiome-Assisted Approaches for Biotic Stress Management

Tina Roy, Pooja Yadav, Anjali Chaudhary, Kanchan Yadav, and Kunal Singh

Abstract The rhizosphere microbiome modulates plant health in many ways and assists in projecting resistance against biotic stress. It harbors phylogenetically diverse microbes, including bacteria, fungi, archaea, and yeast. Many of these microbes carry the potential to promote plant growth and antagonize phytopathogens. Next-generation sequencing (NGS) techniques have been extensively used to study the microbial diversity to gain further insight into the rhizo-microbiome constituents. Application of biocontrol agents (BCAs) obtained from rhizobiome bioprospection with antagonistic activity provides an affordable and environmentally friendly approach to protect plants under biotic stress conditions. Microbe-derived control agents present viable and sustainable paths to avoid use of harmful synthetic fungicides and pesticides having an adverse impact on Earth's atmosphere. Therefore, understanding the advancements and relevant knowledge generation toward microbial bioprospection and diversity characterization simultaneously. Thus, prominent characteristic features of the rhizospheric microbiome, major techniques under application to characterize microbial diversity, and their utilization in disease suppression with their mode of action have been discussed thoroughly utilizing various examples. Analysis of facets of the rhizospheric microbiome and discussion on their application part could provide an updated knowledge framework in the area of the associated rhizospheric microbiome.

Tina Roy and Pooja Yadav contributed equally with all other contributors.

T. Roy
CSIR-Institute of Himalayan Bioresource Technology, Palampur, India

P. Yadav · A. Chaudhary · K. Yadav · K. Singh (✉)
CSIR-Institute of Himalayan Bioresource Technology, Palampur, India

Academy of Scientific and Innovative Research (AcSIR), Ghaziabad, India
e-mail: kunal@ihbt.res.in

P. Singh Chauhan et al. (eds.), *Plant-Microbe Interaction and Stress Management*, Rhizosphere Biology,
https://doi.org/10.1007/978-981-97-4239-4_8

Keywords Microbiome · NGS techniques · Metabolites · Phytopathogens · Biocontrol agents

8.1 Introduction

Plant growth, fitness, and productivity are significantly influenced by the interactions between the soil microbiome and the plant. Studies on plant–microbiome interaction provide insight into the functional and structural diversity of microbial communities related to a particular host plant in a specific environment (Gupta et al. 2021). Microbiome refers to the community of microorganisms that inhabit a particular environment, such as water, air, soil, plants, and living organisms. It is defined as a characteristic microbial community occupying a reasonably well-defined habitat that has distinct physiochemical properties (Whipps et al. 1988). They are incredibly diverse with numerous species of bacteria, fungi, archaea, virus, and protists interacting and coexisting in the environment (Kumawat et al. 2022).

Soil microbiome including rhizospheric ones had effects on the yield of fruits and grains biofuels and other plant products (Oliveira et al. 2022). Various mechanisms, such as the absorption of carbon dioxide from the atmosphere or the storage of carbon in soil and the degradation of organic material in soil, are influenced by the soil microbiomes (Yue et al. 2023). The beneficial interaction between plant root and rhizosphere microbes contributes to various diverse mechanisms that are ultimately profitable to both (Aslam et al. 2022). Prevailing environment could have a significant impact on these interactions, which can increase plant resistance to environmental stress (Trivedi et al. 2022). Microbiome may vary according to the different environmental conditions such as water bodies, volcanic eruption zones, desert, rain forest, etc. The freshwater microbiomes gained exclusive approval for linking the microbial community with the physiochemical properties of the water. The microbial composition of surface sediment communities differed depending on the type of water bodies. For example, lakes had higher concentrations of phototrophic Cyanobacteria lineages, while streams and rivers had higher abundance of Actinobacteria and Acidobacteria (Premke et al. 2022). The volcanic zone has a high abundance of heavy metals and the most abundant phyla reported were Ascomycota, Basidiomycota, Mortierellomycota, and Glomeromycota (Chen et al. 2022a, b). Along the altitudinal gradients of alpine microbiome, bacterial and fungal diversity decreases with elevation, whereas the protistan diversity increases (Fiore-Donno et al. 2024).

In arid ecosystem, microorganisms are important for carbon and nitrogen cycles, and also crucial for soil productivity and other biological processes (Li et al. 2022). Forest ecosystems also inhabit a wide range of soil microbiome dependent on the climate conditions, organic matter content, biotic interactions, nutrient availability, and soil pH (Ahmad et al. 2023). Acidic soils have a low phylogenetic diversity and are dominated by the acidophilic lineages of the microbes (Baldrian et al. 2023; Calvert et al. 2023). While Acidobacteria, Ascomycota, and Basidiomycota were found to be the most abundant phyla in the tropical rainforest region of soil (Deng et al. 2022). Thus, the compositions of microbes may vary according to the habitat.

8.2 An Overview of Rhizosphere Microbiome

The term rhizosphere refers to the region of soil directly influenced by the root secretions and associated with selective soil microorganisms called as rhizospheric microbiome. Therefore, the term rhizosphere describes the phenomenon of plant interaction with such root-associated microbiome (Chen et al. 2023). The importance of these microbes in plant growth through various mechanisms includes nitrogen fixation, phosphorus solubilization, and phytohormone biosynthesis known as plant growth-promoting bacteria (PGPB)/fungi (PGPF) and arbuscular mycorrhizal fungi (AMF) (Yee et al. 2021). Not only mutualistic/beneficial the plant–microbe interaction includes commensalism and antagonism relations that have a huge effect on the plant health and productivity. These biotic interactions induce variations in the microbial diversity and are essential for thriving of plants in any environment (Sannino et al. 2023).

The rhizosphere not only regulates the biology but also the chemistry of the soil. It is a hotspot of biological and chemical activity governed by substances released by both the root and microorganisms (Lippold et al. 2023). Many microorganisms get their nourishment from substances that plant roots release into the soil, including sugars, amino acids, and organic acids (Sokol et al. 2022). Since food availability is in plenty, there is considerably more microbial activity in the rhizosphere than in the non-rhizospheric soil, making rhizosphere the most active habitat zone in soil (Das et al. 2022). In rhizosphere, nutrients and wastes get cycled and recycled converting them into biologically accessible forms, storing them for later use and avoiding leakage into ground and surface waters (Adhikari and Hartemink 2016). Rhizosphere possesses an enormous diversity of microbes contributing to nutrient cycling, decomposition, plant development, protection of soil structure, and disease suppression (Sokol et al. 2022).

8.3 Benefits of Studying Rhizo-Microbial Diversity

The rhizosphere microbial communities associated with plants aid in their growth and resistance to various biotic and abiotic stresses (Dlamini et al. 2022). Plants are directly protected against pathogens by beneficial rhizosphere microorganisms, primarily through niche and resource competition and antagonism. The interaction of plants with microbes results in physiological and biochemical changes via release of various elicitors (Li et al. 2021). Furthermore, recent studies have revealed distinct microbial compositions in different soil types with a high abundance of Proteobacteria in rhizospheric soils and predominance of Actinobacteria and Alphaproteobacteria in bulk soil, indicating selective effects useful for bioremediation (Vu et al. 2022). For example, rhizosphere of healthy *Billardiera scandens* was dominated by Actinobacteria, Proteobacteria, Bacteroidetes, Planctomycetes, Chloroflexi, Cyanobacteria, Acidobacteria, Verrucomicrobia, Gemmatimonadetes, Nitrospirae, and Armatimonadetes (Fuentes et al. 2020). Recent developments in agricultural research have introduced intriguing strategies for plant health. Among these, root microbiome transfer (RMT) is emerging as a noteworthy technique. The RMT model explores the complex interactions between plants and their microbial communities and aims to enhance plant resilience and overall well-being. The consideration of RMT from appropriately resistant donors suggests a promising avenue for potentially modulating protective microbiomes and promoting plant health (Afridi et al. 2022).

8.4 NGS Techniques to Understand Rhizo-Microbial Diversity

Advanced NGS techniques have significantly enhanced the understanding of soil microbiomes, their interactions with plant health under stress, and their roles in ecological processes. The development of molecular ecology has transformed the study of microbial soil diversity, transcending the need to isolate cultivable microorganisms. Modern techniques enable the direct extraction of DNA, RNA, and other biochemicals from the entire rhizospheric microbiome, including nonculturable components. Thus, providing insights into both the phylogenetic relationship and functional capabilities of microbes (Nwachukwu and Babalola 2022). In this context, metagenomics, including shotgun and amplicon sequencing, has been pivotal in exploring microbial diversity in the rhizosphere. The amplicon sequencing focused on specific markers such as 16S rRNA for bacteria and ITS for fungi. It is a cost-effective and sensitive approach but has limitations in terms of taxonomic resolution. The limitations were addressed by third-generation platforms like PacBio and Nanopore (Meslier et al. 2022). In contrast, shotgun sequencing examines the total DNA content of samples, which facilitated the comprehensive microbial identification and functional analysis (Quince et al. 2017). In the advanced landscape of metagenomic research, a sophisticated pipeline incorporates a variety of specialized bioinformatics techniques to unravel the intricate details of soil microbiomes by using commonly utilized tools. The process began with quality control using Trimmomatic and assembly using MetaVelvet-SL, MetaVelvet, IDBA-UD, SOAPdenovo2, Omega, Ray Meta, and MEGAHIT (Afiahayati et al. 2015). The binning was performed using MetaBAT, MaxBin, and CONCOCT and bin refinement using metaWRAP and CheckM. Phylogenetic connections were deduced using PhyloPhlAn, diamond, mafft, and FastTree, while MMseqs2 and eggNOG-mapper aid in protein functional annotation. The prediction of Biosynthetic Gene Clusters (BGCs) utilized antiSMASH and BiG-SCAPE, critical for understanding microbial capabilities. The pipeline may include single-nucleotide variant (SNV) and pangenome analysis utilizing MUMmer and Roary used for investigating genetic variation within species. Additionally, ORF prediction and protein annotation were conducted using Prokka. VirSorter2 and CheckV were used to analyze MAGs and viruses, and extending the understanding of virus–host interactions in microbial communities. This cutting-edge collection of methods represents the latest developments in metagenomic analysis, providing an unprecedented level of insight into the genetic and functional diversity of soil microbiomes (Ma et al. 2023). The database and project related to metagenome like extensive soil metagenomic data from the National Ecological Observatory Network (NEON) and the Earth Microbiome Project 500 (EMP500) offer insights into microbial diversity and potential applications in environmental restoration (Fig. 8.1).

Metagenomics lays the groundwork by providing a comprehensive view of the genetic landscape of microbial communities, capturing the vast array of species present in the soil environment. Advancing from this foundational knowledge, metatranscriptomics takes a crucial step further. The primary aim of metatranscriptome is to explore the active genes and their expression within microbial communities by providing dynamic scenario of gene expression. It enables us to not just identify the microbial constituents but also understand their active roles and responses to environmental stimuli offering a more comprehensive picture of the ecological processes at play. The approach requires the challenges posed by ribosomal RNA (rRNA) dominance in sequencing datasets overcome by robust

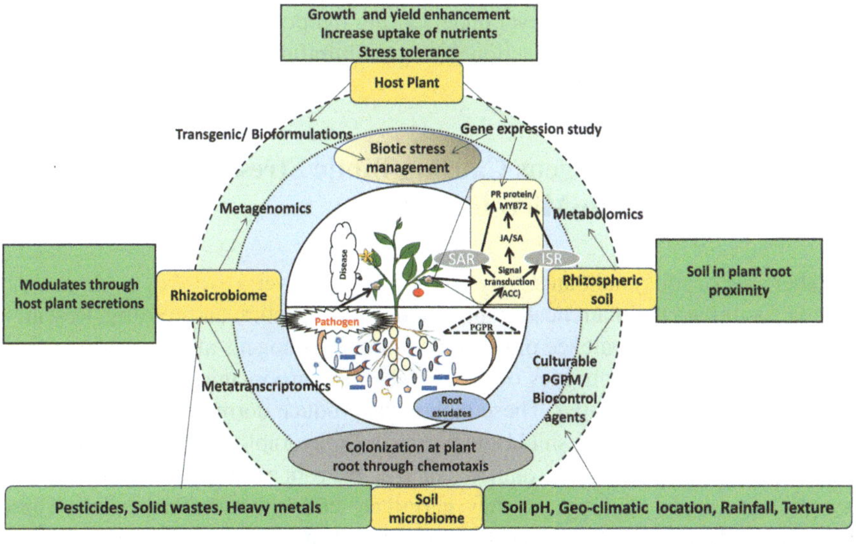

Fig. 8.1 Approaches for the identification of plant organization with beneficial rhizo-microbial composition with their direct and indirect benefits to the host plant

preprocessing tools like FastQC and Trimmomatic for data integrity (Shakya et al. 2019). Metatranscriptome analysis involves software like MetaTrans and COMAN employed for precise mapping, whereas IDBA-MT and MOSCA facilitate de novo assembly (Shakya et al. 2019). Tools such as Kraken2 and Centrifuge recognized for processing longer reads provide enhanced capabilities for microbial sequence analysis. The final stage of metatranscriptomic analysis involves coding transcript annotation by tools like InterProScan, BLAST, RefSeq, Diamond, and HMMER (HMM-based), which were further refined by KEGG, COG, and tools like antiSMASH for detailed annotation (Baltoumas et al. 2023). The set of tools collectively advances the understanding of microbial functional dynamics in diverse environments.

The rigorous utilization of NGS resulted in a very large amount of data that keep getting accumulated with each passing day for which the integration of machine learning (ML) and deep learning (DL) techniques is pivotal. The effectiveness of deep learning techniques, particularly autoencoders in the field of dimensionality reduction, was assessed. This method marks a substantial improvement over traditional techniques like principal component analysis (PCA) and least absolute shrinkage and selection operator (LASSO) providing enhanced accuracy and efficiency in processing extensive microbiome datasets. Such tools demonstrated their proficiency in handling complex and large-scale biological data, thus, setting a new benchmark in the field (Babjac et al. 2022). In metagenome-wide association studies (MWAS) with random forest (RF), an ML algorithm was used to elucidate the relationship between soil microbiome composition and crop productivity. The work identified key microbial taxa associated with nitrogen utility like Bradyrhizodium and Gammaproteobacteria confirmed their varied abundance in soils of differing productivities. The use of RF in this context achieved an impressive prediction accuracy of 0.79 at the order level underscoring the potential of ML in translating complex microbiome data into actionable agricultural insights (Chang et al. 2017).

8.5 Soilborne Pathogens Causing Biotic Stress and Their Effect on Rhizo-Microbiome

Another side of soil inhabitants, they probably cause several diseases in plants. Soil pathogens impact the plant health, influencing soil structure and fertility. Research has shown that the prevalence of certain soilborne pathogens affects the soil structure by disrupting the activities of beneficial soil microorganisms and altering the availability of organic matter. These pathogens produce dormant resting structures in the absence of host and shield them during unfavorable conditions in the soil. When they find a suitable environment, such as the root exudates of specific host or a sufficient supply of nutrients, they germinate and infect the host plant. These pathogens infect the nonhost plant without causing any visual symptoms or disease, for example, *Verticillium* (Katan 2017). The pathogens cause diseases like damping off, root rots, and wilts that lower the effectiveness of water and nutrient uptake, and directly hinder plant growth (Dignam et al. 2022). Various soilborne fungal and bacterial pathogens caused severe loss to the agricultural crops. One such major soilborne pathogens is *Macrophomina phaseolina*, a highly destructive plant pathogen that spreads through the soil and infects a vast range of hosts belonging to ~500 plant species (Khan and Javaid 2023). Another major soil pathogen *Verticillium dahliae* caused high yield of losses of various crop species, including potatoes, lettuce, peppers, strawberries, cotton, cauliflowers, sunflower, aubergines, and olive trees (Bélair et al. 2018; Kowalska 2021).

Rhizoctonia solani was recognized as a significant soilborne fungal pathogen responsible for causing diverse diseases in solanaceous crops, leading to an annual global crop loss ranging from 20% to 40% (Ghosh et al. 2017). There are bacteria of ~7100 plant pathogens worldwide and classified as members of the *Agrobacterium*, *Pseudomonas*, *Erwinia*, *Pectobacterium*, *Ralstonia*, and *Xanthomonas* genera (Aguilar-Marcelino et al. 2020). Numerous bacterial plant pathogens exhibited a broad spectrum of host. For instance, *Ralstonia solanacearum* infect over 200 species across 50 families, including solanaceous crops like tomato, potato, eggplant, pepper, tobacco, and banana. It was considered one of the most dangerous pathogens residing in the soil and reduced crop yield by up to 90–100% (Yanti et al. 2022). Likewise, *Pseudomonas syringae* infected a wide range of crops from potato, olive, banana, tomato, etc. (Mansfield et al. 2012). There are reports that showed that soilborne pathogens altered the rhizospheric microbiome of host plant understood through metagenomics and metatranscriptomics approach. A study highlighted *Tenotrophomonas* spp. and *Buttiauxella* spp. abundance suppress pathogen *Rhizoctonia solani* AG8 disease. The disease occurrence differences in two adjacent wheat fields are revealed through rhizosphere metatranscriptomic data (Hayden et al. 2018). Another research on avocado rhizospheres demonstrated shifts in the microbiome and metabolome due to *Phytophthora cinnamomi* infection and treatment with a biological control agent (Reverchon et al. 2023). Furthermore, an investigation of the colonization behavior of *Burkholderia* strain B23 in citrus, utilizing both metagenomics and metatranscriptomics, shed light on the niche differentiation and differential gene expression patterns critical for nutrient uptake and stress resistance (Wang et al. 2023). In another study, Arecanut (*Areca catechu* L.) had been adversely affected by Yellow Leaf Disease (YLD) by phytoplasmas. NGS data revealed distinct bacterial phyla in the diseased field soil could be indicators of phytoplasmas that will be helpful for future assessment of contaminated fields (Paulraj et al. 2021). Additionally, metagenomic sequencing has provided crucial insights into the rhizosphere microbial community changes due to root-knot nematode (RKN)-Meloidogyne-disease complex in tobacco, identifying the key factors affecting the rhizospheric bacterial communities and offering new perspectives for the biological control (Lu et al. 2023).

8.6 Rhizo-Microbial Biocontrol Agents

Beneficial microbes could gather in the rhizosphere microbiome to directly inhibit pathogens and increase crop productivity. Biocontrol agents (BCAs) release a wide range of antimicrobial polypeptides and secondary metabolites, including enzymes that damage different parts of pathogens (Mullins et al. 2019; Vesga et al. 2020). They inhibit the broad-spectrum pathogen growth through direct biocontrol strategy, including niche competition and antibiotics production. Microbial biocontrol agents competitively excluded pathogens or occupy niches critical for the pathogen survival. This competitive exclusion deprives the phytopathogens of necessary nutrients and space, hindering their proliferation and resulting in pathogen growth inhibition (Wang et al. 2021a, b). Ectomycorrhiza (EcM) fungi provide significant resistance against soilborne pathogens by ensheathing feeder roots and acidifying the soil (Tedersoo et al. 2020). Beneficial rhizospheric bacterial strains like *Pseudomonas*, *Flavobacterium*, *Chitinophaga*, *Chryseobacterium*, *Rhizobium*, *Pedobacter*, and *Variovorax* were attracted by wheat plant and found to antagonize the soilborne pathogen *R. solani* AG8. *Janthinobacterium* exhibited broad antagonistic activity against soilborne pathogens *R. oryzae*, *Pythium ultimum*, and *R. solani* AG8 under in vitro conditions and suppress *R. solani* AG8 in field conditions too (Yin et al. 2021). In order to combat *Fusarium* wilt disease, susceptible cucumber plants gather beneficial microbes (such as Comamonadaceae and Xanthomonadaceae) and secreted organic acids to control the growth of pathogen (Wen et al. 2020). A study confirmed that the biocontrol agent effectively competes against the pathogen's influence on the native plant-associated microorganisms, which is a positive feature. As a result, they might be useful biocontrol agents. Although an endophytic *Acinetobacter* strain obtained from normal stems of the plant *Cinnamomum camphora* has been employed as a biocontrol strain counter to fungal infections (Liu et al. 2007). In another study, the antagonistic capabilities of two *Bacillus* sp. strains capable of degrading methomyl were evaluated against a soilborne pathogen *Alternaria* sp. (Roy et al. 2018). Some of the important BCAs, along with their mechanism of action against several phytopathogens in the last decade, are discussed in Table 8.1.

8.7 Direct Mechanism of Microbial Biocontrol Through Diverse Metabolites

Small antimicrobial molecules known as antibiotics are produced by BCAs to stop pathogen growth or metabolic processes (El-Saadony et al. 2022). Various bacterial genera such as *Pseudomonas*, *Pantoea*, *Streptomyces*, *Agrobacterium*, *Bacillus*, *Serratia*, and *Stenotrophomonas* produced compounds exhibiting broad-spectrum activity. For instance, several species of *Pseudomonas* produced compounds such as pyoluteorin, phenazine-1-carboxamide, 2, 4 diacetylphloroglucinol (DAPG),

Table 8.1 The list of beneficial root microbiota: plant health induction and disease suppression in different crops

Sr. no.	Biocontrol agent	Host plant	Growth-promoting traits	Biocontrol activity	Target pathogen	Location	References
Free-living beneficial rhizosphere inhabiting bacteria							
1.	*Bacillus cereus* Disha A and *Bacillus safensis* Disha B	Lentil (*Lens culinaris*)	IAA, P-solubilization	HCN, siderophore, lytic enzymes	*Alternaria* spp.	West Bengal, India	Roy et al. (2018)
2.	*Pseudomonas syringae*	Chickpea (*Cicer arietinum*)	ACC deaminase	Siderophore	*Fusarium avenaceum*	Saskatoon, Canada	Hynes et al. (2008)
3.	*Bacillus* sp.	Corn mint (*Mentha arvensis*)	IAA, P-solubilization	Siderophore	ND	Lucknow, India	Prakash and Arora (2019)
4.	*Pseudomonas putida*, *Pseudomonas Fluorescens*, and *Burkholderia glumae*	Tomato (*Solanum lycopersicum*)	IAA, P-solubilization, ACC deaminase	HCN, siderophore	ND	Canada	Rashid et al. (2012)
5.	*Pseudomonas* sp. FBJ6	Pea (*Pisum sativum*) and maize (*Zea mays*)	IAA, P solubilization	HCN	ND	Solan, India	Kumar et al. (2012)
6.	*Bacillus siamensis*, LZ8	Tobacco (*Nicotiana tabacum*)	ND	Volatile organic compounds, peroxidase, and polyphenol oxidase	*Alternaria alternata*	Chongqin, China	Xie et al. (2021)
7.	*Paenibacillus polymyxa* GBR-462	Chili pepper (*Capsicum annuum*)	ND	Callose, pectin, phenolic compounds, siderophore	*Phytopthora capsici* (Phytopthora blight)	Seoul, Korea	Kim et al. (2009)
8.	*Azotobacter* sp. FY28	Wheat (*Triticum aestivum*) and barley	IAA, P solubilization	Siderophore, protease	ND	Tabriz, Iran	Farajzadeh et al. (2012)
9.	*Pseudomonas lurida* M2RH3	Wheat (*Triticum aestivum*)	IAA	Siderophore	ND	Uttarakhand, India	Selvakumar et al. (2011)

(continued)

Table 8.1 (continued)

Sr. no.	Biocontrol agent	Host plant	Growth-promoting traits	Biocontrol activity	Target pathogen	Location	References
10.	*Bacillus licheniformis* RCO4	Barley (*Hordeum vulgare*)	N_2-fixation	Starch hydrolysis	ND	Istanbul, Turkey	Canbolat et al. (2006)
11.	*Bacillus subtilis*	Grapevine (*Vitis vinifera*)	ND	Lipopeptides (surfactin and mycosubtilin)	*Botrytis cinerea*	France	Farace et al. (2015)
12.	*Bacillus licheniformis* GL174	Grapevine (*Vitis vinifera* cv. *Glera*)	ND	Lytic enzymes, siderophores, secondary metabolites (lipopeptides)	*Phaeoacremonium leophilum*, *Phaeomoniella* spp., *Botrytis cinerea*, *Sclerotinia sclerotiorum*, and *Phytophthora infestans*	Padova, Italy	Nigris et al. (2018)
13.	*Pseudomonas fluorescens* 2P24	Wheat (*Triticum aestivum*)	ND	HCN, siderophore, 2,4-diacetyl phloroglucinol (2,4-DAPG)	*Ralstonia solanacearum* and *Rhizoctonia solani*	Beijing, China	Wei and Zhang (2006)
14.	*Pseudomonas* sp. NS 1 and *Bacillus* sp. NS 22	Pigeon pea (*Cajanus cajan*)	Enhanced seedling germination and shoot and root elongation	Chitinolytic (endochitinase, exochitinase, chitobiase), lytic enzymes (proteinase, cellulose, amylase, pectinase, lipase), HCN, siderophores, ammonia, phenylalanine ammonia lyase, polyphenol oxidase, and peroxidase	*Fusarium udum*	Abohar, Punjab, India	Dukare and Paul (2021)
15.	*Pseudomonas tolaasii* ACC23	Canola (*Brassica napus*)	IAA, ACC deaminase	Siderophores	ND	Milan, Italy	Dell'Amico et al. (2008)

16.	*Pseudomonas fluorescens* (biovars I, II, III, and V)	Okra (*Abelmoschus esculentu*)	IAA	HCN, siderophore	*Rhizoctonia solani* (root rot)	West Bengal, India	Adhikari et al. (2013)
17.	*Bacillus amyloliquefaciens* DSBA-11 and DSBA-12 b	Tomato (*Solanum lycopersicum*)	IAA, P solubilization	Siderophore, ammonia production	*Ralstonia solanacearum*	New Delhi, India	Singh et al. (2016)
18.	*Bacillus subtilis* (BS2)	*Tomato* (*Solanum lycopersicum*)	ND	Peroxidase, polyphenol oxidase, chitinase, phenylalanine ammonialyase, and phenolics	*Fusarium oxysporum* f.sp. *lycopersici*	Uttar Pradesh, India	Loganathan et al. (2014)
19.	*Burkholderia cepacia* SE4, *Promicromonospora* sp. SE188 and *Acinetobacter calcoaceticus* SE370	Cucumber (*Cucumis sativus*)	Gibberellin (GA_4)	Catalase, peroxidase, polyphenol oxidase, and polyphenol	ND	Korea	Kang et al. (2014)
20.	*Burkholderia contaminans* AY001	Tomato (*Solanum lycopersicum*)	IAA, nitrogen fixation, P solubilization, extracellular protease production, zinc solubilization	Antimicrobial compounds {di(2-ethyl hexyl) phthalate & pyrrolo [1,2-a] pyrazine-1,4-dione, hexahydro-3-(phenylmethyl)}	*Fusarium oxysporum* f.sp. *lycopersici*	Korea	Heo et al. (2022)
21.	*Bacillus velezensis* 5YN8 and DSN01	Pepper (*Piper nigrum*)	ND	Protease, chitinase, siderophore, cellulase, glucans, volatile organic compounds (VOCs)	*Botrytis cinerea* (gray mold)	Nanjing, China	Jiang et al. (2018)

(continued)

Table 8.1 (continued)

Sr. no.	Biocontrol agent	Host plant	Growth-promoting traits	Biocontrol activity	Target pathogen	Location	References
22.	*Bacillus subtilis* LHS1	Canola (*Brassica napus*)	IAA, P solubilization, nitrogenase activity	ND	*Sclerotinia sclerotiorum*	Gansu, China	Sun et al. (2017)
23.	*Bacillus amyloliquefaciens*	Cowpea (*Vigna unguiculata*)	IAA, ACC-deaminase, P-solubilization, zinc solubilization	Siderophores	*Macrophomina phaseolina* (charcoal rot)	Mexico	Rangel-Montoya et al. (2022)
24.	*Pseudomonas* spp. NS-1 and *Bacillus* spp. NS-22	Pigeon pea (*Cajanus cajan*)	ND	Chitinase	*Fusarium udum*	Abohar, Punjab, India	Dukare et al. (2020)
25.	*Pseudomonas aeruginosa* 23_{1-1}	Watermelon (*Citrullus lanatus*)	ND	Peroxidase	*Didymella bryoniae*	Denmark	Nga et al. (2010)
26.	*Bacillus subtilis* var. *amyloliquefaciens* FZB 24	Potato (*Solanum tuberosum*)	ND	Peroxidase, polyphenol oxidase, and phenylalanine ammonia lyase	*Phytophthora infestans* (Late Blight)	Kerala, India	Keerthana et al. (2018)
27.	*Pseudomonas aeruginosa* FG106	Potato (*Solanum tuberosum*), tomato (*Solanum lycopersicum*) (*Solanum lycopersicum*), and taro (*Colocasia esculenta*)	IAA, P solubilization	HCN, siderophore	*Botrytis cinerea*, *Clavibacterm ichiganensis* subsp. *michiganensis*, *Phytophthora colocasiae*, and *Xanthomonas euvesicatoria* pv. *Perforans*	Iran	Ghadamgahi et al. (2022)

28.	*Pseudomonas protegens* MP12	Grapevine (*Vitis vinifera*)	IAA, P solubilization	Antifungal compounds 2,4-diacetyl phloroglucinol (2,4-DAPG), pyoluteorin and pyrrolnitrin	*Botrytis cinerea*, *Alternaria alternata*, *Aspergillus niger*, *Penicillium expansum*, and *Neofusicoccum parvum*	Verona, Italy	Andreolli et al. (2019)
Free-living beneficial rhizosphere inhabiting fungi							
29.	*Aspergillus* sp-01 and sp-03	Wheat (*Triticum aestivum*)	IAA, P solubilization	Siderophore, lipase, amylase	ND	Aligarh, India	Imran et al. (2021)
30.	*Trichoderma asperellum* TS39, TS12, TS42, TS9, TS32, and TS36	Tomato (*Solanum lycopersicum*)	ND	Chitinase and β-1,3-glucanase	*Fusarium oxysporum* f. sp. *lycopersici*	Saudi Arabia	El-Komy et al. (2015)
31.	*Clonostachys rosea* IK726	Barley (*Hordeum vulgare*)	ND	Cell wall-degrading enzymes	*F. Culmorum*	Denmark	Jensen et al. (2007)
32.	*Aspergillus ustus*	*Arabidopsis thaliana* and potato (*Solanum tuberosum*)	IAA and GAs	ND	*Botrytis cinerea* and *Pseudomonas syringae* DC3000	Mexico	Salas-Marina et al. (2011)
33.	*Sphaerodes mycoparasitica*	Wheat (*Triticum aestivum*)	Enhanced seedling germination	Degradation of mycotoxins	*Fusarium graminearum*	Canada	Vujanovic and Goh (2012)
34.	*Penicillium viridicatum* GP15–1	Cucumber (*Cucumis sativus*)	Enhanced root and shoot length and biomass	Secondary metabolites	*Rhizoctonia solani* AG4	Bangladesh	Hossain et al. (2014)
35.	*Trichoderma harzianum* TriH_JSB27	Tomato (*Lycopersicon lycopersicum*)	Enhanced seedling germination and seedling vigor index	Induced PAL, POX, and GLU enzyme activities	*Ralstonia solanacearum*	Kerala, India	Jogaiah et al. (2013)

(continued)

Table 8.1 (continued)

Sr. no.	Biocontrol agent	Host plant	Growth-promoting traits	Biocontrol activity	Target pathogen	Location	References
36.	*Trichoderma harzianum* T22	Maize (*Zea mays*)	Enhanced shoot elongation and area and size of main and secondary roots	Chitinolytic (β-1,3 glucanase, exochitinase, and endochitinase)	*Colletotrichum graminicola*	Ithaca, New York	Harman et al. (2004)
37.	*Aspergillus fumigatus* LH02	Soyabean (*Glycine max*)	GAs Higher proline, salicylic acid (SA) and jasmonic acid (JA)	ND	ND	Texas	Khan et al. (2011)
38.	*Fusarium oxysporum* V5W2	Banana (*Musa* sp.cv. Giant Cavendish and cv. Grand Nain)	Improved yield	ND	*Pratylenchus goodeyi* and *Helicotylenchus multicinctus*	Nairobi, Kenya	Waweru et al. (2014)
39.	*Talaromyces wortmannii* FS2	Field mustard (*Brassica campestris*)	Increased shoot fresh weight	Volatile organic compounds (β-caryophyllene)	*Colletotrichum higginsianum*	Okayama, Japan	Yamagiwa et al. (2011)
40.	*Trichoderma viride*	Sugarcane (*Saccharum officinarum*)	Increased yield and soil organic carbon, N, P, and K content	ND	ND	Lucknow, India	Yadav et al. (2009)

ND not defined

oomycin, phenazine-1-carboxylic acid, and pyrrolnitrin that have antifungal, antiviral, and antibacterial properties (Khabbaz et al. 2015). Siderophores production has been reported as a crucial PGPR mechanism that hinders flourishing of phytopathogens. They are ferric iron chelating agents that prevent phytopathogens from obtaining iron, thus preserving plant health, especially in iron-deficient conditions. The isolates of *Pseudomonas* sp. reduced the proliferation of pathogens present in rhizospheres by reducing the amount of iron surrounding the pathogen (Wilson et al. 2016). Hydrogen cyanide (HCN), a volatile compound, suppressed and inhibited phytopathogens. HCN toxic effect was ascribed to its capacity to inhibit electron transport and cytochrome oxidase, thereby disrupting the energy supply to pathogen cells (Jagadeesh and Kulkarni 2003). In another study, *Pseudomonas protegens* was used against *R. solanacearum*, which inhibited the pathogen through antimicrobial metabolites such as pyoluteorin, diacetyl phloroglucinol, pyrrolnitrin, and HCN production (Rai et al. 2017). *B. subtilis* ATCC6633 inhibited the germination of *Clostridium* sp., causal agent soft rot of potato through production of bacteriocin (Abdelaziz et al. 2023). *Bacillus altitudinis* AMCC1040 was evaluated against bark cracking disease of ginger, inhibited nematode reproduction and root galling, and also reduced the disease severity (Wang et al. 2021a, b). The biocontrol ability was enhanced using a combination of two or more beneficial microbes to increase the reliability and efficiency to control the disease. For instance, *Stenotrophomonas maltophilia* W81 and *Pseudomonas fluorescens* F113 prevent sugar beetroot damping-off disease through extracellular proteolytic activity and 2,4-diacetylphloroglucinol (DAPG), respectively, in field conditions (Dunne et al. 1998; Wang et al. 2021a, b). The plants recruit the beneficial rhizospheric microbes upon pathogen attack via release of specific root exudates that in turn help the plant combat the disease incidence. The barley plants modify the rhizosphere community to favor a particular function like antifungal upon pathogen attack (Dudenhöffer et al. 2016). The invasion of *P. syringae* pv. tomato (Pst) in *Arabidopsis* attracted the advantageous organism *Bacillus subtilis* (Li et al. 2021).

PGPR, as well as biocontrol agent *Paenibacillus* (T22) and *Bacillus subtilis* strains, influenced the rhizosphere dynamics of wheat cultivars. The work unraveled the differential accumulation of lipids, benzenoids, organic acids, and phenylpropanoids in the rhizosphere by using untargeted metabolomics. Benzenoids, involved in the manufacture of cyanogenic glycosides for plant defense, such as 4-(p)-hydroxymandelonitrile, are among the diverse range of chemical metabolites found in the rhizosphere. Benzoxazinoids and its derivatives had antibacterial and allelopathic properties. Microbial growth, metal detoxification, and nutrient solubilization were impacted by abundant organic acids. The presence of 4-fumarylacetoacetate showed the phenylalanine catabolism was mediated by PGPR. Lipids, phenylpropanoids, and polyketides contributed to the chemical space of the rhizosphere, influencing plant–microbe interactions and rhizosphere microbial community. The work emphasized the constitutive character of these metabolites as well as their critical role in plant–microbe interactions (Mashabela et al. 2022). However, in another study, Pseudomonads were reported to generate antibiotics such as DAPG, amphisin, hydrogen cyanide, phenazine, oomycin A, tropolone, pyoluteorin, tensin, pyrrolnitrin, and cyclic lipopeptides. *Streptomyces*, *Bacillus*, and *Stenotrophomonas* spp., on the other hand, produced antibiotics such as kanosamine, oligomycin A, xanthobaccin, and zwittermicin. Antibacterial, antifungal, antiviral, anthelminthic, antimicrobial, cytotoxic, phytotoxic, antioxidant, and antitumor properties were present in these antibiotics (Abbas et al. 2018).

Current investigations revealed *Arabidopsis* can recruit beneficial soil microbes through altered pattern of plant root exudates in response to disease onset (Yuan et al. 2018). Infected plants release microbe-induced plant volatile (MIPVs) (hexenal isomers, 2,3-butanediol, β-caryophyllene), p-coumaric acid, phenolic acid, triterpenes, flavonoid, citric acid, pyruvate acid, succinic acid, fumarate, sugars, organic acids, nucleotides, long-chain fatty acids, short-chain organic acids, and large amount of salicylic acid/jasmonic acid shapes and manipulates the microbial community roughly around the roots (Li et al. 2021). Not only in root disease, but in case of foliar infection of *Arabidopsis* specifically with downy mildew pathogen *Hyaloperonospora arabidopsidis* encourages *Microbacterium*, *Stenotrophomonas*, and *Xanthomonas* sp. in its root proximity (Berendsen et al. 2018). Additionally, lytic enzymes produced by different species of biocontrol agents complement each other; a mixture of these enzymes will be more effective at suppressing multiple pathogens that are present in the host rhizosphere. They secrete cell wall degrading enzymes (CWDEs) as glucanases, cellulases, chitinases, and proteases to disrupt the pathogen cell wall (Santoyo et al. 2021). For instance, the pathogenic fungus *Botrytis cinerea* germination and germ tube elongation were inhibited by *Serratia plymuthica* that secretes chitinase. The *Bacillus* sp. BPR7 demonstrates the inhibitory effects on the growth of fungal pathogens *F. solani*, *F. oxysporum*, and *R. solani* through its chitinase and β-1,4-glucanase activities under in vitro conditions. Similarly, *Paenibacillus ehimensis* KWN38 displays elevated levels of chitinase, cellulase, glucanase protease, and activities against *Phytophthora capsica*, *F. oxysporum* f. sp. lycopersici, and *R. solani* AG-1 (Naing et al. 2014).

8.8 Indirect Mechanism of Microbial Biocontrol Through ISR

In spite of direct methods, microbes inhibit the growth of pathogen through indirect methods specifically through the induction of plant immunity. Biocontrol agents have the ability to produce volatile organic compounds (VOCs), important for plant growth and the development of pathogen-induced systemic resistance (ISR) (Raza et al. 2016). Plants can develop induced systemic resistance (ISR) through the stimulation of nonpathogenic microorganisms, a phenomenon that shares similarities with pathogen-induced systemic acquired resistance (SAR). The reports have showed that rhizospheric microbes provide resistance through induced systemic resistance (ISR). ISR involves the induction of a plant's resistance mechanism through beneficial or nonpathogenic microbes such as plant growth-promoting microbes (PGPM) that suppress or reduce the detrimental effects of plant pathogens (Babalola et al. 2023). *Bacillus* spp., the most explored rhizobacterium, causes ISR. *Paenibacillus* isolate P16-induced systemic resistance in cabbage had been associated to its biocontrol potential against *Xanthomonas campestris* pv. campestris, the causative agent of black rot in cabbage. In the model plant *Arabidopsis*, rhizobacteria-triggered ISR has been shown to confer broad-spectrum resistance against various plant pathogens (Pieterse et al. 1998). This effect is not limited to *Arabidopsis* but extends to food crops such as bean, carnation, cucumber, pepper, radish, and tomato (Salwan et al. 2023). Different bacterial strains have been found to induce varying levels of resistance in different plant species. For instance, *Paenibacillus polymyxa* E681, *Bacillus amyloliquefaciens* EXTN-1, *Bacillus pumilus* SE34, *Pseudomonas fluorescens* 89B61, and *Pseudomonas fluorescens* strain WCS417r are known for their specificity in inducing ISR against broad-spectrum of pathogens (Salwan et al. 2023). Plant growth regulators, including salicylic acid (SA), jasmonic acid (JA), and ethylene (ET), are pivotal in mediating both induced systemic resistance (ISR) and systemic acquired resistance (SAR), as highlighted by Van wees et al. (2000). The transcription factors WRKY11 and WRKY70, regulated by two distinct pathways dependent on NPR1, have been recognized for their involvement in induced systemic resistance (ISR). The biocontrol agents have been found to inhibit the activity of miR472, which is responsible for silencing the nucleotide-binding site (NBS) and leucine-rich repeat (LRR) essential for the immune response. Pathogenesis-related (PR) protein genes such as *CaPR1*, *CaPR4*, *CaPR10, and* MYB72 were also found to activate ISR through rhizobacterial treatments. The induction of induced systemic resistance (ISR) in *Arabidopsis* was observed upon the application of methyl jasmonate (MeJA) or aminocyclopropane-1-carboxylic acid (ACC), akin to the effects observed with jasmonic acid (JA) and ethylene (ET) (Salwan et al. 2023).

8.9 Conclusion

The rhizosphere microbes establish intricate interaction with the host plant by forming a complex web of interaction within the rhizosphere. These microbes provide various benefits to the host and its surrounding in many ways. The host plant releases root exudates that can enhance the suppressive capabilities of beneficial microbes against soilborne pathogens. Beneficial microbes can directly inhibit pathogen growth by competing for nutrients and space, excreting antimicrobial compounds, or antagonizing the pathogen. On the other hand, they can also inhibit pathogen growth indirectly by activating the plant's immune responses and their hormone levels. There are diverse microorganisms present in the rhizosphere, and several techniques like metatranscriptome and metagenome are being used to study the microbial diversity. NGS techniques offer unprecedented opportunities to investigate the interactions between plants and the root microbiome. Biocontrol agents can be an effective and sustainable alternative to chemical bactericides and fungicides in biotic stress mitigation. Thus, further research is required to modify the composition of the rhizospheric microbiome to create a biocontrol-friendly environment.

References

Abbas T, Zahir ZA, Naveed M, Kremer RJ (2018) Limitations of existing weed control practices necessitate development of alternative techniques based on biological approaches. Adv Agron 147:239–280

Abdelaziz AM, Hashem AH, El-Sayyad GS, El-Wakil DA, Selim S, Alkhalifah DH, Attia MS (2023) Biocontrol of soil-borne diseases by plant growth-promoting rhizobacteria. Trop Plant Pathol 48(2):105–127

Adhikari K, Hartemink AE (2016) Linking soils to ecosystem services-a global review. Geoderma 262:101–111

Adhikari A, Dutta S, Nandi S, Bhattacharya I, De Roy M, Sarkar G, Mandal T (2013) Antagonistic potentiality of native rhizobacterial isolates against root rot disease of okra, incited by *Rhizoctonia solani*. Afr J Agric Res 8(4):405–412

Afiahayati, Sato K, Sakakibara Y (2015) MetaVelvet-SL: an extension of the velvet assembler to a de novo metagenomic assembler utilizing supervised learning. DNA Res 22(1):69–77

Afridi MS, Fakhar A, Kumar A, Ali S, Medeiros FHV, Muneer MA, Ali H, Saleem M (2022) Harnessing microbial multitrophic interactions for rhizosphere microbiome engineering. Microbiol Res 265:127–199. https://doi.org/10.1016/j.micres.2022.127199

Aguilar-Marcelino L, Mendoza-de-Gives P, Al-Ani LKT, López-Arellano ME, Gómez-Rodríguez O, Villar-Luna E, Reyes-Guerrero DE (2020) Using molecular techniques applied to beneficial microorganisms as biotechnological tools for controlling agricultural plant pathogens and pests. In: Molecular Aspects of Plant Beneficial Microbes in Agriculture, Academic Press, 333–349

Ahmad MF, Abdullah H, Hassan MN, Jamaludin MI, Sivam A, Komatsu K, Yaacob NS (2023) Topographically distinguished microbiome taxonomy and stress-response genes of Royal Belum Rainforest and Raja Muda Musa peat swamp revealed through metagenomic inquisition. Int J Mol Sci 24(1):872

Andreolli M, Zapparoli G, Angelini E, Lucchetta G, Lampis S, Vallini GS (2019) Pseudomonas protegens MP12: a plant growth-promoting endophytic bacterium with broad-spectrum antifungal activity against grapevine phytopathogens. Microbiol Res 219:123–131

Aslam MM, Idris AL, Zhang QIAN, Weifeng XU, Karanja JK, Wei YUAN (2022) Rhizosphere microbiomes can regulate plant drought tolerance. Pedosphere 32(1):61–74

Babalola OO, Olowe OM, Ayangbenro AS (2023) Shotgun metagenomics dataset of *Striga hermonthica*-infested maize (*Zea mays* L.) rhizospheric soil microbiome. Data Brief 48:109–132

Babjac A, Royalty T, Steen AD, Emrich SJ (2022) A comparison of dimensionality reduction methods for large biological data. In: Proceedings of the 13th ACM international conference on bioinformatics, computational biology and health informatics 1-7

Baldrian P, López-Mondéjar R, Kohout P (2023) Forest microbiome and global change. Nat Rev Microbiol 21:487–501. https://doi.org/10.1038/s41579-023-00876-4

Baltoumas FA, Karatzas E, Paez-Espino D, Venetsianou NK, Aplakidou E, Oulas A, Finn RD, Ovchinnikov S, Pafilis E, Kyrpides NC, Pavlopoulos GA (2023) Exploring microbial functional biodiversity at the protein family level-from metagenomic sequence reads to annotated protein clusters. Front Bioinform 3:1157956

Bélair G, Jean Coulombe J, Dauphinais N (2018) Management of *Pratylenchus penetrans* and *Verticillium* symptoms in strawberry. Phytoprotection 98(1):1–3

Berendsen RL, Vismans G, Yu K, Song Y, de Jonge R, Burgman WP, Pieterse CM (2018) Disease-induced assemblage of a plant-beneficial bacterial consortium. ISME J 12(6):1496–1507

Calvert J, McTaggart A, Carvalhais LC, Rensink S, Dennis PG, Drenth A, Shivas R (2023) Divergent rainforest tree microbiomes between phases of the monsoon cycle, host plants and tissues. Plant Biol 25:860–870. https://doi.org/10.1111/plb.13569

Canbolat MY, Barik K, Çakmakçi R, Şahin F (2006) Effects of mineral and biofertilizers on barley growth on compacted soil. Acta Agric Scand Sect B Soil Plant Sci 56(4):324–332

Chang HX, Haudenshield JS, Bowen CR, Hartman GL (2017) Metagenome-wide association study and machine learning prediction of bulk soil microbiome and crop productivity. Front Microbiol 8:519

Chen J, Wu Y, Yang H, Wang S, Wu W, Lyu L et al (2022a) Long-term cultivation drives dynamic changes in the rhizosphere microbial community of blueberry. Front Plant Sci 13:962759

Chen Y, Xi J, Xiao M, Wang S, Chen W, Liu F (2022b) Soil fungal communities show more specificity than bacteria for plant species composition in a temperate forest in China. BMC Microbiol 22:208. https://doi.org/10.1186/s12866-022-02591-1

Chen P, Yu K, He Y (2023) The dynamics and transmission of antibiotic resistance associated with plantmicrobiomes. Environ Int 176:107986. https://doi.org/10.1016/j.envint.2023.107986

Das PP, Singh KR, Nagpure G, Mansoori A, Singh RP, Ghazi IA, Singh J (2022) Plant-soil-microbes: a tripartite interaction for nutrient acquisition and better plant growth for sustainable agricultural practices. Environ Res 214:113821

Dell'Amico E, Cavalca L, Andreoni V (2008) Improvement of *Brassica napus* growth under cadmium stress by cadmium-resistant rhizobacteria. Soil Biol Biochem 40(1):74–84

Deng X, Zhang N, Li Y, Zhu C, Qu B, Liu H, Falcao Salles J (2022) Bio-organic soil amendment promotes the suppression of *Ralstonia solanacearum* by inducing changes in the functionality and composition of rhizosphere bacterial communities. New Phytol 235(4):1558–1574

Dignam BE, Marshall SD, Wall AJ, Mtandavari YF, Gerard EM, Hicks E, Bell NL (2022) Impacts of soil-borne disease on plant yield and farm profit in dairying soils. J Sustain Agric Environ 1(1):16–29

Dlamini SP, Akanmu AO, Babalola OO (2022) Rhizospheric microorganisms: the gateway to a sustainable plant health. Front Sustain Food Syst 6:925802. https://doi.org/10.3389/fsufs.2022.925802

Dudenhöffer JH, Scheu S, Jousset A (2016) Systemic enrichment of antifungal traits in the rhizosphere microbiome after pathogen attack. J Ecol 104:1566–1575. https://doi.org/10.1111/1365-2745.12626

Dukare A, Paul S (2021) Biological control of fusarium wilt and growth promotion in pigeon pea (*Cajanus cajan*) by antagonistic rhizobacteria, displaying multiple modes of pathogen inhibition. Rhizosphere 17:100278

Dukare A, Paul S, Arambam A (2020) Isolation and efficacy of native chitinolytic rhizobacteria for biocontrol activities against fusarium wilt and plant growth promotion in pigeon pea (*Cajanus cajan* L.). Egypt. J Biol Pest Control 30(1):1–12

Dunne C, Moenne-Loccoz Y, McCarthy J, Higgins P, Powell J, Dowling DN, O'Gara F (1998) Combining proteolytic and phloroglucinol-producing bacteria for improved biocontrol of Pythium-mediated damping-off of sugar beet. Plant Pathol 47:299–307

El-Komy MH, Saleh AA, Eranthodi A, Molan YY (2015) Characterization of novel *Trichoderma asperellum* isolates to select effective biocontrol agents against tomato fusarium wilt. Plant Pathol 31(1):50

El-Saadony MT, Saad AM, Soliman SM, Salem HM, Ahmed AI, Mahmood M, El-Tahan AM, Ebrahim AAM, Abd El-Mageed TA, Negm SH, Selim S, Babalghith AO, Elrys AS, El-Tarabily KA, AbuQamar SF (2022) Plant growth-promoting microorganisms as biocontrol agents of plant diseases: mechanisms, challenges and future perspectives. Front Plant Sci 13:923880. https://doi.org/10.3389/fpls.2022.923880

Farace G, Fernandez O, Jacquens L, Coutte F, Krier F, Jacques P, Dorey S (2015) Cyclic lipopeptides from *Bacillus subtilis* activate distinct patterns of defence responses in grapevine. Mol Plant Pathol 16(2):177–187

Farajzadeh D, Yakhchali B, Aliasgharzad N, Sokhandan-Bashir N, Farajzadeh M (2012) Plant growth promoting characterization of indigenous Azotobacteria isolated from soils in Iran. Curr Microbiol 64:397–403

Fiore-Donno AM, Freudenthal J, Dahl MB, Rixen C, Urich T, Bonkowski M (2024) Biotic interactions explainseasonal dynamics of the alpine soil microbiome. ISME commun 4(1):ycae028.10.1093/ismeco/ycae028

Fuentes A, Herrera H, Charles TC, Arriagada C (2020) Fungal and bacterial microbiome associated with therhizosphere of native plants from the Atacama Desert. Microorganisms 8(2):209. https://doi.org/10.3390/microorganisms8020209

Ghadamgahi F, Tarighi S, Taheri P, Saripella GV, Anzalone A, Kalyandurg PB, Vetukuri RR (2022) Plant growth-promoting activity of *Pseudomonas aeruginosa* FG106 and its ability to act as a biocontrol agent against potato, tomato and taro pathogens. Biology 11(1):140

Ghosh S, Kanwar P, Jha G (2017) Alterations in rice chloroplast integrity, photosynthesis and metabolome associated with pathogenesis of *Rhizoctonia solani*. Sci Rep 7:41610

Gupta R, Anand G, Gaur R, Yadav D (2021) Plant–microbiome interactions for sustainable agriculture: a review. Physiol Mol Biol Plants 27:165–179

Harman GE, Petzoldt R, Comis A, Chen J (2004) Interactions between *Trichoderma harzianum* strain T22 and maize inbred line Mo17 and effects of these interactions on diseases caused by *Pythium ultimum* and *Colletotrichum graminicola*. Phytopathology 94(2):147–153

Hayden HL, Savin KW, Wadeson J, Gupta VV, Mele PM (2018) Comparative metatranscriptomics of wheat rhizosphere microbiomes in disease suppressive and non-suppressive soils for *Rhizoctonia solani* AG8. Front Microbiol 9:85

Heo AY, Koo YM, Choi HW (2022) Biological control activity of plant growth promoting *rhizobacteria Burkholderia* contaminans AY001 against tomato Fusarium wilt and bacterial speck diseases. Biology 11(4):619

Hossain MM, Sultana F, Miyazawa M, Hyakumachi M (2014) The plant growth-promoting fungus *Penicillium* spp. GP15-1 enhances growth and confers protection against damping-off and anthracnose in the cucumber. J Oleo Sci 63(4):391–400

Hynes RK, Leung GC, Hirkala DL, Nelson LM (2008) Isolation, selection, and characterization of beneficial rhizobacteria from pea, lentil, and chickpea grown in western Canada. Can J Microbiol 54:248–258

Imran M, Abulreesh HH, Monjed MK, Elbanna K, Samreen AI (2021) Multifarious functional traits of free-living rhizospheric fungi, with special reference to *Aspergillus* spp. isolated from north Indian soil, and their inoculation effect on plant growth. Ann Microbiol 71:1–17

Jagadeesh K, Kulkarni J (2003) Mechanisms of biocontrol in Rhizobacteria of tomato antagonistic to *Ralstonia solanacearum* EF smith causing bacterial wilt in tomato. Proceedings of the sixth international PGPR workshop on plant growth promoting Rhizobacteria, Calicut, India 5–10

Jensen DF, Knudsen IM, Lübeck M, Mamarabadi M, Hockenhull J, Jensen B (2007) Development of a biocontrol agent for plant disease control with special emphasis on the near commercial fungal antagonist *Clonostachys rosea* strain 'IK726'. Australas Plant Pathol 36:95–101

Jiang CH, Liao MJ, Wang HK, Zheng MZ, Xu JJ, Guo JH (2018) *Bacillus velezensis*, a potential and efficient biocontrol agent in control of pepper gray mold caused by *Botrytis cinerea*. Biol Control 126:147–157

Jogaiah S, Abdelrahman M, Tran LS, Shin-Ichi I (2013) Characterization of rhizosphere fungi that mediate resistance in tomato against bacterial wilt disease. J Exp Bot 64(12):3829–3842

Kang SM, Latif Khan A, You YH, Kim JG, Kamran M, Lee IJ (2014) Gibberellin production by newly isolated strain *Leifsonia soli* SE134 and its potential to promote plant growth. J Microbiol Biotechnol 24:106–112

Katan J (2017) Diseases caused by soilborne pathogens: biology, management and challenges. J Plant Pathol:305–315

Keerthana U, Nagendran K, Raguchander T, Prabakar K, Rajendran L, Karthikeyan G (2018) Deciphering the role of *Bacillus subtilis* var. *amyloliquefaciens* in the management of late blight pathogen of potato, *Phytophthora infestans*. Proc Natl Acad Sci India Sect B Biol Sci 88(3):1071–1080

Khabbaz S, Zhang L, Cáceres L, Sumarah M, Wang A, Abbasi P (2015) Characterisation of antagonistic *Bacillus* and *Pseudomonas* strains for biocontrol potential and suppression of damping-off and root rot diseases. Ann Appl Biol 166:456–471

Khan IH, Javaid A (2023) *Macrophomina phaseolina* causing various diseases in different crops. In: Macrophomina Phaseolina 21–31

Khan AL, Hamayun M, Kim YH, Kang SM, Lee JH, Lee IJ (2011) Gibberellins producing endophytic *Aspergillus fumigatus* sp. LH02 influenced endogenous phytohormonal levels, isoflavonoids production and plant growth in salinity stress. Process Biochem 46(2):440–447

Kim SG, Khan Z, Jeon YH, Kim YH (2009) Inhibitory effect of *Paenibacillus polymyxa* GBR-462 on *Phytophthora capsici* causing Phytophthora blight in chili pepper. J Phytopathol 157:329–337

Kowalska B (2021) Management of the soil-borne fungal pathogen–*Verticillium dahliae* Kleb. causing vascular wilt diseases. J Plant Pathol 103(4):1185–1194

Kumar A, Kumar A, Devi S, Patil S, Payal C, Negi S (2012) Isolation, screening, and characterization of bacteria from Rhizospheric soils for different plant growth promotion (PGP) activities: an *in vitro* study. Recent Res Sci Technol 4(1):1–5

Kumawat KC, Razdan N, Saharan K (2022) Rhizospheric microbiome: bio-based emerging strategies for sustainable agriculture development and future perspectives. Microbiol Res 254:126901

Li J, Wang C, Liang W, Liu S (2021) Rhizosphere microbiome: the emerging barrier in plant-pathogen interactions. Front Microbiol 12:772420. https://doi.org/10.3389/fmicb.2021.772420

Li W, Li Y, Lv J, He X, Wang J, Teng D, Jiang L, Wang H, Lv G (2022) Rhizosphere effect alters the soil microbiome composition and C, N transformation in an arid ecosystem. Appl Soil Ecol 170:104296. https://doi.org/10.1016/j.apsoil.2021.104296

Lippold E, Schlüter S, Mueller CW, Höschen C, Harrington G, Kilian R, Vetterlein D (2023) Correlative imaging of the rhizosphere— a multimethod workflow for targeted mapping of chemical gradients. Environ Sci Technol 57(3):1538–1549

Liu CH, Chen X, Liu TT, Lian B, Gu Y, Caer V, Xue YR, Wang BT (2007) Study of the antifungal activity of Acinetobacter baumannii LCH001 in vitro and identification of its antifungal components. Appl Microbiol Biotechnol 76:459–466. https://doi.org/10.1007/s00253-007-1010-0

Loganathan M, Garg R, Venkataravanappa V, Saha S, Rai AB (2014) Plant growth promoting rhizobacteria (PGPR) induces resistance against fusarium wilt and improves lycopene content and texture in tomato. Afr J Microbiol Res 8(11):1105–1111

Lu P, Shi H, Tao J, Jin J, Wang S, Zheng Q, Liu P, Xiang B, Chen Q, Xu Y, Cao P (2023) Metagenomic insights into the changes in the rhizosphere microbial community caused by the root-knot nematode *Meloidogyne incognita* in tobacco. Environ Res 216:114848

Ma B, Lu C, Wang Y, Yu J, Zhao K, Xue R, Ren H, Lv X, Pan R, Zhang J, Zhu Y, Xu J (2023) A genomic catalogue of soil microbiomes boosts mining of biodiversity and genetic resources. Nat Commun 14(1):7318

Mansfield J, Genin S, Magori S, Citovsky V, Sriariyanum M, Ronald P, Dow M, Verdier V, Beer SV, Machado MA (2012) Top 10 plant pathogenic bacteria in molecular plant pathology. Mol Plant Pathol 13:614–629

Mashabela MD, Tugizimana F, Steenkamp PA, Piater LA, Dubery IA, Mhlongo MI (2022) Untargeted metabolite profiling to elucidate rhizosphere and leaf metabolome changes of wheat cultivars (*Triticum aestivum* L.) treated with the plant growth-promoting *rhizobacteria Paenibacillus alvei* (T22) and *Bacillus subtilis*. Front Microbiol 13:971836

Meslier V, Quinquis B, Da Silva K (2022) Benchmarking second and third-generation sequencing platforms for microbial metagenomics. Sci Data 9:694. https://doi.org/10.1038/s41597-022-01762-z

Mullins AJ, Murray JAH, Bull MJ, Jenner M, Jones C, Webster G (2019) Genome mining identifies cepacin as a plant-protective metabolite of the biopesticidal bacterium *Burkholderia ambifaria*. Nat Microbiol 4:996–1005. https://doi.org/10.1038/s41564-019-0383-z

Naing KW, Anees M, Kim SJ, Nam Y, Kim YC, Kim KY (2014) Characterisation of antifungal activity of *Paenibacillus ehimensis* KWN38 against soilborne phytopathogenic fungi belonging to various taxonomic groups. Ann Microbiol 64:55–63

Nga NT, Giau NT, Long NT, Lübeck M, Shetty NP, De Neergaard E, Thuy TT, Kim PV, Jørgensen HJ (2010) Rhizobacterially induced protection of watermelon against *Didymella bryoniae*. J Appl Microbiol 109(2):567–582

Nigris S, Baldan E, Tondello A, Zanella F, Vitulo N, Favaro G, Guidolin V, Bordin N, Telatin A, Barizza E, Marcato S, Zottini M, Squartini A, Valle G, Baldan B (2018) Biocontrol traits of *Bacillus licheniformis* GL174, a Culturable endophyte of *Vitis vinifera* cv. Glera. BMC Microbiol 18(1):133

Nwachukwu BC, Babalola OO (2022) Metagenomics: a tool for exploring key microbiome with the potentials for improving sustainable agriculture. Front Sustain Food Syst 6:886987

Oliveira VM, Andreote FD, Cortelo PC, Castro-Gamboa I, Costa-Lotufo LV, Polizeli MDL, Setubal JC (2022) Microorganisms: the secret agents of the biosphere, and their key roles in biotechnology. Biota Neotropica (Edicao em Iingles) 22

Paulraj S, Bhat R, Rajesh MK, Ramesh SV, Priya UK, Pandian RTP, Hegde V, Chowdappa P (2021) Data of 16S rRNA gene amplicon-based metagenomic signatures of arecanut rhizosphere soils in yellow leaf disease (YLD) endemic region of India. Data Brief 38:107443

Pieterse CM, Van Wees SC, Van Pelt JA, Knoester M, Laan R, Gerrits H, Weisbeek PJ, Van Loon LC (1998) A novel signaling pathway controlling induced systemic resistance in Arabidopsis. Plant Cell 10(9):1571–1580

Prakash J, Arora NK (2019) Phosphate-Solubilizing *Bacillus* sp Enhances Growth, Phosphorus Uptake, and Oil Yield of *Mentha arvensis* L. 3 Biotech 9(4):126

Premke K, Wurzbacher C, Felsmann K, Fabian J, Taube R, Bodmer P, Hölker F (2022) Large-scale sampling of the freshwater microbiome suggests pollution-driven ecosystem changes. Environ Pollut 308:119627

Quince C, Walker AW, Simpson JT, Loman NJ, Segata N (2017) Shotgun metagenomics, from sampling to analysis. Nat Biotechnol 35(9):833–844

Rai R, Srinivasamurthy R, Dash PK, Gupta P (2017) Isolation, characterization and evaluation of the biocontrol potential of *Pseudomonas protegens* RS-9 against *Ralstonia solanacearum* in tomato. Indian J Exp Biol 55:595–603

Rangel-Montoya EA, Delgado-Ramírez CS, Sepulveda E, Hernández-Martínez R (2022) Biocontrol of *Macrophomina phaseolina* using *bacillus amyloliquefaciens* strains in Cowpea (*Vigna unguiculata* L.). Agronomy 12(3):676

Rashid S, Charles TC, Glick BR (2012) Isolation and characterization of new plant growth-promoting bacterial endophytes. Appl Soil Ecol 61:217–224

Raza W, Ling N, Yang L, Huang Q, Shen Q (2016) Response of tomato wilt pathogen *Ralstonia solanacearum* to the volatile organic compounds produced by a biocontrol strain *Bacillus amyloliquefaciens* SQR-9. Sci Rep 6:24856. https://doi.org/10.1038/srep24856

Reverchon F, García-Meléndez M, Guevara-Avendaño E, Mora-Chávez O, Solís-García IA, Dáttilo W, Guererro-Analco JA, Méndez-Bravo A, Monribot-Villanueva JL, Patiño-Conde V, Pineda-García F, Rebollar EA (2023) Shifts in the rhizosphere microbiome and exudation profile of avocado (*Persea americana* Mill.) during infection by *Phytophthora cinnamomi* and in presence of a biocontrol bacterial strain. CABI Agric Biosci 4(1):23

Roy T, Bandopadhyay A, Sonawane PJ, Majumdar S, Mahapatra NR, Alam S, Das N (2018) Bio-effective disease control and plant growth promotion in lentil by two pesticide degrading strains of Bacillus sp. Biol Con 127:55–63. https://doi.org/10.1016/j.biocontrol.2018.08.018

Salas-Marina MA, Silva-Flores MA, Cervantes-Badillo MG, Rosales-Saavedra MT, Islas-Osuna MA, Casas-Flores S (2011) The plant growth-promoting fungus *Aspergillus ustus* promotes growth and induces resistance against different lifestyle pathogens in *Arabidopsis thaliana*. J Microbiol Biotechnol 21:686–696

Salwan R, Sharma M, Sharma A, Sharma V (2023) Insights into plant beneficial microorganism-triggered induced systemic resistance. Plant Stress 100140

Sannino C, Qi W, Rüthi J, Stierli B, Frey B (2023) Distinct taxonomic and functional profiles of high Arctic and alpine permafrost-affected soil microbiomes. Environ Microbiome 18(1):1–22

Santoyo G, Urtis-Flores CA, Loeza-Lara PD, Orozco-Mosqueda MDC, Glick BR (2021) Rhizosphere colonization determinants by plant growth-promoting rhizobacteria (PGPR). Biology 10:475

Selvakumar G, Joshi P, Suyal P, Mishra PK, Joshi GK, Bisht JK, Gupta HS (2011) *Pseudomonas lurida* M2RH3 (MTCC 9245), a *psychrotolerant bacterium* from the Uttarakhand Himalayas, solubilizes phosphate and promotes wheat seedling growth. World J Microbiol Biotechnol 27:1129–1135

Shakya M, Lo CC, Chain PS (2019) Advances and challenges in metatranscriptomic analysis. Front Genet 10:904

Singh D, Yadav DK, Chaudhary G, Rana VS, Sharma RK (2016) Potential of *Bacillus amyloliquefaciens* for biocontrol of bacterial wilt of tomato incited by *Ralstonia solanacearum*. J Plant Pathol Microbiol 7(327):1–6

Sokol NW, Slessarev E, Marschmann GL, Nicolas A, Blazewicz SJ, Brodie EL, Pett-Ridge J (2022) Life and death in the soil microbiome: how ecological processes influence biogeochemistry. Nat Rev Microbiol 20(7):415–430

Sun G, Yao T, Feng C, Chen L, Li J, Wang L (2017) Identification and biocontrol potential of antagonistic bacteria strains against *Sclerotinia sclerotiorum* and their growth-promoting effects on *Brassica napus*. Biol Control 104:35–43

Tedersoo L, Bahram M, Zobel M (2020) How mycorrhizal associations drive plant population and community biology. Science 367:1223. https://doi.org/10.1126/science.aba1223

Trivedi P, Batista BD, Bazany KE, Singh BK (2022) Plant–microbiome interactions under a changing world: responses, consequences and perspectives. New Phytol 234(6):1951–1959

Van Wees SC, De Swart EA, Van Pelt JA, Van Loon LC, Pieterse CM (2000) Enhancement of induced disease resistance by simultaneous activation of salicylate-and jasmonate-dependent defense pathways in *Arabidopsis thaliana*. Proc Natl Acad Sci 97(15):8711–8716

Vesga P, Flury P, Vacheron J, Keel C, Croll D, Maurhofer M (2020) Transcriptome plasticity underlying plant root colonization and insect invasion by *Pseudomonas protegens*. ISME J 14:2766–2782. SS

Vu MT, Geraldi A, Do HDK, Luqman A, Nguyen HD, Fauzia FN, Amalludin FI, Sadila AY, Wijaya NH, Santoso H, Manuhara YSW, Wibowo AT (2022) Soil mineral composition and salinity are the main factors regulating the bacterial community associated with the roots of coastal sand dune halophytes. Biology 11(5):695

Vujanovic V, Goh YK (2012) qPCR quantification of *Sphaerodes mycoparasitica* biotrophic mycoparasite interaction with *Fusarium graminearum*: *in vitro* and *in planta* assays. Arch Microbiol 194:707–717

Wang H, Liu R, You MP, Barbetti MJ, Chen Y (2021a) Pathogen biocontrol using plant growth-promoting bacteria (PGPR): role of bacterial diversity. Microorganisms 9:1988. https://doi.org/10.3390/microorganisms9091988

Wang JY, Guo C, Zhao P, Yu FY, Su Y, Qu JP, Zhou B (2021b) Biocontrol potential of *Bacillus altitudinis* AMCC1040 against root-knot nematode disease of ginger and its impact on rhizosphere microbial community. Biol Control 158:104598

Wang Y, Duan S, Xu J, Long Y, Zhang S, Li S, Wu L, Zhang Y (2023) Comparison of the colonization ability of *Burkholderia* strain B23 in the citrus rhizoplane and rhizosphere and assessment of the underlying mechanisms using full-length 16S rDNA amplicon and metatranscriptomic analyses. Microb Biotechnol 16(8):1657–1670

Waweru B, Turoop L, Kahangi E, Coyne D, Dubois T (2014) Non-pathogenic *Fusarium oxysporum* endophytes provide field control of nematodes, improving yield of banana (*Musa sp.*). Biol Control 74:82–88

Wei HL, Zhang LQ (2006) Quorum-sensing system influences root colonization and biological control ability in *Pseudomonas fluorescens* 2P24. Antonie Van Leeuwenhoek 89(2):267–280

Wen T, Yuan J, He X, Lin Y, Huang Q, Shen Q (2020) Enrichment of beneficial cucumber rhizosphere microbes mediated by organic acid secretion. Hortic Res 7:154. https://doi.org/10.1038/s41438-020-00380-3

Whipps JM, Karen L, Cooke RC (1988) Mycoparasitism and plant disease control. In: Burge NM (ed) Fungi in biological control systems. Manchester University Press, Manchester, pp 161–187. https://doi.org/10.1186/s40101-015-0061-7

Wilson BR, Bogdan AR, Miyazawa M, Hashimoto K, Tsuji Y (2016) Siderophores in iron metabolism: from mechanism to therapy potential. Trends Mol Med 22(12):1077–1090

Xie Z, Li M, Wang D, Wang F, Shen H, Sun G, Feng C, Wang X, Chen D, Sun X (2021) Biocontrol efficacy of *Bacillus siamensis* LZ88 against Brown spot disease of tobacco caused by *Alternaria alternata*. Biol Control 154:104508

Yadav RL, Shukla SK, Suman A, Singh PN (2009) *Trichoderma* inoculation and trash management effects on soil microbial biomass, soil respiration, nutrient uptake and yield of ratoon sugarcane under subtropical conditions. Biol Fertil Soils 45:461–468

Yamagiwa Y, Inagaki Y, Ichinose Y, Toyoda K, Hyakumachi M, Shiraishi T (2011) *Talaromyces wortmannii* FS2 emits β-Caryophyllene, which promotes plant growth and induces resistance. J Gen Plant Pathol 77:336–341

Yanti Y, Hamid H, Reflin R, Yaherwandi Y, Suhendra D, Hariandi D, Suriani NL (2022) Evaluation of the effect of PGPR strains on tomato growth and suppression of *Ralstonia* wilt disease. In: KnE Life Sciences, The First Asian PGPR Indonesian Chapter International e-Conference 664–671

Yee MO, Kim P, Li Y, Singh AK, Northen TR, Chakraborty R (2021) Specialized plant growth chamber designs to study complex rhizosphere interactions. Front Microbiol 12:507

Yin C, Casa Vargas JM, Schlatter DC, Hagerty CH, Hulbert SH, Paulitz TC (2021) Rhizosphere community selection reveals bacteria associated with reduced root disease. Microbiome 9:86

Yuan TT, Xu HH, Zhang Q, Zhang LY, Lu YT (2018) The COP1 target SHI-RELATED SEQUENCE5 directly activates photomorphogenesis-promoting genes. The Plant Cell 30(10): 2368-2382. https://doi.org/https://doi.org/10.1105/tpc.18.00455

Yue H, Yue W, Jiao S, Kim H, Lee YH, Wei G, Shu D (2023) Plant domestication shapes rhizosphere microbiome assembly and metabolic functions. Microbiome 11(1):1–19

Chapter 9
Sustainable Management of Economically Important Plant Viral Pathogens by Plant Growth-Promoting Rhizobacteria

Archana Rathore, Tanu Vishwakarma, Dilip Kumar Maurya, Prabhat Kumar Maurya, S. Vijay Anand Raj, Puneet Singh Chauhan, S. K. Raj, and Susheel Kumar

Abstract Plant viruses are a category of viruses that infect plants, predominantly affecting vascular plants or higher plants. Collectively, these viruses contribute about $600 million in global crop yield losses. Functioning as obligate intracellular parasites, plant viruses lack the necessary molecular machinery for replication independent of a host, and therefore, they are required to highjack host machinery. The major class of these viruses have small single-stranded ssRNA genome, although some exhibit double-stranded RNA followed by ssDNA or double-stranded DNA also. Viruses of plant are less comprehensively understood unlike animal counterparts; the one prominent example is the tobacco mosaic virus (TMV). As of now, plant viruses are categorized into 73 genera and 49 families; however, these classifications pertain mainly to cultivated plants, representing only a fraction of the vast variety of plant species. This chapter focuses and provides insight into the top 10 plant viruses ranked based on their prevalence and the economic damages they inflict on their host plants and their management using growth-promoting rhizobacteria (PGPR). The top 10 (in rank order) list includes (1) Tobacco mosaic virus, (2) Tomato spotted wilt virus, (3) Tomato yellow leaf curl virus, (4) Cucumber mosaic virus, (5) Potato virus Y, (6) Cauliflower mosaic virus, (7) African cassava mosaic virus, (8) Plum pox virus, (9) Brome mosaic virus, and (10) Potato virus X. Unlike to fungi and bacteria pathogens, plant viruses lack chemical control. However, the utilization of PGPR stands out as an environmentally friendly mean for the manage-

A. Rathore · T. Vishwakarma · D. K. Maurya · P. K. Maurya · S. V. A. Raj · S. Kumar (✉)
Plant Molecular Virology Laboratory, Molecular Biology and Biotechnology Division, CSIR-National Botanical Research Institute, Lucknow, Uttar Pradesh, India
e-mail: susheel.kr79@nbri.res.in

P. Singh Chauhan
Microbial Technology Division, CSIR-National Botanical Research Institute, Lucknow, Uttar Pradesh, India

S. K. Raj
Sector-G, Janki Puram, Lucknow, Uttar Pradesh, India

P. Singh Chauhan et al. (eds.), *Plant-Microbe Interaction and Stress Management*, Rhizosphere Biology,
https://doi.org/10.1007/978-981-97-4239-4_9

ment of viruses and enhances crop yield. In recent times, there has been a notable shift towards leveraging PGPR-induced systemic resistance (ISR) for viral disease management, although previous research predominantly focused on fungal and bacterial pathogens. This chapter will explore the range of PGPR-induced ISR focused on combating the top 10 plant viruses.

Keywords DNA · RNA · Viral structure · Vector · Replication · Virus transmission

9.1 Introduction

Adolf Eduard Mayer, a German agricultural biologist from the Netherlands, is credited with the discovery of "plant virus" that causes disease. Mayer's experiments involved mechanical sap transmission from the diseased tobacco leaves exhibiting mosaic symptom into healthy plants, resulting in the development of typical "mosaic" as were observed in the original host. Mayer also established that infection was rendered ineffective when the sap was boiled. Initially attributing the cause to bacteria, Mayer conducted larger inoculations with a substantial quantity of bacteria but failed to induce typical "mosaic" symptoms, leading him to reconsider the nature of the causal agent that was being considered a bacterium. Further, Martinus Beijerinck in 1898 introduced the idea that "viruses were minute entities." Beijerinck observed that disease (mosaic) retained its infectious nature after passing from Chamberland filter-candle, distinguishing it from bacterium. He therefore coined the "contagium vivum fluidum" term for the infectious filtrate, marking the inception of the contemporary term "virus" (in Latin meaning "poison, slime, venom") (Beijerinck 1898).

Following recognition of viral concept was necessary to categorize additional viral diseases according to their modes of transmission arose. In 1939, Holmes, using this information on 129 plant viruses, compiled data on classification. Over time, this list expanded significantly, and International Committee on Taxonomy of Viruses (ICTV) tenth report (2011) enlists 2284 species of virus and viroid scattered within 349 genera, 19 subfamilies, 87 families, and 6 orders (https://ictv.global/report). Here, we provide insights into the top 10 plant viruses, ranked depending on their prevalence along with socioeconomic damages they inflict on their host plants (Table 9.1). The top 10 (in rank order) list includes (1) Tobacco mosaic virus (TMV), (2) Tomato spotted wilt virus (TSWV), (3) Tomato yellow leaf curl virus (TYLCV), (4) Cucumber mosaic virus (CMV), (5) Potato virus Y (PVY), (6) Cauliflower mosaic virus (CaMV), (7) African cassava mosaic virus (ACMV), (8) Plum pox virus (PPV), (9) Brome mosaic virus (BMV) and (10) Potato virus X (PVX) (Scholthof et al. 2011).

Pertinent to mention that within top 10 viruses, most belong to the category of ss(+)RNA viruses. However, representation of other nucleic acid genome forms is also there, such as dsDNA exemplified by CaMV, and ssDNA viruses illustrated by two begomovirus, namely TYLCV and ACMV. The ICTV committee nominated

Table 9.1 List of top ten plant viruses. Table represents the ranked list of plant viruses suggested by the respective authorities

Rank	Virus	Abbreviation	Genome type and Strandness	Family/Genus
1	Tobacco mosaic virus	TMV	RNA, positive-sense single-stranded	Virgaviridae/*Tobamovirus*
2	Tomato spotted wilt virus	TSWV	RNA, negative and ambisense single-stranded	Tospoviridae/*Orthotospovi rus*
3	Tomato yellow leaf curl virus	TYLCV	DNA, single-stranded	Geminiviridae/*Begomovirus*
4	Cucumber mosaic virus	CMV	RNA, positive-sense single-stranded	Bromoviridae/*Cucumovirus*
5	Potato virus Y	PVY	RNA, positive-sense single-stranded	Potyviridae/*Potyvirus*
6	Cauliflower mosaic virus	CaMV	DNA, double-stranded	Caulimoviridae/*Caulimovirus*
7	African cassava mosaic virus	ACMV	DNA, single-stranded	Geminiviridae/*Begomovirus*
8	Plum pox virus	PPV	RNA, positive-sense single-stranded	Potyviridae/*Potyvirus*
9	Brome mosaic virus	BMV	RNA, positive-sense single-stranded	Bromoviridae/*Bromovirus*
10	Potato virus X	PVX	RNA, positive-sense single-stranded	Alphaflexiviridae /*Potexvirus*

TMV at position 1 in the top 10, which is globally the most studied virus. TSWV is at second position encompassing ambisense ssRNAs as genome. Occupying fifth and eighth positions are PVY and PPV, respectively, both belonging to one of the largest families, Potyviridae. With a global presence, both PVY and PPV are adeptly transmitted by aphids, posing challenges for effective control measures. The economic impact of BMV and PVX, though lying in the bottom of the list, are prevalent due to their wide hot range and crop losses that they do. Now, we will discuss in detail about the genome organization, symptoms, and mode of transmission to replication for all aforementioned viruses. This chapter will also explore the range of plant growth-promoting rhizobacteria (PGPR)-modulated induced resistance against various plant viruses.

9.2 Tobacco Mosaic Virus

Wendell Stanley (1935) conducted the initial purification (crystallization) of TMV (genus *Tobamovirus*) for which he received the Nobel Prize in Chemistry in 1946, though he was not able to identify RNA as the infectious material at that time. In the 1950s, simultaneous discoveries by two laboratories revealed RNA capability of TMV to be infectious as virus. Since then, TMV is the most studied virus.

9.2.1 Morphology

It is the first virus to be discovered, purified, characterized through X-ray crystallography, and sequenced among plant viruses (Wyckoff and Corey 1936). TMV is characterized by rigid rod-shaped particles measuring 300 × 18 nm (Stanley and Anderson 1941). These particles consist of coat protein molecules arranged helically around a single strand of RNA, which serves as its genome (Fig. 9.1). A solitary TMV particle comprises 2130 units of envelope protein (CP) enveloping its about 6400 nucleotides long genetic RNA (Lewis 2004). The genome encodes two replicase (Rep)-associated proteins and a movement protein and a CP, translated through subgenomic RNA segment (Anderer et al. 1960).

9.2.2 Etiology of Disease and Mode of Infection

Tobacco mosaic disease by TMV stands as a widespread viral ailment, impacting over 200 species of herbaceous plants on a global scale and, to a lesser extent, in woody plants. TMV mostly induces mosaic patterns, mottling, necrosis, and stunting (Fig. 9.1). Typical leaf symptoms comprise light and dark mosaic, cupping, elongation, and stunting, often resulting in smaller-than-normal size. Primary hosts include tobacco, pepper, tomato, cucumber, melon, squash, begonia, coleus, geranium, impatiens, million bells, and petunia. Particularly significant in vegetables, the disease can significantly diminish yield and compromise quality, rendering commercial crops unsuitable for market. Fruits in TMV-infected plants may exhibit blotchy coloration, uneven ripening, deformities, or an off flavor. TMV infiltrates plants through wounds, exploiting compromised plant cell membranes that allow infectious particles to enter the cytoplasm. TMV is transmitted very easily making control difficult. Infected leaves rubbing against healthy plants, contaminated tools, or worker's hands (which can become contaminated with virus from cigarettes) can all spread virus. TMV is not transmitted by insects (Creager 2022).

9.2.3 Replication and Movement

Upon infiltrating the cell, TMV undergoes a coordinated disassembly, revealing its RNA to the host system. The virus co-opts the host's machinery to produce virus proteins and replicates its RNA extensively (Buck 1999). TMV RNA functions directly as messenger RNA (mRNA), undergoing translation utilizing the plant ribosomal machinery. The translation process of 126- and 183-kDa Rep proteins initiates shortly after infection. Once these proteins are available in host, the replicase bind to 3′ end of the +sense RNA for –ve sense RNA form production. This antiparallel form is used as template for generating both complete genomic +ve

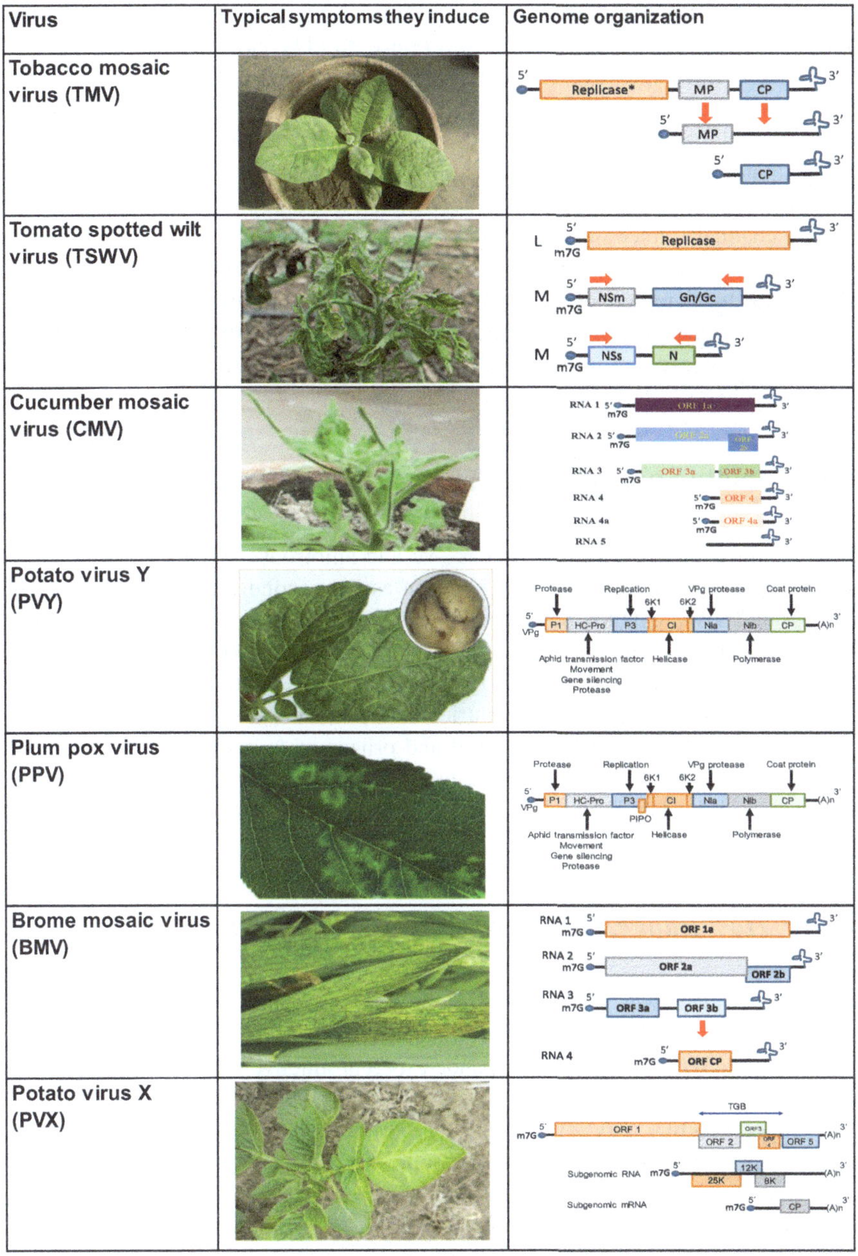

Fig. 9.1 Depiction of single stranded type species of RNA viruses having positive, negative and ambience coding capacity. Standard abbreviations used for virus nomenclature. In the middle panel, the plants exhibiting typical symptoms induced by their respective viruses on their systemic host plants is shown. The genome size represents the standard proteins that virus encodes

sense RNA and subgenomic RNAs (sgRNAs) (Fig. 9.1). Then, host ribosomal machinery translates these sgRNAs to produce 30 kDa and 17.5 kDa MP and CP, respectively. Subsequently, CP interacts with the newly formed +ve sense RNA, facilitating the formation of new virus particles, virions. Alternatively, +ve sense RNA is enveloped in MP, forming a complex capable of infecting the nearby cells. To facilitate vascular movement of TMV within the host cell, virus utilizes MP allowing it to propagate from one to another cell through plasmodesmata, an intercellular channel connecting plant cells (Liu and Nelson 2013), and penetrate plant's system (specifically the veins), facilitating fast systemic dissemination through the phloem to distal region of plant. MP expands plasmodesmal openings enabling peripheral movement of RNA, where it releases MP and associated host proteins, initiating a new cycle of infection (Benitez-Alfonso et al. 2010).

9.3 Tomato Spotted Wilt Virus

TSWV, which (genus *Tospovirus*) has –ve and ambisense ssRNA genome and is thrips-transmitted, holds a notable position as the second in the top 10 list. The interest in TSWV extends beyond its economic impact to its fascinating biology being replicate both in host and vector (thrips). The description of TSWV causing disease dates back to the early twentieth century (Brittlebank 1919), likely entrenched in all Mediterranean countries. In the Mediterranean region, TSWV initially posed challenges in horticultural and ornamental crops in Italy, Spain, and France. Subsequently, its impact was felt in Portugal, Slovenia, Albania, Greece, Turkey, Israel, Jordan, Egypt, Tunisia, and Montenegro (Zindović et al. 2014).

9.3.1 Host Range, Symptoms, Signs, and Epidemiology

It is distinguished by its global prevalence and extensive host range across economically vital crops. TSWV emerged as a significant threat to European horticulture when *Frankliniella occidentalis* expanded its reproductive range globally from its origin in the western US during 1980s. Notably, TSWV is distinctive for its broad host range, infecting approximately 150 reported species in 1968 (Best 1968) and reaching around 1100 in 2003 (Parrella et al. 2003). TSWV stands out as the most economically significant tospovirus in the Mediterranean basin, particularly posing a major threat, and, in some specific situations, the primary threat to tomato and pepper crops causing stunting and ringspots (Fig. 9.1). Apart from its broad list of hosts, TSWV is a devastating pathogen of artichoke, lettuce, celery, basil, eggplant, and chicory, too (Vovlas and Lafortezza 1992; Parisi et al. 1998; Moreno et al. 2004; Testa et al. 2011). Over 10 thrips species transmit tospoviruses, with *F. occidentalis* considered a key vector due to global distribution and compatibility with most tospoviruses. The broad host range and multiple thrips species transmitting TSWV make eliminating primary inoculum sources impractical (Sherwood et al. 2003).

9.3.2 Genome Organization

TSWV virions exhibit a complexity uncommon in many viruses of plant. The viral genome is tri-partite RNAs, each individually encapsidated and enveloped by a host-origin membrane (Walter and Barr 2011). The roughly spherical virions measure between 80 and 110 nm (in dia).

The out envelope harbors two glycoproteins (GPs), namely glycoprotein C (GC) and glycoprotein N (GN), both processed to contain sugars during replication, dispersed across its surface with slight size variations (Fig. 9.1). Units of CP interact each of the three viral RNAs labeled as Large (L), Middle (M), and Small (S). Additionally, multiple units of Rep protein, crucial for initiating replication of virus in a host, are also present (Oliver and Whitfield 2016). This intricate virion structure sets TSWV apart from the majority of other plant viruses. The envelope's glycoproteins aid in assembly of virions and acquiring TSWV in thrips. The L RNA encodes an RNA-dependent RNA polymerase (RdRp). The M RNA has information for a nonstructural protein (NSm), facilitating tospovirus movement between plant cells via plasmodesmata, overcoming the limitation by forming tubules to aid pass of nucleocapsids from one cell to another. Moreover, the S RNA translates nonstructural protein (NSs) that exhibits RNA silencing suppressor activity, potentially influencing PTGS or RNA metabolism in plant.

9.4 Tomato Yellow Leaf Curl Virus

Tomato crops in Israel during the late 1950s faced the pandemic of an unidentified disease, coinciding with noticeable hike of whitefly populations. The TYLCV was responsible for the outbreak subsequently (genus *Begomovirus*, family Geminiviridae) (Anfoka et al. 2009). Its presence spans all five continents. Infected tomato plants reveal stunting, leaf curling, and yellowing symptoms, resulting in significant production losses in tomato cultivation (Fig. 9.2). Follow-up laboratory experiments, devised to manage transmission, validated the suspicion that whiteflies (*Bemisia tabaci*) acted as vectors for a viral disease (Czosnek and Laterrot 1997). Since the initial report in 1931 from Middle East, TYLCV is persistently occurring in warmer regions and becomes a major pathogen of tomato crop worldwide.

<table>
<tr><th>Virus</th><th>Tomato yellow leaf curl virus (TYLCV)</th><th>Cauliflower mosaic virus</th><th>African cassava mosaic virus</th></tr>
<tr><td>Typical symptoms they induce</td><td></td><td></td><td></td></tr>
<tr><td>Genome organization</td><td>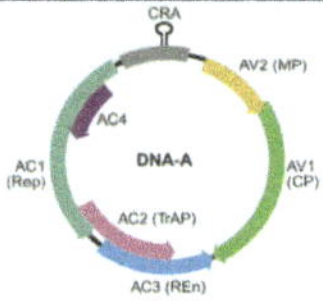
</td><td>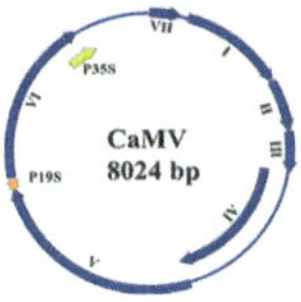
</td><td>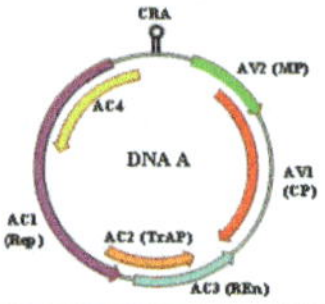
</td></tr>
</table>

Fig. 9.2 Depiction of single stranded DNA (Begomovirus, type species TYLCV and ACMV) and double stranded DNA (Caulimovirus, type species Cauliflower mosaic virus) virus and the typical symptoms they induce on their systemic host plant. The genome size represents the standard proteins that virus encodes

9.4.1 Host Range, Symptoms, and Epidemiology

Tomato plants infected with the TYLCV exhibit stunted and upright growth, with more pronounced stunting when infection occurs early. The leaves in infected plants become smaller, curl upward, and show significant crumpling (Poston and Anderson 1997). Yellowing occurs along the veins and margins. Infected plants also undergo a reduction in internode length, giving them a bushy appearance often compared to "bonsai-like" growth (Fig. 9.2). Flowers on infected plants commonly fail to develop and typically detach (abscise) leading to reduced fruit production and sometimes suffer losses of up to 100%. Apart from tomatoes, plants including pepper, common bean, eustoma (*Eustoma grandiflora*), and cucurbit (*Cucumis* sp.) have also been reported as hosts for TYLCV. Studies have delved into the complex relationships among TYLCV, whitefly (vector), and tomato (host) plant (De Barro et al. 2011). The virus has not only spread significantly over the past decades but also exhibits a wide distribution range for its insect vector. *B. tabaci*, responsible for transmitting TYLCV, is found across all five continents (Brown 2010).

9.4.2 Genome Organization

TYLCV has ssDNA genome of about 2800 bases and is enclosed in geminate structure measuring about 25–30 nm (in dia) (Czosnek et al. 2001). Genome span in six partially overlapping open reading frame (ORF)s organized into two the virion and complementary sense, spaced by 300 nucleotides long noncoding intergenic region (IR) with an origin of replication (Díaz-Pendón et al. 2010). Within this IR, a reserved stem-loop structure with regulatory elements like nonanucleotide sequence

(TAATATTAC), TATA box and iterons (attachment sites for the Rep) are present (Hanley-Bowdoin et al. 1999). There are two genes on viral (sense) strand: V1 and V2. V1 encodes CP, which acts as a nuclear shuttle protein facilitating vector transmission. On the other hand, V2 encodes a protein responsible for viral movement as well as suppressor of host-silencing machinery (Fig. 9.2). CP is responsible for vector-mediated virus transmission (Noris et al. 1998). On antiparallel strand, four genes, namely C1, C2, C3, and C4, are present. C1 encodes Rep protein responsible for initiating viral replication, C2 serves as a transcriptional activator (TrAP) that also disrupts transcriptional and PTGS, C3 encodes a replication enhancer (REn), and C4 is engaged in silencing suppression, also influencing symptom development (Caciagli et al. 2009). Within plants, TYLCV replicates in the infected nucleus through rolling-circle replication model (Gutierrez 1999).

9.4.3 Vector-Mediated TYLCV Transmission

Whitefly transmits TYLCV in circulative persistent manner. Although extensive researches were made on acquisition and transmission parameters of TYLCV (Zeidan and Czosnek 1991), the relationship between TYLCV and *B. tabaci* remains poorly understood. *B. tabaci* is known to be complex comprising undistinguishable 24 biotypes known so far. However, two most invasive and destructive whiteflies, *B. tabaci* biotype B (originating Asia) and *B. tabaci* biotype Q (originating in Mediterranean basin), exhibit notable distinctions (De Barro et al. 2011).). Two biotypes exhibit variations in host range, feeding habits, endosymbionts, resistance to insecticides, and efficiency in transmitting viruses. Both B and Q cause considerable harm to plants through their phloem-feeding activity and transmission of plant viruses.

9.5 Cucumber Mosaic Virus

The Bromoviridae family encompasses five genera: Alfamo, Ilar, Bromo, Cucumo, and Oleavirus. Among these, Cucumoviruses stand out as the most widely distributed plant viruses globally, infecting over 1000 plant species, including chrysanthemum cultivars (Codoner et al. 2005). ICTV endorsed the classification of CMV, TAV, and PSV as type species in *Cucumovirus* genus (Fauquet et al. 2005). The initial report of CMV dates back to 1916 when Doolittle identified it in cucumber and muskmelon, resulting in its name, CMV (Doolittle 1916). CMV, a functionally tripartite virus, possesses a ss(+)sense RNA and has a global distribution causing significant penalty both in quantity and in quality. CMV infects over 1200 host species across 85 plant families including ornamental, vegetable, and other plants (Bouwen and van Zaayen 1995).

9.5.1 Symptom and Epidemiology

CMV is widespread in temperate regions, existing in numerous strains that vary in host preference, induced symptoms, mode of transmission, and other properties (Francki et al. 1979). CMV induces varying symptoms that very widely in nature according to strain type (Doolittle 1916). Severe leaf symptoms include mottling, flecking, fern leaf or shoestring, and stunting or yellowing of entire plants, whereas mild symptoms include ringspot and hardness of fruit (Raj et al. 2008b). Ornamental hosts susceptible to CMV encompass aster, salvia, chrysanthemum, gladiolus, amaranths, delphinium, geranium, gilia, periwinkle, phlox, petunia, tulip, zinnia, and so forth (Agrios 1978; Kaper and Waterworth 1981). Among vegetable hosts are squash, broad bean, turnip, pumpkin, onion, eggplant, peppers, potato, carrot, parsley, loofah, and artichoke (Huang et al. 1987). CMV is also reported on jatropha plants (Raj et al. 2008c), *Rauvolfia serpentine* (Raj et al. 2007b), *Cymbopogon citratus* (Raj et al. 2007c), and *Catharanthus roseus* (Samad et al. 2008). Coinfection of CMV with some other nonrelated virus produces severe synergistic effects (Xi et al. 2007). CMV-associated RNA 5 (CARNA5) are satellite RNAs associated with CMV and have regulatory effect on symptom expression. Satellite RNAs may exacerbate or cause attenuation in disease symptom (Waterworth et al. 1979). Virions are distributed throughout the host plant's cytoplasm. In infected cells, there are inclusions in the cytoplasm, which are often crystalline and may appear as rhomboidal, hexagonal, or roughly spherical structures, sometimes resembling solid hollow formations, and contain virions (Brunt et al. 1996).

9.5.2 Genome Organization

Virions have 18.2% RNA and exhibit icosahedral symmetry with T = 3 quasisymmetry, measuring 29 nm (in dia) (Brunt et al. 1996; Roossinck 2001). The discovery of the multicomponent nature of the CMV genome was initially made by Kaper et al. (1965) through the physicochemical characterization of a cucumovirus. Lot et al. (1974) further demonstrated that CMV genome has tripartite ss(+)sense RNA, comprising RNA-1, RNA-2, RNA-3, and CP expressing subgenomic RNA-4 (Ding et al. 1994). These RNA segments have 7-methyl guanosine cap and t-RNA like structure at their 5′ and 3′ ends, respectively, crucial for viral RNA replication (Gould and Symons 1977; Kataoka et al. 1990) (Fig. 9.1). On occasion, satellite RNA of less than 400 nucleotides, a fifth RNA, is sometime associated with genome (Lot et al. 1977; Peden and Symons 1973). Extensive investigation of CMV genome has been undergone for 1a (111 kD) and 2a (94–97 kD) proteins coding for replicase functions by RNA-1 and RNA-2, respectively (Palukaitis and Garcia-Arenal 2003). CMV genome replication requires the creation of a replicase complex which consists of viral encoded 1a and 2a proteins, in addition to certain unidentified cellular host proteins (Sivakumaran et al. 2000). RNA3 is dicistronic where 5′ ORF codes

Table 9.2 Detail of ORF and their encoded protein with their function

ORF	Protein designation	Protein function
ORF I	P1	Movement protein.
ORF II	P2	Aphid/insect transmission factor.
ORF III	P3	Virion-associated protein (VAP): Structural protein, DNA-binding capabilities
ORF IV	P4	Capsid protein (CP)
ORF V	P5	Pro-pol: Protease, bifunctional reverse transcriptase and RNaseH
ORF VI	P6	Transactivator/viroplasmin: Inclusion body formation/ trafficking; possibly other functions
ORF VII/ VIII	–	Unknown (appears note to be required for infection, contains a tRNA-Met binding site

for 30 kDa 3a MP while 3′ end for CP of about 24 kDa, expressed through the subgenomic RNA-4 (Gould and Symons 1978; Cillo et al. 2002). MP facilitates adjacent cell movement (Taliansky and Garcia-Arenal 1995) (Table 9.2).

9.5.3 *Vector-Mediated Virus Transmission*

A prominent feature facilitating the natural dissemination of CMV is its transmission by various species of aphids (Palukaitis et al. 1992). Over 80 distinct species of aphids, notably *Aphis gossypii, Myzus persicae,* and *Macrosiphoniella sanborni* (chrysanthemum aphid), are identified to transmit CMV (Roossinck 2002) in a non-persistent manner (Palukaitis et al. 1992). In hosts on which aphid colonies are present, the infected source plants, test plants, aphid species, and strains are the factors influencing virus transmission efficacy (Hamilton 1985). Although seed transmission has been documented in various hosts by Bhattiprolu (1991), the dynamics of seed transmission remains not fully understood.

9.6 Potato Virus Y

PVY of genus *Potyvirus* (family Potyviridae) belongs to large plant virus families and stands at positions 5 among agriculturally important viruses. PVY was initially identified as a factor linked to 80% yield deterioration of potatoes (*Solanum tuberosum*) in 1931 (Khurana and Garg 2000). It is present in various groups, namely PVY-O, PVY-N (having NA-N and EU-N), PVY-C, and PVY-O5 (non-recombinant types). Additionally, nine strains are recombinant, including PVY-N:O, PVY-N-Wi, PVY-NTN-NW, PVY-NTNa, PVY-NTNb, PVY-NE11, PVY-E, PVY-SYR-I, PVY-SYR-II, and PVY-SYR-III. These strains contain segments of

PVY-O and PVY-N sequences, serving as contributors (Glasa et al. 2021). Despite potatoes being a significant concern and leading to crop losses, the impact extends to tobacco, tomato, brinjal, and chilli (Quenouille et al. 2013).

9.6.1 Symptomatology and Host Range of PVY

Plants of Solanaceae family are favorable hosts of PVY, leading to diverse symptoms ranging from asymptomatic to lesions, mosaic, stunting, and death of plant when severe. PVY infection has been documented in about 500 host plants of various families (Kerlan 2006). It infects a diverse variety of hosts including food crops such as potato, tobacco, pepper, and tomato; flower plants like Geranium, Dahlia, and Petunia; and weeds such as Datura, Physalis, and *Solanum nigrum* (Chikh-Ali et al. 2008). Leaf symptoms may encompass rough texture, veinal necrosis, mosaic patterns, and, occasionally, leaf drop but varies significantly from cultivar to cultivar. For instance, in *S. nigrum*, it triggers vein clearing, mottling, and leaf asymmetry (Eskarous et al. 1983), whereas in *Nicotiana tabacum*, it causes vein clearing and vein banding (Rich 1983).

9.6.2 Vector-Mediated Virus Transmission

The main methods of PVY transmission involve vector-mediated transmission and the propagation of infected material through cuttings, although transmission via contact is less common. About 40 aphid species nonpersistently transmit PVY (Kerlan 2006). Just a brief moment is enough for virus acquisition or inoculation while aphids probe/feed. *M. persicae* exhibits highest transmission efficiency with its widespread geographical presence, which substantially aids in its dissemination.

9.6.3 Genome Organization

PVY exhibits filamentous and flexuous morphology characterized by helical symmetry measuring 730–740 × 12 nm. Its genome is made up of ss(+ve)-sense RNA of about 9.7 kilobases with attached genome-linked viral protein (VPg) at the 5′ end and a polyadenosine (polyA) tail at the 3′ end (Fig. 9.1) coding for one long ORF surrounded by UTRs and is translated into a polyprotein consisting of approximately 3062 aa residues. Three viral proteases cleave this polyprotein in to 10 proteins (Carrington and Freed 1990). Another protein is synthesized via an overlapping ORF following a +2 frame shift of the P3 cistron, which is fused to the amino (N)-terminal segment of P3 (P3N-PIPO) (Chung et al. 2008). P1 interacts with both ss and dsRNAs, contributing to genome amplification and augmenting the suppression of plant defenses via PTGS mediated by RNA silencing. HC-Pro participates in aphid transmission, virus replication, cell-to-cell and systemic spread, and influencing the severity of symptoms. It is suggested that HC-Pro functions as a link connecting the virion CP and an unidentified receptor within the aphid's stylet. The proteins P3 and 6 K2 constitute the two potyviral membrane proteins. P3, characterized by two hydrophobic domains at its N- and C-termini, has been associated with virus replication, systemic infection, pathogenicity, and movement. A translational frameshift within the P3 cistron results in the synthesis of P3N-PIPO, which localizes to plasmodesmata and collaborates with the CI protein in facilitating viral movement to adjacent cells (Wei et al. 2010b). However, 6K1 protein function is still unclear. Membrane-bound 6K2 protein is crucial for replication of virus (Wei et al. 2010a), as it secures connection of viral replication complex with smooth and rough endoplasmic reticulum. The first nuclear inclusion protein (NIa) contains VPg and a protease domain that cleave proteins within the precursor polyprotein (Carrington and Dougherty 1987). The interaction between VPg and eIF4E of host is a critical factor in the multiplication of the virus, involving translation and/or replication processes (Robaglia and Caranta 2006; Jiang and Laliberte 2011). The second nuclear inclusion protein (NIb) acts as the RNA-dependent RNA polymerase (RdRp), facilitating the high copy number formation of viral RNAs. The CP facilitates virion assembly, peripheral and systemic movement (Rojas et al. 1997; Dolja et al. 1995), and vector (aphid) transmission (Blanc et al. 1997).

9.7 Cauliflower Mosaic Virus

CaMV genus of *Caulimovirus* is one of the sixth genera in the Caulimoviridae family. These genera consist of pararetroviruses specifically targeting plants. The pararetroviruses have DNA and therefore employ reverse transcription for replication, a process similar to retroviruses (Rothnie et al. 1994).

9.7.1 Hosts, Symptoms, Transmission, and Disease Etiology

CaMV elicits array of phenotypes including mild to severe chlorosis, dark–light mosaic patches, vein clearing, and plant stunting (Fig. 9.2). Mainly radish, turnip, canola, mustard, cauliflower, broccoli, and cabbage, plants belonging to the Brassicaceae family, are vulnerable to CaMV infection. Of all documented cases, the D4 and W260 strains of CaMV exhibit the capability to cause disease in plants of the Solanaceae family such as Datura and tobacco, displaying diverse symptoms, levels of virulence, and rates of transmission (Scholelz and Shepherd 1988; Yasaka et al. 2014; Sukal et al. 2018). CaMV is spread through a noncirculatory method by

M. persicae, and when introduced inside a plant host cell, the virions move towards the nuclear envelope of the plant cell. (Brault et al. 2010).

9.7.2 Morphology and Genome Organization

CaMV particles are icosahedron and have about 52 nm diameter. Consisting of 420 CP units adjusted in $T = 7$ triangulation pattern, it encloses a central cavity filled with solvent (Cheng et al. 1992). It houses a genome consisting of a circular dsDNA of about 8.0 kb, featuring interruptions caused by single-stranded nicks by RNAse-H originating from Met-tRNA and two RNA primers employed in the process during reverse transcription (Fig. 9.2). After entering the host cell, these cuts undergo repair, leading to the formation of histone-bound supercoiled DNA which is then transcribed into complete 35S RNA and subgenomic 19S RNA. The 35S RNA-associated promoter is a robust constitutive promoter responsible for transcribing the entire CaMV genome. It is well-known for its effectiveness in plant transformation, particularly for achieving high levels of gene expression in dicot plants rather than monocots. Its utility is especially pronounced in cereals. The 35S RNA associated with this promoter is notably intricate, featuring a structured leader sequence spanning 600 nucleotides with 6–8 small ORFs (Pooggin et al. 1998). Following CaMV replication, seven elongated ORFs are closely positioned coding for its proteins. Of these, protein of ORF VI governs translational reinitiation of major ORFs (Park et al. 2001). This process, typically exclusive to bacterial mRNAs, adds a unique aspect to the regulation of protein expression in CaMV. P6 facilitate assembly of virions and transmission by aphids by communicating with P2 and P3. It also alters expression of host NON-EXPRESSOR OF PATHOGENESIS RELATED 1, NPR1 protein during infection (Love et al. 2012). Furthermore, there is evidence of P6 binding to P7. Exploring the interactions between these two proteins may provide insights into the yet undiscovered function of P7 (Lutz et al. 2012).

9.7.3 Replication

The place of CaMV genome replication is the cytoplasm of infected cells involving reverse transcription with an RNA template, beginning with the entry of viral particles into a plant cell, where they undergo unencapsulation (Laliberté and Sanfaçon 2010). Nevertheless, RNA synthesis takes place within the nucleus and is facilitated by the transcription of supercoiled DNA in the form of episomal minichromosome. The DNA assembles as three fragments: α on the (−ve) strand, and β and γ on the (+ve) strand into a circular genome with three gaps (D1, D2, and D3). After entering nucleus, gaps are filled before DNA binds to host histone proteins, resulting in the formation of a minichromosome. Within the nucleus, host DNA-dependent RNA polymerase transcribes the viral genome starting from the 35S promoter, proceeding beyond it to initiate transcription at the 19S promoter. Subsequently, these transcripts move into the cytoplasm for translation. Synthesized virus genome packaged to form virions and subsequently passaged using MP to adjacent uninfected tissue. CaMV, with its unique translation strategies, utilizes reverse transcription as a crucial element in its replication process.

9.8 African Cassava Mosaic Virus

Initial documentation of whitefly transmitted cassava mosaic disease (CMD) in 1894, also known as "Kräuselkrankheit or rippling/crinkling illness" traces back to northeast Tanzania (Warburg 1894). The virus is later called as ACMV and underwent several physicochemical to molecular characterization. ACMV inflicts substantial economic losses that range between USD 1.9 and 2.7 billion (Rey and Vanderschuren 2017), further intensified by its effective transmission through whitefly vectors. Pandemic of CMD in the East-Central Africa, primarily attributable to ACMV, has led to significant adversity and challenges.

9.8.1 Disease Etiology, Transmission, and Distribution

CMD manifests diverse foliar symptoms encompassing mosaic patterns, mottling, deformities, and the twisting of leaflets, resulting in plant size (Fig. 9.2). Cassava plants affected by CMD exhibit minimal tuber production. The extent of tuber yield depends on disease severity, plant's age while infection, and whitefly intensity (Palaniswami et al. 1996). Initially, it was believed that CMD exhibited geographic structuring, with ACMV primarily observed in West and African countries. However, subsequent reports have indicated the dispersal of CMD to various regions across sub-Saharan Africa, spanning the entire subcontinent to the tropical and subtropical region across the globe (Patil and Fauquet 2009).

9.8.2 Morphology and Genome Organization

ACMV exhibits geminate particles measuring 30 × 20 nm and possesses circular, ssDNA genome encapsulated by 30 kDa CP units (Stanley et al. 2005). DNA-A has two overlapping virion-sense ORFs (AV2 and AV1) and four ORFs, AC1, AC2, AC3, and AC4 which are overlapping and in complementary sense (Fig. 9.2). The CP encoded from AV1 gene serves as factor for encapsidation of genetic material and facilitating insect-vector transmission (Harrison et al. 2002). Movement protein is encoded from AV2. The AC1 to AC4 genes, both independently and collectively, play an essential role in replication inside host. AC1, AC2, and AC3 encodes Rep protein, TrA protein, and Ren protein, respectively. The TrA protein coded by AC2 gene has been demonstrated to serve as a trans-activator for various host genes (Trinks et al. 2005), besides functioning as trans-activator for AV1 and BV1, the late viral genes. In addition, TrA protein suppresses the host PTGS machinery for virus survival. Furthermore, AC4 functions as a host activation protein, serving as a crucial determinant of symptoms besides modulating the Rep protein expression in host (Stanley et al. 2005). The AC5, hypothetical ORF containing the coat protein sequence, has been identified in some viruses. However, its confirmation for transcription and translation in complementary sense remains unconfirmed.

9.8.3 Replication and Movement in Planta

For initiation of replication, viral DNA must undergo conversion into a double-stranded intermediate which is subsequently replicated through rolling-circle replication generating multiple copies of original viral genome. During replication, five putative replication intermediates were detected including hybrid of –ve and +ve strand DNA, full-length sense and antisense strand DNA, concatemeric of ssDNA, dsDNA, and partially ssDNA, which suggested its rolling circle replication mechanism in host (Paszkowski et al. 1993). The replication of viral DNA requires only Rep protein which is necessary for reprogramming the cell-cycle, recruiting the DNA replication machinery of host to the viral genome, and facilitating the nicking and rejoining events required for replication initiation and for releasing new DNA after construction. There is quite a possibility that other proteins of viral origin may support replication (Etessami et al. 1991).

9.9 Plum Pox Virus

PPV stands at position 8 in the top 10 entries and is from genus *Potyvirus* and family Potyviridae (Dunez and Sutic 1988). PPV is globally distributed and is efficiently spread by aphids, posing challenges in its control. The Sharka disease (skarka means pox in Bulgarian) of stone fruit was noticed first on plum cv. Kjustendil in Bulgari (Atanasoff 1935), later identified to be caused by PPV, is a great threat and requires expenditure of huge money to control. Interestingly, PPV is naturally restricted to stone fruits and ornamental *Prunus* sp. PPV and is extensively distributed across regions of Europe, the Mediterranean region, parts of Asia, and North and South America, albeit with varying incidence levels among these countries (Cervera et al. 1993). On the basis of biological, serological and molecular characters of PVP, the M, D, Rec, C, EA, and W strains/groups are known (Candresse et al. 1995).

9.9.1 Host Range, Symptomatology, and Transmission

Identification of natural host range is limited to genus *Prunus* including *Prunus domestica* (cultivated plum), *Prunus salicina* (Japanese plum), *Prunus aarmenica* (apricot), *Prunus persica* (nectarine), and *Prunus amygdalus* (almond). Besides plum as host, PVP includeS peach, apricot, and on sour and sweet cherry as host. Major leaf symptoms include pale green chlorotic rings or spots (Fig. 9.1) and chlorotic rings or arabesque depressions and gumming on fruit including premature fruit drop (Badenes et al. 1996). The fruits are small and deformed, and seeds of infected fruits also showed discolored rings. Symptom types and intensity depends on the type of infecting isolate, host age, and physiological status including surrounding climate conditions. Co-occurrence of other viruses such as prunus necrotic spot virus (PNRSV) or prune dwarf virus (PDV) exacerbates the symptom intensity. PVP is not seed transmitted but nonpersistently by 20 aphid species in nature besides grafting and vegetative mode (Atreya et al. 1991; Gaborjanyi and Basky 1995).

9.9.2 Virus Structure and Genome

PVP shape is flexuous filamentous having size of about 750 nm × 15 nm. Virion encapsulate ss(+ve) sense RNA genome of about 9741–9795 nucleotides covering uniform subunits of coat protein. RNA genome is covalently bound with virus-encoded protein (VPg) at its 5′ end and poly A tail at 3′ end. PVP genomic organization is typical of potyviruses featuring a single ORF encoding a large polyprotein precursor spanning 3125 to 3143 amino acids, producing a 355 kDa protein. This precursor is then processed by virus-encoded proteinases P1, HC-Pro, and NlaPro, resulting in the generation of 10 mature functional proteins (García and Laín 1991). The function of proteins are the same as described for the PPY.

9.10 Brome Mosaic Virus

BMV, the representative species in genus *Bromovirus* of Bromoviridae family, stands as a valuable model for investigating various processes associated with +ve-strand RNA viruses. These cumulatively perform RNA replication, host interactions, recombination, and encapsidation (Bockstahler and Kaesberg 1965). Operating as an alphavirus-like virus with a well-characterized presence in cereal plants, BMV contributed in advancing our understanding towards fundamental mechanisms in positive-strand RNA virus biology. Notably, it has contributed insights into translation, replication, and assembly, leveraging its tripartite RNA genome and icosahedral capsid structure.

9.10.1 Host Range and Symptomatology

The transmission of BMV in plants lacks comprehensive characterization. In field conditions, human activities related to agricultural production, including machinery trampling, can contribute to BMV transmission. However, they may be mechanically inoculated from either using inoculum from virus-infected tissue or virion RNAs and *Nicotiania benthamiana* and *Vigna unguiculata* used as laboratory host for BMV (Trzmiel et al. 2016). To date, the insect vector/s responsible for this virus transmission in natural settings remain unidentified. However, laboratory studies have shown that spotted cucumber beetles and nematodes have the potential to induce disease in test plants (Maekawa et al. 1985). BMV predominantly targets monocots like barley, maize, rice, wheat, and sorghum, along with *Brachypodium distachyon* and *Setaria viridis*. BMV coupled with other viruses infecting wheat is also reported (Ranabhat et al. 2023).

9.10.2 Virion Structure

BMV particles consist of approximately 22% nucleic acid and 78% protein. These non-enveloped virions produced by BMV have a diameter of 28 nm and lack polyamines. The virion coat is formed by arranging 180 units of CP in T¼ 3 quasi-icosahedral symmetry (Rao 2006). Inside the virion, RNA is arranged as an internal subshell within the capsid, forming a hollow core. The pivotal N-terminal 26 amino acids of the CP, crucial for RNA packaging, interact with the RNA to neutralize its charge (Okuno and Furusawa 1979). Both cryo-electron microscopy reconstruction and X-ray crystallography studies have shown that the capsid structure of BMV closely resembles that of cowpea chlorotic mottle virus. (Siegel et al. 1999). The distinctive characteristics include orientation of core b-barrel which positioned almost perpendicular to surface and forming distinct hexameric and pentameric capsomeres (Ahlquist 1992). These features suggest a common ancestry between CP of BMV to CPs of polyoma- and papilloma-viruses.

9.10.3 Genome Organization

The BMV genome, spanning approximately 8.2 kb, was fully characterized in 1984. It has four ssRNAs with varying molecular weights of 1.1×10^6 (RNA-1), 1.0×10^6 (RNA-2), 0.7×10^6 (RNA-3), and 0.3×10^6 (RNA-4). Each of these RNAs has cap like structure at 5′ termini and 200 nucleotides long tRNA like structure at 3′ prime (Fig. 9.1). For replication of BMV inside the host cell, it requires copy of minus-strand, plus-strand, and subgenomic plus-strand RNAs which attach to the replicase protein to initiate replication. The RNA1 and RNA2 encode 1a and 2a proteins of109 and 94 kDa, respectively. The 1a protein, encoded by RNA-1 with 961 aa spanning 3234 nucleotides, functions in RNA capping and facilitates the formation of membrane-associated replication structures after its binding to viral RNAs (Schwartz et al. 2004). The 1a protein also has helicase-like motifs whose function is yet to be explored. The 2a protein of 822 aa encoded by RNA-2 segment contributes as RNA-dependent RNA polymerase for nucleotide polymerization. The RNA 3 segment is dicistronic and directly translates code 3a protein of 35 kDa by 5′-half known as MP. However, to encode the CP of 20.3 kDa BY 3′ end, it requires synthesis of subgenomic RNA4 (Shih and Kaesberg 1976). RNA4, derived from RNA3, produces CP indicating a sophisticated gene regulation system through translational competition. In early infection stages with low viral RNA concentrations, translation occurs for structural and nonstructural RNA replication factors. As virion RNAs amplify, there is preferential translation of CP to encapsulate these RNAs (Miller et al. 1985). It is interesting to know that the 1a and 2a proteins responsible for synthesis of Replicase protein share a very high sequence similarity with diverse plant and animal positive-strand RNA viruses (O'Reilly and Kao 1998).

9.10.4 BMV Spread

Virus encapsulates the newly synthesized RNAs within sap transmitted tissue across various plants, yet its systemic infection of entire plants is confined to a narrow test range (Choi and Rao 2003). Although BMV 3a (MP) and CP proteins are not crucial for RNA replication, they play essential roles in cell-to-cell and long-distance viral spread as modifications in 3a gene limit or abolish systemic movement, thereby restricting infection to site of inoculations. The MP collaborates in binding with ssRNA, localizes this complex to plasmodesmata connections, and triggers the formation of virion-containing tubules protruding from infected protoplasts. Disruption of CP halts virus spread to nearby tissues, whereas local cell-to-cell spread without CP relies on the presence of the 3a allele and characteristics of host. Adaptation of MP in hosts predominantly impacts the spread of infection, as evidenced by a few amino-acid alterations enabling an expansion of the host range from monocotyledonous to dicotyledonous plants.

9.11 Potato Virus X

PVX, genus *Potexvirus* and family Alphaflexiviridae, stands as the earliest identified pathogen of potato, a substantial risk to the fourth most vital food commodity globally (Scholthof et al. 2011). Its presence in scientific literature dates back to 1938 with the seminal work of Loughnane and Murphy. Known for inducing a mild mosaic disease, PVX affects various solanaceous plants, including potatoes, tomatoes, and tobacco. Within the genus *Potexvirus*, which comprises 48 species, many members are responsible for causing diseases in their respective hosts, as outlined by Martelli et al. (2008).

9.11.1 Virion Morphology, Genome Organization, and Replication

PVX genome comprises an ss(+ve)-sense RNA molecule of about 6.4 kb. Like all members of Alphaflexiviridae, the PVX RNA genome features 5′ mG7-cap and pilyadenylated at 3′ (Fig. 9.1) (Kreuze et al. 2020). The virions, characterized as relatively short flexuous filaments, measure between 470 and 580 nm in length. A left-handed helical virion is formed by 1300 copies of the CP (Grinzato et al. 2020). The genome includes five ORFs coding indispensable RNA-dependent RNA polymerase (RdRp), triple gene block (TGB)-complex of group of proteins essential for virus cell-to-cell movement, and CP gene (Morozov and Solovyev 2015). Once inside a cell, PVX virions undergo unpacking facilitated by the host cellular enzymes, a process aided by phosphorylation of the exposed N-terminus of the CP. Released genomic RNA then undergoes translation, makes a copy of RdRp of 166 kDa from ORF1, which leads to the formation and replication of genomic and subgenomic RNAs (Huisman et al. 1988; Samuels et al. 2007). RdRp has N-terminal methyltransferase-like domain, an NTP-binding/helicase-like domain, and a C-terminal RNA-dependent polymerase (Linnik et al. 2013). The TGB2 and TGB3 proteins are essential for movement facilitation as they recruit TGB1 and CP to virus replication complex, enabling the transportation of RNA through plasmodesmata to adjacent cells (Mukhamedzhanova et al. 2009; Tilsner et al. 2013; Park et al. 2014). TGB1 also functions as a suppressor of host PTGS, thereby disrupting SGS3, which is a cofactor for RDR6 and essential for siRNAs biogenesis (Voinnet et al. 2000; Linnik et al. 2013). TGB2 modulates and checks the function of ER tubules and is, therefore, responsible for membrane curvature (Ju et al. 2005; Lazareva et al. 2021).

The protein of TGB3 is responsible for unfolded protein response (UPR), initiating ER-to-nuclear signaling and transcriptional changes through transcription factors bZIP60, bZIP28, and bZIP17 (Gayral et al. 2020).

9.12 Management of Plant Viruses by PGPR

Contrary to the pests, fungi, and bacteria, there are no direct control methods available against viruses. The management of plant viral diseases primarily relies on the genetic resistance of host plants, environmental factors, and the efficacy of synthetic pesticides in controlling vectors, which is a crucial strategy. Although effective pesticides exist for plant viral diseases, their long-term utility is hindered by residual poisoning, posing environmental and public health risks. Consequently, there's a call for new approaches to complement existing strategies for better and more sustainable viral disease control. One approach is the use of bioinoculants, which can protect crops by reducing viral infections and enhancing plant growth, leading to significant economic benefits for growers (Sofy et al. 2019). In recent years, there has been a growing trend towards utilizing PGPR for systemic resistance against plant viruses. Although much of the research on PGPR-mediated induced systemic resistance (ISR) has focused on fungal and bacterial pathogens, this chapter will explore the potential of PGPR-mediated ISR against various plant viruses. Here, we will discuss about the recent advances in controlling this notorious pathogen using the sustainable PGPR-mediated system.

9.12.1 Management of TMV

An isolate of bacteria named PM1, extracted from the rhizosphere of healthy plants, was identified as *Pseudomonas aeruginosa* through biochemical characteristics and the sequencing of its 16S rRNA gene. This strain demonstrated the ability to trigger resistance in *N. tabacum* cv. Xanthi-nc against TMV and in *Cyamopsis tetragonoloba* against Sunn-hemp Rosette Virus (SRV) (Gupta et al. 2022). When applied to the foliage, isolate PM1 successfully reduced TMV accumulation in susceptible *N. tabacum*. Foliage and soil drench treatments with PM1 led to disease indices of 48% and 86%, respectively, with control rates of 48.9% and 8.5%. PM1 exhibited various plant growth-promoting traits, including phosphate solubilization, siderophore production, auxin secretion, hydrogen cyanide (HCN) production, and ammonia release. Foliar application of PM1 enhanced tobacco growth, whereas its volatiles notably stimulated seedling growth in *C. tetragonoloba*. Many metabolites produced by PM1 are recognized for their roles in inducing systemic resistance, exhibiting antibiosis, and promoting plant growth.

9.12.2 Management of PVY

PVY is a widespread plant virus that causes significant crop losses in Solanaceae crops. Recent research has identified *Serratia marcescens*-S3, as a PGPR strain, capable of inhibiting PVY replication in *N. benthamiana* (Ge et al. 2022). In this study, authors have uncovered that the ubiquitination of NbHsc70–2 plays a crucial role in how *S. marcescens*-S3 induces systemic resistance. Following treatment with *S. marcescens*-S3, there was a notable reduction in the protein level of NbHsc70–2. Inhibiting ubiquitination led to increased accumulation of NbHsc70–2 in plants, consequently reducing *S. marcescens*-S3-mediated resistance to PVY. Additionally, transgenic *N. benthamiana* lines were engineered, including NbHsc70–2 knockout (NbHsc70–2KO) and NbHsc70–2 ubiquitination site mutation (NbHsc70–2USM), using CRISPR-Cas9 technology. *S. marcescens*-S3 was found to significantly diminish the inhibition of NbHsc70–2 protein accumulation in both NbHsc70–2KO and NbHsc70–2USM lines. Moreover, PVY exhibited increased virulence in NbHsc70–2USM compared to wild-type plants. These findings highlight that *S. marcescens*-S3 enhances the ubiquitination of NbHsc70–2, thereby impeding the recruitment of the molecular chaperone NbHsc70–2 and reducing its ability to support PVY replication and infection. However, the precise mechanism behind this inhibition has not been thoroughly explored.

A research conducted by the Beris et al. (2018) demonstrated that drenching with *Bacillus amyloliquefaciens* strain MBI600 led to a delay in the systemic accumulation of PVY. Transcriptomic analysis revealed the induction of the salicylic acid (SA) signaling pathway in tomato plants following MBI600 treatment, along with distinct gene expression patterns associated with defense mechanisms against PVY infections.

9.12.3 Management of TSWV

This research focused on evaluating the antiviral capabilities of *B. amyloliquefaciens* strain MBI600, the active ingredient in the biological fungicide Serifel® (BASF SE), when applied to tomato plants. Through drenching, foliar spraying, or soil amendment with MBI600, a reduction of up to 80% in TSWV incidence was observed across two different environmental conditions. Furthermore, drenching with MBI600 led to a delay in the systemic accumulation of PVY. Transcriptomic analysis revealed the induction of the salicylic acid (SA) signaling pathway in tomato plants following MBI600 treatment, along with distinct gene expression patterns associated with defense mechanisms against TSWV infections (Beris et al. 2018).

9.12.4 Management of CMV

Earlier studies on a PGPR *Paenibacillus lentimorbus* B-30488 have shown its capacity to improve plant quality under both normal growth conditions and stress environments. Kumar et al. (2016) explored its potential as BCA against CMV in tobacco cultivar White Burley plants, elucidating the biophysical and molecular alterations resulting from the interactions between B-30488, tobacco, and CMV. Earthen pot treatment with PGPR not only boosted plant health but also efficiently reduced RNA accumulation by 12-fold in apex leaves of virus-infected plants than in control plants along with reduction in virulence. Histological analysis of sectioned leaves of B-30488-treated tobacco plants revealed enhanced tissue health and reduced signs of aging regardless of CMV-infection status. These plants exhibited fewer intercellular spaces between collenchyma cells, decreased levels of xyloglucans and pectins in connecting primary cells, and increased polyphenol accumulation in the hypodermis layer extending to collenchyma cells. B-30488 inoculation positively modulated essential biophysical parameters (ion leakage and photosynthetic efficiency) and biochemical attributes (sugar, proline, chlorophyll, malondialdehyde, acid phosphatase, and alkaline phosphatase) of tobacco plants, thereby alleviating virus-induced stress. Furthermore, B-30488 inoculation mitigated the activities of defense-related enzymes (ascorbate peroxidase, guaiacol peroxidase, superoxide dismutase, and catalase) induced by CMV infection, indicating protection against CMV through systemic induced resistance in tobacco. Quantitative RT-PCR analysis of genes associated with normal plant development, stress responses, and pathogenesis aligned well with the biochemical findings, revealing the regulation (either upregulation or downregulation) of these genes in favor of plants combating CMV-induced stress. These enhancements resulted in increased flower and seed production in tobacco plants without any adverse effects on plant health. The findings of this study advocate for the potential applicability of B-30488 for improving crop yields in virus-infested areas.

In another study, Murphy et al. (2003) assessed combinations of two strains of PGPR, formulated with the carrier chitosan, for their potential to promote the growth of tomato plants and confer resistance to infection by CMV. Each combination consisted of GB03 (*Bacillus subtilis*) paired with one of the following PGPR strains: SE34 (*Bacillus pumilus*), IN937a (*B. amyloliquefaciens*), IN937b (*B. subtilis*), INR7 (*B. pumilus*), or T4 (*B. pumilus*). These combinations, formulated with chitosan, were termed biopreparations. Upon challenge with CMV, all plants treated with biopreparations and the older control exhibited significantly greater height, fresh weight, and numbers of flowers and fruits compared to plants in the same-age control treatment inoculated with CMV. Ratings of CMV disease severity were notably lower for tomato plants treated with biopreparations and the older control than for same-age control plants at 14 and 28 dpi. Moreover, CMV accumulation in young, non-inoculated leaves was significantly reduced in all biopreparation-treated plants, in the older control compared to same-age control plants at 14 dpi, and in four out of the five biopreparation treatments at 28 dpi. Among infected tomato plants, the amount of CMV in non-inoculated leaves was significantly lower in three of the biopreparation treatments and the older control at 14 dpi. Additionally, in the biopreparation G/INR7 treatment, CMV accumulation was significantly lower at 28 dpi compared to the control treatment. These findings demonstrate that treatment of tomato plants with biopreparations leads to significant improvements in growth and provides protection against CMV infection.

Ryu et al. (2007) conducted an assessment of a commercially available biopreparation comprising PGPR strains *B. subtilis* GB03 and *B. amyloliquefaciens* IN937a, formulated with chitosan as BioYield. The biopreparation demonstrated the ability to enhance plant growth across various Arabidopsis hormonal mutants, including those insensitive to auxin, gibberellic acid, ethylene, jasmonate, salicylic acid, and brassinosteroids, as well as in wild-type plants. Although the biopreparation provided protection against CMV based on reduced disease severity in wild-type plants, there was no observed decrease in virus titer in either control or treated plants, indicating the induction of tolerance rather than resistance against CMV. Notably, the biopreparation induced resistance against CMV in NahG plants, as evidenced by reduced disease severity and virus titer. Furthermore, it triggered induced resistance against *Pseudomonas syringae* pv. tomato in wild-type plants but not in NahG transgenic plants, which degrade endogenous salicylic acid, suggesting the involvement of salicylic acid signaling. These findings suggest that such a PGPR strains can promote plant growth through mechanisms distinct from known hormonal signaling pathways, and the elicitation of induced resistance by PGPR may vary depending on the pathogen.

9.13 Management of Other Economically Important Viruses

9.13.1 Groundnut Bud Necrosis Virus (GBNV)

GBNV infects a number of plant species and is notorious for causing necrosis disease in chilli, posing a significant threat to its production. Authors demonstrated microbe-associated molecular pattern (MAMP) triggered immunity through *Agrobacterium*-mediated ectopical expression of flagellin protein sourced from *B. amyloliquefaciens* CRN9 in chilli. This expression upregulated EAS1 gene synthesizing epi-aristolochene synthase, enzyme involved in capsidiol phytoalexin synthesis. Furthermore, chilli plants challenged with GBNV exhibited upregulation of PDF and LOX genes of jasmonate (JA)-pathway. The analysis showed a considerable increase in the expression of these genes during treatment with CRN9 and benzothiadiazole (BTH), where transcript level of defense-responsive and PR proteins remaining significantly elevated even at 3 dpi as compared to the control group. Hence, flagellin expression effectively triggers innate immunity and impedes GBNV proliferation in host through ISR activated via both the SA and JA/ET signaling pathways. Similarly, Vanthana et al. (2019) demonstrated the induction of flagellin (Flg) and elongation factor-TU (EF-Tu) mediated obtained from *B. amyloliquefaciens* (VB7) against GBNV in tomato.

9.13.2 Bean Common Mosaic Virus (BCMV)

The study conducted by Udaya Shankar et al. (2009) investigated the impact of seed treatments with seven different Bacillus species as PGPR on inducing disease resistance in cowpea against mosaic disease caused by BCMV. First, remarkable increase in seedling vigor after seed germination was observed in all PGPR treatments with 7 PGPR strains, whereas *B. pumilus* (T4) and *B. subtilis* (GBO3) strains exhibited particularly significant results. Treatment of cowpea seeds with T4 and GBO3 strains demonstrated disease severity rates of 42% and 41%, respectively, against

BCMV. In field conditions, strain GBO3 provided 34% protection against BCMV. Evaluation of PGPR strain protection against BCMV was conducted using indirect ELISA, with cowpea seeds treated with GBO3 and T4 strains exhibiting the lowest immunoreactive values compared to the non-bacterized control. It is documented that combination of strains produce higher quantity and in inducing resistance than individual strains as seeds treated with GBO3 + T4 showed maximum level of protection against BCMV.

9.13.3 *Tomato Mottle Virus (ToMV)*

ToMV is a serious problem of tomato cultivation. In a study to assess induced resistance, the tomato seeds/plants treated by seed treatment or spore preparation mixed with potting medium (powder) or a combined seed-powder treatment revealed varying level of reduction in virus titer (Murphy et al. 2000). For this, PGPR strains including *B. amyloliquefaciens* 937a, *B. subtilis* 937b, and *B. pumilus* SE34 were chosen based on their previous success in inducing resistance. The highest protection was observed in a trial where PGPR powder-based treatments showed significantly lower ToMV disease severity compared to seed or control treatments. Although few PGPR treatments showed decreased symptom severity and reduced incidence of viral DNA, there was no clear corresponding improvement in yield responses.

However, little or no resistance was observed among treatments at 80 days after transplant. These findings indicate that, under natural conditions with high vector-virus pressure, certain PGPR treatments can reduce ToMV incidence and disease severity, potentially leading to increased fruit yield, thus suggesting the potential integration of PGPR in managing this virus in tomatoes.

9.13.4 *Alfalfa Mosaic Virus (AMV)*

AMV has an extensive host range leading to crop losses. Abdelkhalek et al. (2020) revealed the effectiveness of *Bacillus licheniformis* strain POT1 against AMV-infecting potato strain. Most successful treatment involved foliar application of culture filtrate (CF) resulting in 86.79% decrease in virus concentration along with enhancements in various growth parameters. Furthermore, high-performance liquid chromatography analysis revealed the accumulation of 7218.86 mg/kg of 20 polyphenolic compounds in PGPR-treated potato which was comparable to 1606.49 mg/kg in control potato plants. Transcriptomic analysis demonstrated the upregulation of 13 genes involved in phenylpropanoid, chlorogenic acid, and flavonoid biosynthetic pathways following POT1 treatments. Particularly noteworthy was the significant increase of 588.133-fold in expression levels of a key enzyme in flavonoid biosynthesis F3H and AN2, an anthocyanin 2 transcription factor (97.005-fold), suggesting that flavonoid accumulation, may exert crucial roles in plant defense. GC-MS identified pyrrolo[1,2-a]pyrazine-1,4-dione as main compound from CF, indicating its potential role as an elicitor molecule for inducing SAR in potato.

9.14 Perspectives

Plant viruses poses significant threat to crop production globally coupled with climate and ecological niche. As viruses replicate in hosts, they highjack cellular processes and host physiology, leading to onset of diseases. Typically, these reactions manifest in plants lacking specific defensive responses against virulence factors produced by the virus. However, instances exist where plants, in both natural and cultivated populations, possess inherent disease resistance but remain vulnerable to virus infections. Effective disease management strategies focus on preventing or reducing virus infections, requiring information on their survival and dissemination. Due to the distinct epidemiological characteristics of different diseases, no universal approach to management is available. The deep information on their dissemination, replication, transmission, and information on interaction with their host and vectors will certainly improve our understanding towards facile and efficient virus management. Also, the producers and farmers need to be introduced and made aware with the life cycle and biology of the virus and for the independent or integrated approaches for virus management in their field conditions. Although, at present, focus is being given to study the plant viruses in India and abroad, much focus is desired, as the case with animal viruses, so that more effective and robust tools can be developed to combat with plant viruses.

Acknowledgments The authors wish to thank the Director of CSIR-National Botanical Research Institute (NBRI), Lucknow for providing necessary facilities. AR is thankful to DBT for granting Junior Research Fellowship (JRF). TV and DKM are thankful to UGC-CSIR for granting JRF.

References

Abdelkhalek A, Al-Askar AA, Behiry SI (2020) *Bacillus licheniformis* strain POT1 mediated polyphenol biosynthetic pathways genes activation and systemic resistance in potato plants against *Alfalfa mosaic virus*. Sci Rep 10(1):16120. https://doi.org/10.1038/s41598-020-72676-2

Agrios GN (1978) Plant pathology, 2nd edn. Academic Press, San Diego, CA, pp 466–470

Ahlquist P (1992) Bromovirus RNA replication and transcription. Curr Opin Genet Dev 2:71–76. https://doi.org/10.1016/S0959-437X(05)80325-9

Anderer FA, Uhlig H, Weber E, Schramm G (1960) Primary structure of the protein of tobacco mosaic virus. Nature 186:922–925. https://doi.org/10.1038/186922a0

Anfoka G, Haj Ahmad F, Abhary M, Hussein A (2009) Detection and molecular characterization of viruses associated with tomato yellow leaf curl disease in cucurbit crops in Jordan. Plant Pathol 58:754–762. https://doi.org/10.1111/j.1365-3059.2009.02060.x

Atanasoff D (1935) Mosaic of stone fruits. Phytopathol Z 8:259–284

Atreya PL, Atreya CD, Pirone TP (1991) Amino acid substitutions in the coat protein result in loss of insect transmissibility of a plant virus. Proc Natl Acad Sci U S A 88:7887–7891. https://doi.org/10.1073/pnas.88.17.7887

Badenes ML, Asins MJ, Carbonell EA, Llácer G (1996) Genetic diversity in apricot, *Prunus armeniaca*, aimed at improving resistance to plum pox virus. Plant Breed 115:133–139. https://doi.org/10.1111/j.1439-0523.1996.tb00888.x

Beijerinck MW (1898) Über ein Contagium vivum fluidum als Ursache der Fleckenkrankheit der Tabaks-blätter. Verh K Akad Weterschappen Amst Afd Natuurkunde 6:3–21

Benitez-Alfonso Y, Faulkner C, Ritzenthaler C, Maule A (2010) Plasmodesmata: gateways to local and systemic virus infection. Mol Plant Microbe Interact 23:1403–1412. https://doi.org/10.1094/MPMI-05-10-0116

Beris D, Theologidis I, Skandalis N et al (2018) Bacillus amyloliquefaciens strain MBI600 induces salicylic acid dependent resistance in tomato plants against tomato spotted wilt virus and potato virus Y. Sci Rep 8:10320. https://doi.org/10.1038/s41598-018-28677-3

Best RJ (1968) Tomato spotted wilt virus. Adv Virus Res 13:66–146. https://doi.org/10.1016/S0065-3527(08)60251-1

Bhattiprolu SL (1991) Seed borne cucumber mosaic virus infecting French bean (*Phaseolus vulagaris*) in India. Indian J Virol 7:66–67

Blanc S, López-Moya JJ, Wang R, García-Lampasona S, Thornbury DW, Pirone TP (1997) A specific interaction between coat protein and helper component correlates with aphid transmission of a potyvirus. Virology 231:141–147. https://doi.org/10.1006/viro.1997.8521

Bockstahler LE, Kaesberg P (1965) Isolation and purification of RNA from bromegrass mosaic virus. J Mol Biol 13:127–137. https://doi.org/10.1016/S0022-2836(65)80084-5

Bouwen I, van Zaayen AZ (1995) Chrysanthemum. In: Loebenstein G, Lawwon RH, Brunt AA (eds) Virus and virus-like diseases of bulb and flower crops. Wiley, New York, pp 396–408

Brault V, Uzest M, Monsion B, Jacquot E, Blanc S (2010) Aphids as transport devices for plant viruses. C R Biol 333(6–7):524–538. https://doi.org/10.1016/j.crvi.2010.04.001

Brittlebank CC (1919) Tomato diseases. J Agric Victoria 27:213–235

Brown JK (2010) Phylogenetic biology of the *Bemisia tabaci* sibling species group. In: Stansley PA, Naranjo SE (eds) Bemisia-bionomics and management of a global pest. Springer, New York, pp 31–67. https://doi.org/10.1007/978-90-481-2460-2_2

Brunt AA, Crabtree K, Dallwitz MJ, Gibbs AJ, Watson L, Zurcher EJ (1996) Plant viruses online: descriptions and lists from the VIDE database. http://biology.anu.edu.au/groups/MES/vide

Buck KW (1999) Replication of tobacco mosaic virus RNA. Philos Trans R Soc Lond B Biol Sci 354(1383):613–627. https://doi.org/10.1098/rstb.1999.0413

Caciagli P, Medina V, Marian D, Vecchiati M, Masenga V, Falcioni T, Noris E (2009) Virion stability is important for the circulative transmission of tomato yellow leaf curl Sardinia virus by *Bemisia tabaci*, but virion access to salivary glands does not guarantee transmissibility. J Virol 83:5784–5795. https://doi.org/10.1128/JVI.02267-08

Candresse T, Macquaire G, Lanneau M, Bousalem M, Quiot-Douine L, Quiot JB, Dunez J (1995) Analysis of plum pox virus variability and development of strain-specific PCR assays. Acta Hort 386:357–369

Carrington JC, Dougherty WG (1987) Small nuclear inclusion protein encoded by plant potyvirus genome is a protease. J Virol 61:2540–2548. https://doi.org/10.1128/jvi.61.8.2540-2548.1987

Carrington JC, Freed DD (1990) Cap-independent enhancement of translation by a plant potyvirus 5′ nontranslated region. J Virol 64:1590–1597. https://doi.org/10.1128/JVI.64.4.1590-1597.1990

Cervera MT, Riechmann JL, Martín MT, García JA (1993) 3′-terminal sequence of the plum pox virus PS and o6 isolates: evidence for RNA recombination within the potyvirus group. J Gen Virol 74:329–334. https://doi.org/10.1099/0022-1317-74-3-329

Cheng RH, Olson NH, Baker TS (1992) Cauliflower mosaic virus: a 420 subunit (T=7), multilayer structure. Virology 186(2):655–668. https://doi.org/10.1016/0042-6822(92)90032-k

Chikh-Ali M, Katayama K, Maoka T, Natsuaki K (2008) Significance of weed hosts for potato virus Y protection in Syria. Bull OEPP/EPPO Bull 38:226–232. https://doi.org/10.1111/j.1365-2338.2008.01209.x

Choi YG, Rao AL (2003) Packaging of brome mosaic virus RNA3 is mediated through a bipartite signal. J Virol 77:9750–9757. https://doi.org/10.1128/JVI.77.18.9750-9757.2003

Chung BYW, Miller WA, Atkins JF, Firth AE (2008) An overlapping essential gene in the Potyviridae. Proc Natl Acad Sci U S A 105:5897–5902. https://doi.org/10.1073/pnas.0800468105

Cillo F, Roberts IM, Palukaitis P (2002) In situ localization and tissue distribution of the replication-associated proteins of cucumber mosaic virus in tobacco and cucumber. J Virol 76:19654–19664. https://doi.org/10.1128/jvi.76.21.10654-10664.2002

Codoner FM et al (2005) Molecular evolution of the plant virus family Bromoviridae based on RNA3-encoded proteins. J Mol Evol 61:697–705. https://doi.org/10.1007/s00239-005-0021-7

Creager ANH (2022) Tobacco mosaic virus and the history of molecular biology. Annu Rev Virol 9:39–45. https://doi.org/10.1146/annurev-virology-100520-014520

Czosnek H, Laterrot H (1997) A worldwide survey of tomato yellow leaf curl viruses. Arch Virol 142:1391–1406. https://doi.org/10.1007/s007050050168

Czosnek H, Ghanim M, Rubinstein G, Morin S, Fridman V, Zeidan M (2001) Whiteflies: vectors, and victims(?), of geminiviruses. Adv Virus Res 57:291–322. https://doi.org/10.1016/S0065-3527(01)57006-2

De Barro PJ, Liu SS, Boykin LM, Dinsdale AB (2011) Bemisia tabaci: a statement of species status. Annu Rev Entomol 56:1–19. https://doi.org/10.1146/annurev-ento-112408-085504

Díaz-Pendón JA, Cañizares MC, Moriones E, Bejarano ER, Czosnek H, Navas-Castillo J (2010) Tomato yellow leaf curl viruses: ménage à trois between the virus complex, the plant, and the whitefly vector. Mol Plant Pathol 11:441–450. https://doi.org/10.1111/j.1364-3703.2010.00618.x

Ding SW, Anderson BJ, Haase HR, Symons RH (1994) New overlapping gene encoded by the cucumber mosaic virus genome. Virology 198:593–601. https://doi.org/10.1038/313068a0

Dolja VV, Haldeman-Cahill R, Montgomery AE, Vandenbosch KA, Carrington JC (1995) Capsid protein determinants involved in cell-to-cell and long distance movement of tobacco etch potyvirus. Virology 206:1007–1016. https://doi.org/10.1006/viro.1995.1023

Doolittle SP (1916) A new infectious mosaic disease of cucumber. Phytopathology 6:145–147

Dunez J, Sutic D (1988) Plum pox virus. In: Smith IM, Dunez J, Elliot RA, Phillips DH, Arches SA (eds) European handbook of plant diseases. Blackwell, London, pp 44–46. https://doi.org/10.1002/9781444314199

Eskarous JK, Habib HM, Kishtah AA, Ismail MH (1983) A strain of potato virus Y isolated from Solanum nigrum var. judaicum in Egypt. Phytopatol Mediterr 22:53–58

Etessami P, Saunders K, Watts J, Stanley J (1991) Mutational analysis of complementary-sense of genes of African cassava mosaic virus DNA a. J Gen Virol 72:1005–1012. https://doi.org/10.1099/0022-1317-72-5-1005

Fauquet CM, Mayo MA, Malinoff J, Desselberger U, Ball LA (2005) Virus taxonomy, VIIIth report of the ICTV. Elsevier/Academic Press, London

Francki RIB, Mossop DW, Hatta T (1979) Cucumber mosaic virus. In: CMI/AAB descriptions of plant viruses no. 213. Commonwealth Agricultural Bureau/Association of Applied Biologists, Kew

Gaborjanyi R, Basky S (1995) Correlation between migration of aphid vector and natural spread of plum pox virus. Acta Hort 386:376–378. https://doi.org/10.17660/ActaHortic.1995.386.26

García JA, Laín S (1991) Proteolytic activity of plum pox virus-tobacco etch virus chimeric proteins. FEBS Lett 281:67–72. https://doi.org/10.1016/0014-5793(91)80360-F

Gayral M, Gaguancela OA, Vasquez E, Herath V, Flores FJ, Dickman MB, Verchot J (2020) Multiple ER-to-nucleus stress signaling pathways are activated during Plantago asiatica mosaic virus and turnip mosaic virus infection in *Arabidopsis thaliana*. Plant J 103:1233–1245. https://doi.org/10.1111/tpj.14798

Ge M, Gong M, Jiao Y, Li Y, Shen L, Li B, Wang Y, Wang F, Zhang S, Yang J (2022) Serratia marcescens-S3 inhibits *potato virus Y* by activating ubiquitination of molecular chaperone proteins NbHsc70-2 in *Nicotiana benthamiana*. J Microbial Biotechnol 15(4):1178–1188. https://doi.org/10.1111/1751-7915.13964

Glasa M, Hančinský R, Šoltys K, Predajňa L, Tomašechová J, Hauptvogel P, Mrkvová M, Mihálik D, Candresse T (2021) Molecular characterization of potato virus Y (PVY) using high-throughput sequencing: constraints on full genome reconstructions imposed by mixed infection involving recombinant PVY strains. Plan Theory 10(4):753. https://doi.org/10.3390/plants10040753

Gould AR, Symons RH (1977) Determination of sequence homology between the four RNA species of cucumber mosaic virus by hybridization analysis with complementary DNA. Nucleic Acids Res 4:3787–3802. https://doi.org/10.1093/nar/4.11.3787

Gould AR, Symons RH (1978) Alfalfa mosaic virus RNA. Determination of sequence homology between the four RNA species of and comparison with the four RNA species of cucumber mosaic virus. Eur J Biochem 91:269–278. https://doi.org/10.1111/j.1432-1033.1978.tb20962.x

Grinzato A, Kandiah E, Lico C, Betti C, Baschieri S, Zanotti G (2020) Atomic structure of potato virus X, the prototype of the Alphaflexiviridae family. Nat Chem Biol 16:564–569. https://doi.org/10.1038/s41589-020-0502-4

Gupta AK, Jyoti V, Aparana S, Shalini S, Vivek P (2022) Pseudomonas aeruginosa isolate PM1 effectively controls virus infection and promotes growth in plants. Arch Microbiol 204(8):494. https://doi.org/10.1007/s00203-022-03105-3

Gutierrez C (1999) Geminivirus DNA replication. Cell Mol Life Sci 56:313–329. https://doi.org/10.1007/s000180050433

Hamilton RI (1985) Virus transmission. In: Francki RIB (ed) The plant viruses, Polyhedral Virions with tripartite genomes, vol 1. Plenum press, New York, London, pp 245–267. https://doi.org/10.1007/978-1-4684-4937-2

Hanley-Bowdoin L, Settlage SB, Orozco BM, Nagar S, Robertson D (1999) Geminiviruses: models for plant DNA replication, transcription, and cell cycle regulation. CRC Crit Rev Plant Sci 18:71–106. https://doi.org/10.1016/S0735-2689(99)00383-4

Harrison BD, Swanson MM, Fargette D (2002) Begomovirus coat protein: serology, variations and functions. Physiol Mol Plant Pathol 60:257–271. https://doi.org/10.1006/pmpp.2002.0404

Huang CH, Chao YJ, Chang CA, Hseu SH, Hsaio CH (1987) Identification and comparison of different viruses on symptom expression in loofah. J Agric Res China 36:413–420

Huisman MJ, Linthorst HJ, Bol JF, Cornelissen JC (1988) The complete nucleotide sequence of potato virus X and its homologies at the amino acid level with various plus-stranded RNA viruses. J Gen Virol 69:1789–1798. https://doi.org/10.1099/0022-1317-69-8-1789

Jiang J, Laliberte J-F (2011) The genome-linked protein VPg of plant viruses-a protein with many partners. Curr Opin Virol 1:347–354. https://doi.org/10.1016/j.coviro.2011.09.010

Ju H-J, Samuels TD, Wang Y-S, Blancaflor E, Payton M, Mitra R et al (2005) The potato virus X TGBp2 movement protein associates with endoplasmic reticulum-derived vesicles during virus infection. Plant Physiol 138:1877–1895. https://doi.org/10.1104/pp.105.066019

Kaper JM, Waterworth HE (1981) Cucumoviruses. In: Kurstak E (ed) Handbook of plant virus infections and comparative diagnosis. Elsevier/North-Holland Biomedical Press, Amsterdam/New York, pp 257–332

Kaper JM, Diener TO, Scott HA (1965) Some physical and chemical properties of cucumber mosaic virus (strain Y) and of its isolated ribonucleic acid. Virology 27:54–72. https://doi.org/10.1016/0923-2516(90)90058-q

Kataoka J, Masuta C, Takami Y (1990) Complete nucleotide sequence of RNA 1 of cucumber mosaic virus Y-strain and evolutionary relationships among genome RNAs of the virus strains. Ann Phytopathol Soc Japan 56:501–507

Kerlan C (2006) Potato virus Y. AAB/CMI Descriptions of Plant Viruses 414. http://www.dpvweb.net/dpv/showdpv.php?dpvno=414

Khurana SMP, Garg ID (2000) Potato mosaic. In: Khurana SMP (ed) Diseases and pests of potato—a manual, pp 36–40

Kreuze JF, Vaira AM, Menzel W, Candresse T, Zavriev SK, Hammond J et al (2020) ICTV virus taxonomy profile: Alphaflexiviridae. J Gen Virol 101:699–700. https://doi.org/10.1099/jgv.0.001436

Kumar S, Chauhan PS, Agrawal L, Raj R, Srivastava A, Gupta S, Mishra SK, Yadav S, Singh PC, Raj SK, Nautiyal CS (2016) *Paenibacillus lentimorbus* inoculation enhances tobacco growth and extenuates the virulence of cucumber mosaic virus. PLoS One 11(3):e0149980. https://doi.org/10.1371/journal.pone.0149980

Laliberté JF, Sanfaçon H (2010) Cellular remodeling during plant virus infection. Annu Rev Phytopathol 48:69–91. https://doi.org/10.1146/annurev-phyto073009-114239

Lazareva EA, Lezzhov AA, Chergintsev DA, Golyshev SA, Dolja VV, Morozov SY et al (2021) Reticulon-like properties of a plant virus-encoded movement protein. New Phytol 229:1052–1066. https://doi.org/10.1111/nph.16905

Lewis J (2004) From virus research to molecular biology: tobacco mosaic virus in Germany, 1936–1956. J Hist Biol 37:259–301. https://doi.org/10.1023/b:hist.0000038206.09052.7a

Linnik O, Liesche J, Tilsner J, Oparka KJ (2013) Unraveling the structure of viral replication complexes at super-resolution. Front Plant Sci 4:6. https://doi.org/10.3389/fpls.2013.00006

Liu C, Nelson RS (2013) The cell biology of tobacco mosaic virus replication and movement. Front Plant Sci 4:12. https://doi.org/10.3389/fpls.2013.00012

Lot H, Marrou J, Kaper JM, West CK, van Vloten-Doting L, Hull R (1974) Evidence for threee functional RNA species in several strains of cucumber mosaic virus. J Gen Virol 22:81–93. https://doi.org/10.1099/0022-1317-22-1-81

Lot H, Jonard G, Richard K (1977) Cucumber mosaic virus RNA 5. Partial characterization and evidence for no large sequence homologies with genomic RNAs. FEBS Lett 80:395–400. https://doi.org/10.1016/0014-5793(77)80484-5

Loughnane J, Murphy P (1938) Mode of dissemination of potato virus X. Nature 141:120–121. https://doi.org/10.1038/141120a0

Love AJ, Geri C, Laird J, Carr C, Yun BW, Loake GJ, Tada Y, Sadanandom A, Milner JJ (2012) Cauliflower mosaic virus protein P6 inhibits signaling responses to salicylic acid and regulates innate immunity. PLoS One 7(10):e47535. https://doi.org/10.1371/journal.pone.0047535

Lutz LM, Raikhy G, Leisner SM (2012) Cauliflower mosaic virus major inclusion body protein interacts with the aphid transmission factor, the virion-associated protein, and gene VII product. Virus Res 170(1–2):150–153. https://doi.org/10.1016/j.virusres.2012.08.017

Maekawa K, Horikoshi M, Furusawa I (1985) Modes of multiplication of brome mosaic virus in protoplasts from host and non-host plants. Ann Phytopathol Soc Japan 51:227–230

Martelli GP, Adams MJ, Kreuze JF, Dolja VV (2008) Family Flexiviridae: a case study in virion and genome plasticity. Annu Rev Phytopathol 45:73–100. https://doi.org/10.1146/annurev.phyto.45.062806.094401

Miller WA, Dreher TW, Hall TC (1985) Synthesis of brome mosaic virus subgenomic RNA in vitro by internal initiation on (−)-sense genomic RNA. Nature (London) 313:68–70. https://doi.org/10.1038/313068a0

Moreno A, De Blas C, Biurrun R, Nebreda M, Palacios I, Duque M, Fereres A (2004) The incidence and distribution of viruses infecting lettuce, cultivated brassica and associated natural vegetation in Spain. Ann Appl Biol 144(3):339–346. https://doi.org/10.1111/j.1744-7348.2004.tb00349.x

Morozov SY, Solovyev AG (2015) Phylogenetic relationship of some "accessory" helicases of plant positive-stranded RNA viruses: toward understanding the evolution of triple gene block. Front Microbiol 6:1–8. https://doi.org/10.3389/fmicb.2015.00508

Mukhamedzhanova A, Karpova O, Rodionova N, Atabekov I (2009) Nonspecific activation of translation of encapsidated potexviral RNA with involvement of potato virus X movement protein TGB1. Dokl Biochem Biophys 428:239–241. https://doi.org/10.1134/S1607672909050044

Murphy JF, Zehnder GW, Schuster DJ, Sikora EJ, Polston JE, Kloepper JW (2000) Plant growth-promoting rhizobacterial mediated protection in tomato against tomato mottle virus. Plant Dis 84(7):779–784. https://doi.org/10.1094/PDIS.2000.84.7.779

Murphy JF, Reddy MS, Ryu C-M, Kloepper JW, Li R (2003) Rhizobacteria-mediated growth promotion of tomato leads to protection against cucumber mosaic virus. Phytopathology 93(10):1301–1307. https://doi.org/10.1094/PHYTO.2003.93.10.1301

Noris E, Vaira AM, Caciagli P, Masenga V, Gronemborn B, Accotto GP (1998) Amino acids in the capsid protein of tomato yellow leaf curl virus that are crucial for systemic infection, particle formation, and insect transmission. J Virol 72:10050–10057. https://doi.org/10.1128/JVI.72.12.10050-10057.1998

O'Reilly EK, Kao CC (1998) Analysis of RNA-dependent RNA polymerase structure and function as guided by known polymerase structures and computer predictions of secondary structure. Virology 252:287–303. https://doi.org/10.1006/viro.1998.9463

Okuno T, Furusawa I (1979) RNA polymerase activity and protein synthesis in brome mosaic virus-infected protoplasts. Virology 99:218–225. https://doi.org/10.1016/0042-6822(79)90002-3

Oliver JE, Whitfield AE (2016) The genus Tospovirus: emerging Bunyaviruses that threaten food security. Annu Rev Virol 3:101–124. https://doi.org/10.1146/annurev-virology-100114-055036

Palaniswami MS, Nair RR, Pillai KS, Thankappan M (1996) Whiteflies on cassava and its role as vector of cassava mosaic in India. J Root Crops 22:1–8

Palukaitis P, Garcia-Arenal F (2003) Cucumoviruses. Adv Virus Res 62:241–323. https://doi.org/10.1016/s0065-3527(03)62005-1

Palukaitis P, Roossinck MJ, Dietzgen RG, Francki RIB (1992) Cucumber mosaic virus. Adv Virus Res 41:281–348. https://doi.org/10.1016/s0065-3527(08)60039-1

Parisi B, Roggero P, Grassi G, Masenga V (1998) Tospovirus (TSWV) on aubergine in protected cultivation. L'Informatore Agrario 54(23):45–48

Park HS, Himmelbach A, Browning KS, Hohn T, Ryabova LA (2001) A plant viral reinitiation factor interacts with the host translational machinery. Cell 106(6):723–733. https://doi.org/10.1016/S0092-8674(01)00487-1

Park MR, Jeong RD, Kim KH (2014) Understanding the intracellular trafficking and intercellular transport of potexviruses in their host plants. Front Plant Sci 5:60. https://doi.org/10.3389/fpls.2014.00060

Parrella G, Gognalons P, Gebre-Selassiè K, Vovlas C, Marchoux G (2003) An update of the host range of tomato spotted wilt virus. J Plant Pathol 85:227–264

Paszkowski U, Zhang S-B, Potrykus I, Paszkowski J (1993) Replication of the DNA a component of African cassava mosaic virus in a heterologous system. J Gen Virol 74(12):2725–2272. https://doi.org/10.1099/0022-1317-74-12-2725

Patil BL, Fauquet CM (2009) Cassava mosaic geminiviruses: actual knowledge and perspectives. Mol Plant Pathol 10:685–701. https://doi.org/10.1111/j.1364-3703.2009.00559.x

Peden K, Symons R (1973) Cucumber mosaic virus contains a functionally divided genome. Virology 53:487–492. https://doi.org/10.1016/0042-6822(73)90232-8

Pooggin MM, Hohn T, Fütterer J (1998) Forced evolution reveals the importance of short open reading frame a and secondary structure in the cauliflower mosaic virus 35S RNA leader. J Virol 72(5):4157–4169. https://doi.org/10.1128/JVI.72.5.4157-4169

Poston JE, Anderson PK (1997) The emergency of whitefly-transmitted Geminiviruses in tomato in the Western hemisphere. Plant Dis 81(12):1358–1369. https://doi.org/10.1094/pdis.1997.81.12.1358

Quenouille J, Vassilakos N, Moury B (2013) Potato virus Y. Mol Plant Pathol 14:439–452. https://doi.org/10.1111/mpp.12024

Raj SK, Kumar S, Pratap D, Vishnoi R, Snehi SK (2007b) Natural occurrence of cucumber mosaic virus on Rauvolfia serpentine, a new record. Plant Dis 91:322. https://doi.org/10.1094/PDIS-91-3-0322C

Raj SK, Kumar S, Pratap D, Vishnoi R, Choudhary S, Chandra S (2007c) Natural occurrence of cucumber mosaic virus on lemongrass (*Cymbopogon citratus*), a new record. Aus Plant Dis Notes 2:95–96. https://doi.org/10.1071/DN07039

Raj SK, Khan MS, Kumar S, Pratap D, Vishnoi R (2008b) Cucumber mosaic virus infecting vegetable and pulse crops in India. In: Rao GP, Kumar PL, Holguin-Pena RJ (eds) Characterization, diagnosis & management of plant viruses: vegetable and pulse crops, vol 3. Studium Press LLC, Houston, TX, pp 39–62

Raj SK, Kumar S, Snehi SK, Pathre U (2008c) First report of cucumber mosaic virus on Jatropha curcas in India. Plant Dis 92:171

Ranabhat NB, Fellers JP, Bruce MA, Rupp JLS (2023) Brome mosaic virus detected in Kansas wheat co-infected with other common wheat viruses. Front Plant Sci 14:1096249. https://doi.org/10.3389/fpls.2023.1096249

Rao ALN (2006) Genome packaging by spherical plant RNA viruses. Annu Rev Phytopathol 44:61–87. https://doi.org/10.1146/annurev.phyto.44.070505.143334

Rey C, Vanderschuren H (2017) Cassava mosaic and Brown streak diseases: current perspectives and beyond. Annu Rev Virol 4(1):429–452. https://doi.org/10.1146/annurev-virology-101416-041913

Rich AE (1983) Potato diseases. Academic Press, New York, p 238

Robaglia C, Caranta C (2006) Translation initiation factors: a weak link in plant RNA virus infection. Trends Plant Sci 11:40–45. https://doi.org/10.1016/j.tplants.2005.11.004

Rojas MR, Zerbini FM, Allison RF, Gilbertson RL, Lucas WJ (1997) Capsid protein and helper component-proteinase function as potyvirus cell-to-cell movement proteins. Virology 237:283–295. https://doi.org/10.1006/viro.1997.8777

Roossinck MJ (2001) Cucumber mosaic virus, a model for RNA virus evolution. Mol Plant Pathol 2:59–63. https://doi.org/10.1046/j.1364-3703.2001.00058.x

Roossinck MJ (2002) Evolutionary history of cucumber mosaic virus deduced by phylogenetic analyses. J Virol 76:3382–3387. https://doi.org/10.1128/jvi.76.7.3382-3387.2002

Rothnie HM, Chapdelaine Y, Hohn T (1994) Pararetroviruses and retroviruses: a comparative review of viral structure and gene expression strategies, pp 1–67. https://doi.org/10.1016/s0065-3527(08)60327-9

Ryu C-M, Murphy JF, Reddy MS, Kloepper JW (2007) A two-strain mixture of rhizobacteria elicits induction of systemic resistance against pseudomonas syringae and cucumber mosaic virus coupled to promotion of plant growth on Arabidopsis thaliana. J Microbiol Biotechnol 17(2):280–286

Samad A, Ajay Kumar PV, Gupta MK, Shukla AK, Darokar MP, Somkuwar B, Alam M (2008) Natural infection of periwinkle (*Catharanthus roseus*), with cucumber mosaic virus subgroup Ib. Aus Plant Dis notes 3:30–34. https://doi.org/10.1007/BF03211229

Samuels TD, Ye C-M, Ju H-J, Motes CM, Blancaflor EB, Verchot-Lubicz J (2007) Potato virus X TGBp1 protein accumulates independently of TGBp2 and TGBp3 to promote virus cell-to-cell movement. Virology 367:375–389. https://doi.org/10.1016/j.virol.2007.05.022

Scholelz JE, Shepherd RJ (1988) Host range control of cauliflower mosaic virus. Virology 162(1):30–37. https://doi.org/10.1016/0042-6822(88)90391-1

Scholthof KB, Adkins S, Czosnek H, Palukaitis P, Jacquot E, Hohn T, Hohn B, Saunders K, Candresse T, Ahlquist P, Hemenway C, Foster GD (2011) Top 10 plant viruses in molecular plant pathology. Mol Plant Pathol 12(9):938–954. https://doi.org/10.1111/j.1364-3703.2011.00752.x

Schwartz M, Chen J, Lee WM, Janda M, Ahlquist P (2004) Alternate, virus-induced membrane rearrangements support positive-strand RNA virus genome replication. Proc Natl Acad Sci U S A 101:11263–11268. https://doi.org/10.1073/pnas.0404157101

Sherwood JL, German TL, Moyer JW, Ullman DE (2003) Tomato spotted wilt virus: tomato spotted wilt. Plant Health Instruc. https://doi.org/10.1094/PHI-I-2003-0613-02

Shih DS, Kaesberg P (1976) Translation of the RNAs of brome mosaic virus: the monocistronic nature of RNA1 and RNA2. J Mol Biol 103:77–88. https://doi.org/10.1016/0022-2836(76)90053-X

Siegel R, Bellon L, Beigelman L, Kao CC (1999) Use of DNA, RNA and chimeric templates by a viral RNA-dependent RNA polymerase: evolutionary implications for the transition from the RNA to the DNA world. J Virol 73:6424–6429. https://doi.org/10.1128/jvi.73.8.6424-6429.1999

Sivakumaran K, Bao Y, Roossinck MJ, Kao CC (2000) Recognition of the core RNA promoter for minus-strand RNA synthesis by the replicases of brome mosaic virus and cucumber mosaic virus. J Virol 74:10323–10331. https://doi.org/10.1128/jvi.74.22.10323-10331.2000

Sofy AR, Sofy MR, Hmed AA, El-Dougdoug NK (2019) Potential effect of plant growth-promoting rhizobacteria (PGPR) on enhancing protection against viral diseases. In: Maheshwari D, Dheeman S (eds) Field crops: sustainable management by PGPR. Sustainable development and biodiversity, vol 23. Springer, Cham. https://doi.org/10.1007/978-3-030-30926-8_15

Stanley WM (1935) Isolation of a crystalline protein possessing the properties of tobacco-mosaic virus. Science 81:644–645. https://doi.org/10.1126/science.81.2113.644

Stanley WM, Anderson TF (1941) A study of purified viruses with the electron microscope. J Biol Chem 139:325–338. https://doi.org/10.1016/S0021-9258(19)51389-8

Stanley J, Bisaro DM, Briddon RW, Brown JK, Fauquet CM, Harrison BD, Rybicki EP, Stenger DC (2005) Geminiviridae. In: Fauquet CM, Mayo MA, Maniloff J, Desselberger U, Ball LA (eds) Virus taxonomy, VIIIth report of the ICTV. Elsevier/Academic Press, London, pp 301–326

Sukal AC, Kidanemariam DB, Dale JL, Harding RM, James AP (2018) Characterization of a novel member of the family Caulimoviridae infecting *Dioscorea nummularia* in the Pacific, which may represent a new genus of dsDNA plant viruses. PLoS One 13:e0203038. https://doi.org/10.1371/journal.pone.0203038

Taliansky ME, Garcia-Arenal F (1995) Role of cucumovirus capsid protein in long distance movement within the infected plant. J Virol 69:916–922. https://doi.org/10.1128/jvi.69.2.916-922.1995

Testa M, Sanna D, Pintore R, Marongiu G, Marras PM (2011) A two-year survey of TSWV occurrence in globe artichokes in Sardinia (Italy). Acta Hortic 917:297–302. https://doi.org/10.17660/ActaHortic.2011.917.42

Tilsner J, Linnik O, Louveaux M, Roberts IM, Chapman SN, Oparka KJ (2013) Replication and trafficking of a plant virus are coupled at the entrances of plasmodesmata. J Cell Biol 201(7):981–995. https://doi.org/10.1083/jcb.201304003

Trinks D, Rajeswaran R, Shivaprasad PV, Akbergenov R, Oakeley EJ, Veluthambi K, Hohn T, Pooggin MM (2005) Suppression of RNA silencing by a geminivirus nuclear protein, AC2, correlates with transactivation of host genes. J Virol 79:2517–2527. https://doi.org/10.1128/JVI.79.4.2517-2527.2005

Trzmiel K, Zarzyńska-Nowak A, Lewandowska M et al (2016) Identification of new brome mosaic virus (BMV) isolates systemically infecting *Vigna unguiculata* L. Eur J Plant Pathol 145:233–238. https://doi.org/10.1007/s10658-015-0830-5

Udaya Shankar AC, Nayaka SC, Niranjan-Raj S, Kumar HB, Reddy MS, Siddapura RN, Prakash HS (2009) Rhizobacteria-mediated resistance against the blackeye cowpea mosaic strain of bean common mosaic virus in cowpea (*Vigna unguiculata*). Pest Manag Sci 65(10):1059–1064. https://doi.org/10.1002/ps.1791

Vanthana M, Nakkeeran S, Malathi VG, Renukadevi P, Vinodkumar S (2019) Induction of in planta resistance by flagellin (Flg) and elongation factor-TU (EF-TU) of bacillus amyloliquefaciens (VB7) against groundnut bud necrosis virus in tomato. Microb Pathog 137:103757. https://doi.org/10.1016/j.micpath.2019.103757

Voinnet O, Lederer C, Baulcombe DC (2000) A viral movement protein prevents systemic spread of the gene silencing signal in *Nicotiana benthamiana*. Cell 103:157–167. https://doi.org/10.1016/S0092-8674(00)00095-7

Vovlas C, Lafortezza R (1992) TSWV infection of chicory crops. L'Informatore Agrario 48:63–64. (in Italian)

Walter CT, Barr JN (2011) Recent advances in the molecular and cellular biology of bunyaviruses. J Gen Virol 92:2467–2484. https://doi.org/10.1099/vir.0.035105-0

Warburg O (1894) Die Kulturpflantzen Usambaras. Mittcilungen aus den Deutschen Schutzgebieten 7:131

Waterworth HE, Kaper JM, Tousignant ME (1979) CARNA 5, the small cucumber mosaic virus dependent replicating RNA regulates disease expression. Science 204:845–847. https://doi.org/10.1126/science.204.4395.84

Wei T, Huang T-S, McNeil J, Laliberté J-F, Hong J, Nelson RS, Wang A (2010a) Sequential recruitment of the endoplasmic reticulum and chloroplasts for plant potyvirus replication. J Virol 84:799–809. https://doi.org/10.1128/jvi.01824-09

Wei T, Zhang C, Hong J, Xiong R, Kasschau KD, Zhou X, Carrington JC, Wang A (2010b) Formation of complexes at plasmodesmata for potyvirus intercellular movement is mediated by the viral protein P3N-PIPO. PLoS Pathog 6:e1000962. https://doi.org/10.1371/journal.ppat.1000962

Wyckoff RWG, Corey RB (1936) The ultracentrifugal crystallization of tobacco mosaic virus protein. Science 84:513. https://doi.org/10.1126/science.84.2188.513.a
Xi D, Feng H, Lan L, Du J, Wang J, Zhang Z, Xue L, Xu W, Lin H (2007) Characterization of synergy between cucumber mosaic virus and tobacco necrosis virus in *Nicotiana benthamiana*. J Phytopathol 155:570–573. https://doi.org/10.1111/j.1439-0434.2007.01279.x
Yasaka R, Nguyen HD, Ho SYW, Duchêne S, Korkmaz S, Katis N et al (2014) The temporal evolution and global spread of cauliflower mosaic virus, a plant Pararetrovirus. PLoS One 9(1):e85641. https://doi.org/10.1371/journal.pone.0085641
Zeidan M, Czosnek H (1991) Acquisition of tomato yellow leaf curl virus by the whitefly *Bemisia tabaci*. J Gen Virol 72:2607–2614. https://doi.org/10.1099/0022-1317-72-11-2607
Zindović J, Marina C, Massimo T (2014) Molecular characterization of tomato spotted wilt virus in Montenegro. J Plant Pathol 96:201–205. https://doi.org/10.4454/JPP.V96I1.001

Chapter 10
Insects as Biotic Stress in Agriculture

Vinay Kumar Dhiman, Devendra Singh, Vivek Kumar Dhiman, Minakshi Pandey, Avinash Sharma, Himanshu Pandey, and Sunil Kumar Verma

Abstract Plants are surrounded by the biotic factors around their ecosystem. Their interaction with insects is one such important factor that shapes the habitat around them. Insects may be beneficial or pests depending on their activity in the microclimate they live with the flora around them. The plants are provided with physical strength, chemical responses, and physiological changes with time to adapt to the herbivory nature of insects. Some insects live in mutual and take food and shelter for survival, providing plants with defense and predatory response to pest populations. Others cause tissue damage, triggering responses like chemical secretion that either attract pests for toxic feeders or repel them. The physical architecture of tissue structures also provides the advantage to plants to cope with damage and regeneration. Some defense mechanism varies with species of plants and are specific based on their interaction with insects.

V. K. Dhiman
Department of Basic Sciences, College of Forestry, Dr. Yashwant Singh Parmar University of Horticulture and Forestry, Nauni, Solan, India

D. Singh (✉) · S. K. Verma
Faculty of Biotechnology, Institute of Biosciences and Technology, Shri Ramswaroop Memorial University, Barabanki, Uttar Pradesh, India
e-mail: Devendrasingh.ibst@srmu.ac.in

V. K. Dhiman
Department of Biotechnology, College of Horticulture, Dr. Yashwant Singh Parmar University of Horticulture and Forestry, Nauni, Solan, India

M. Pandey
Faculty of Biosciences, Institute of Biosciences and Technology, Shri Ramswaroop Memorial University, Barabanki, Uttar Pradesh, India

A. Sharma
Faculty of Agricultural Sciences, Arunachal University of Studies, Namsai, Arunachal Pradesh, India

H. Pandey
PG Department of Agriculture, Khalsa College, Amritsar, Punjab, India

P. Singh Chauhan et al. (eds.), *Plant-Microbe Interaction and Stress Management*, Rhizosphere Biology,
https://doi.org/10.1007/978-981-97-4239-4_10

Keywords Insects · Plants · Defense · Pests · Biotic · Herbivory

10.1 Introduction

Agriculture is the most important part of human lives, and its sustainability is crucial for our survival. Green plants that are agriculturally important crops or economically valuable face both abiotic and biotic stresses. Abiotic stress from climate change and human anthropogenic activities like mining and dumping of chemicals provide challenges for their survival (Singh et al. 2022a). Biotic stress, like insects and pests, has also caused a detrimental effect on crop health and yield, thus making them agents of biotic stress. They cause major economic losses by reducing yield and quality in agriculture and forestry. Globally, the majority of food crops have been affected by the attack of pathogens, insect pests, and weeds, causing loss in average yield losses ranging in major crops such as rice, wheat, and potato (Tudi et al. 2021; Sunya et al. 2022; Sangma et al. 2022; Yadav et al. 2023).

Insects have the most species of any organism on Earth, with green plants making up the majority of land biomass. Nearly half of all insect species feed on living plants as herbivorous living on around vascular plant species (Paganizza 2017). Beneficial insects can be used to prosper agriculture hand to hand by regulating the ecosystem, suppressing and controlling pests, and serving as natural alternatives for pesticides for ensuring organically rich nutritive food, making it available to feed for healthy beings. It is potentially challenging to introduce the beneficial insects to the environment as their survival and success of incorporation into the ecosystem is not sure with climate and surrounding habitat (Getanjaly et al. 2015). Moreover, the food chain is more than capable of introducing natural enemies against these beneficial insects, making their survival a bit harder. Figure 10.1 represents the different abiotic and biotic stress.

Herbivory is not uniform across all insect groups, with some orders being almost exclusively herbivorous, whereas others are less frequent or absent (Belsky 1986). Some common herbivores include Lepidoptera, Hemiptera, Orthoptera, Thysanoptera, and Phasmida. Coleoptera, Hymenoptera, and Diptera are the most common orders, including numerous predatory and parasitic species (Denno and Kaplan 2007; Wardhaugh 2015; Shikano 2017). Terrestrial tracheophytes, including the tree and ferns, also harbor herbivore members. Previously, it was believed that ancient plants, such as the maidenhair tree and ferns, were devoid of insect consumers (War et al. 2020; Del-Claro and Torezan-Silingardi 2021). Plants have evolved a range of responses to insect stress, like physical defenses, including thorns and tough leaves, and chemical defenses, such as the production of toxins or deterrents. Some plants even have the ability to "call for help" by releasing volatile compounds that attract insect predators (Strauss and Zangerl 2002; Schoonhoven et al. 2004).

Insects are a significant cause of stress for plants. They can cause physical damage, spread diseases, and even alter plant physiology. Insects interact with plants in numerous ways—they feed on plant tissues, lay eggs, and can even act as vectors for

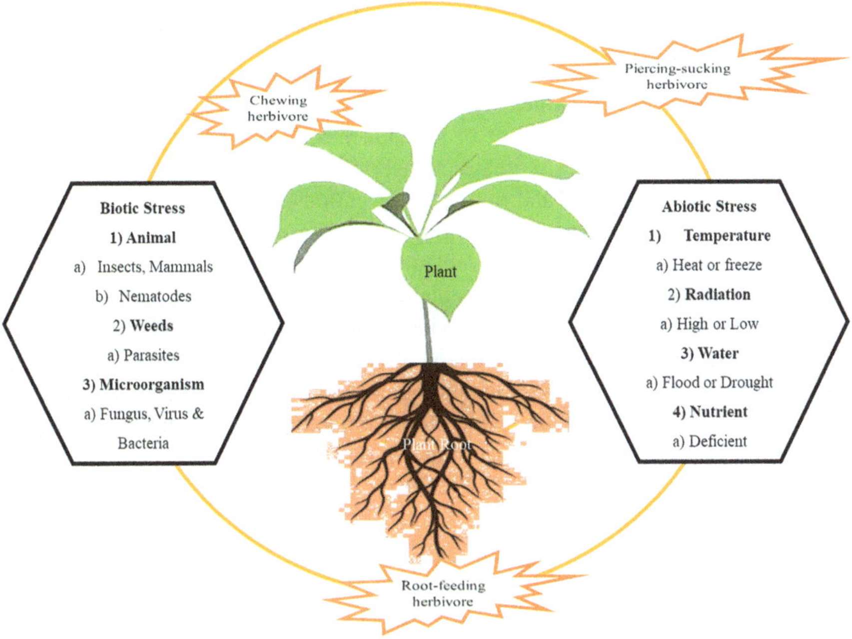

Fig. 10.1 Shows various probable stress that affects plants

plant diseases. These interactions can lead to substantial physical damage and physiological changes in plants, thereby inducing stress. This stress is not just limited to the loss of plant biomass (Sarwar and Salman 2015; Whitehill et al. 2023). It extends to a more complex level where the plant's growth, development, and productivity are hampered. Insects like locusts, aphids, beetles, and caterpillars are some common pests that cause extensive damage to agricultural crops (Stotz et al. 1999; Denno et al. 1995). Their impact is not localized but spreads across fields, affecting a wide range of crops, including grains, pulses, fruits, and vegetables. In tropical countries, these pests are believed to cause losses approaching more than half of the stored products. In response to this, there has been an increased focus on introducing new high-yielding varieties and hybrids with disease resistance towards insect pests (Manosathiyadevan et al. 2017). Control methods such as the use of antifeedants, chemosterilants, and irradiation techniques are commonly used. These approaches have led to the adoption of integrated pest management in crops. Efficient pest surveillance services and quick forecasting of pests have become essential tools to bring down the pest population before any significant damage is done (Chauhan et al. 2023).

Understanding the role of insects as a source of biotic stress in agriculture is crucial for developing effective pest management strategies. It helps in devising methods to reduce crop loss and ensure food security. This topic will delve into the various aspects of insect-induced stress in agriculture—the types of insects involved, the kind of stress they cause, plant responses to this stress, and potential mitigation strategies.

Moreover, the build-up of biotic stresses in nature is a continuous process, particularly in favorable environments. Several factors are responsible for the buildup of biotic stresses, such as the shift of secondary causal agents to primary status, injudicious use of pesticides, climate change, resistant pest populations, invasive pests, biotype development, and human interventions. As a result, it is critical to assess the effects of these newly emerging biotic stresses and to create innovative technologies for their management (Venkateswarlu et al. 2011; Marwal et al. 2020). Research innovations such as understanding the fundamental/molecular mechanisms of host–pathogen/insect interactions, endophytic mechanisms of plant protection, nanotechnology in pest management, host resistance strengthening through gene cloning, recombinant DNA technologies, and employing gene editing technologies like CRISPR/Cas9 are needed to be implemented in order to develop an efficient preventive and eradicative strategy for containing these biotic stresses (Tyagi and Keval 2020; Kumar et al. 2021; Kumari et al. 2022, Chauhan et al. 2022; Singh et al. 2022d).

10.2 Insect–Plant Interactions

Insect–plant interactions are increasingly important for biologists due to their quantitative importance in understanding the earth's biosphere. The plant kingdom and insects represent extensive taxa of living organisms, with green plants being the most abundant and insects leading in species. Insect–plant interactions are complex and multilayered relationships that involve various aspects influencing the performance and fitness of both participants (Andrew and Terblanche 2013; Karthik et al. 2021). In herbivory, insects use particular chemical cues to identify their host plants, and plants respond chemically and morphologically to ward off insect attacks. These interactions are frequently mediated by parasitoids, competitors, predators, and related microbiota that engage in facultative or obligatory interactions (Sharma et al. 2021). The perception, nutrition, metabolism, and reproduction of their host insects all of which together govern the insects' survival and fitness on the plant—are greatly influenced by the endosymbionts of insects. On the other hand, by strengthening chemical and morphological defense, plant-associated microbes aid in the induction of systemic plant resistance. These interactions affect both the ecosystem as a whole and high trophic levels (Clark et al. 2010).

Together, insects and plants have evolved defense mechanisms that allow them to outcompete one another in a series of coevolutionary events involving interactions. For defense against insect attacks, plants generate a variety of chemical compounds, and herbivorous insects have evolved strategies to get around constitutive and induced plant changes. These multilayered relationships determine the outcome of insects feeding on plants. Insects interact with plants in numerous ways, from feeding on plant tissues to laying eggs on plant surfaces (Mello and Silva-Filho 2002).

These interactions can cause significant stress to the plant, leading to a variety of responses. For instance, when insects feed on a plant, they can cause physical damage that disrupts the plant's ability to photosynthesize and transport nutrients, and they can also act as vectors for plant diseases. Plant interactions with insect herbivores are complex and multifaceted. These interactions can be either negative (competitive) or positive (facilitative) (Bendix and Lewis 2018) (Table 10.1).

10.2.1 Host-Plant Specialization

Insect–plant relationships are characterized by a high degree of food specialization among herbivorous insects. Monophagous insects, such as lepidopterous larvae, hemipterans, and coleopterans, feed only some closely related plant species (Chapman 2009). Oligophagous insects, like the cabbage white butterfly and Colorado potato beetle, feed on multiple plant species within the same plant family. Polyphagous insects, like the green peach aphid, accept many plants from different plant families, such as the peach tree or closely related rosaceous plants. These relationships are based on the degree of dietary specialization or generalization shown by herbivores (Sablon et al. 2012). Herbivores with narrow host-plant ranges prefer young growing leaves, while polyphagous species larvae prefer mature leaves. Young leaves are rich in nutrients but contain higher levels of toxic secondary plant substances, necessitating the use of specialized detoxification systems to compensate for this disadvantage (Lampert 2012).

10.2.2 Insect Herbivory

Insect herbivores can cause significant damage to plants, affecting their growth, development, and productivity. Damage caused by insect bites can induce intricate and dynamic defense systems in plants. These defense responses include calcium ion fluxes and phosphorylation cascades, as well as systemic and jasmonate

Table 10.1 Biotic stress of common insect pests in major crops

Common insect pests, along with major crop damage		
Common name	Botanical name	Insect stress crops
Locust	*Locusta migratoria*	Grasses
Tomato fruitworm	*Heliothis armigera*	Cotton, tomato
Cotton bollworm	*Helicoveroa zea*	Cotton
Brown plant hooper	*Nilaparvata lugens*	Rice
Colorado beetle	*Leptinotarsa decemlineata*	Potato
European corn borer	*Ostrinia nubilalis*	Maize
Cotton leafworm	*Spodoptera littoralis*	Maize, Rice, cotton, tobacco

signaling (Tlak Gajger and Dar 2021; Singh et al. 2022b). In some instances, when insect herbivores were present, every individual of a particular plant species displayed shoot damage; however, when these herbivores were excluded, no damage was observed. Under herbivory, these species' roots underwent morphological changes that resulted in thinner roots with an increased specific root surface area. Additionally, their root biomass increased (Heinze et al. 2019). Insect pests are a major factor in biotic stress that can cause significant losses in crop production worldwide. The damage caused by these pests is one of the primary reasons for reduced crop yield. They can cause harm through direct feeding and transmission of plant diseases. Moreover, at higher elevations, natural herbivore populations and plant damage from herbivory were significantly higher compared to their low-elevation counterparts. This suggests that changes in climate and elevation can also impact the level of damage caused by insect herbivores.

Insect herbivory can be categorized into two types based on the plant parts they target: above-ground herbivory and below-ground herbivory (Blossey and Hunt-Joshi 2003; Pitteloud et al. 2020).

1. Above-ground herbivory refers to the consumption of plant tissues that are located above the ground, such as leaves, stems, flowers, and fruits. Insects that engage in above-ground herbivory can cause significant damage to plants by feeding on these plant parts. This can result in reduced photosynthesis, decreased yield, and compromised plant quality.
2. Below-ground herbivory involves insects feeding on plant tissues that are located below the ground, such as roots, bulbs, and tubers. Below-ground herbivores can affect plants by disrupting nutrient uptake and transport systems, reducing root growth, and increasing susceptibility to diseases. Some examples of below-ground herbivores include plant-parasitic nematodes and fossorial insects.

10.2.3 Host Plant as Food and Shelter

The host plant is a crucial habitat for insects, providing them with biotic and abiotic factors beyond food. Insects living on a plant face various cohabitants, including competitors, natural enemies, a specific microclimate, and pathogen effects. For example, larvae of *Platyprepia virginalis* collected from hemlock were parasitized by a tachinid fly in most cases, compared to only half of cases of caterpillars from lupin. If the parasitoid's heavy toll is not compensated by physiological or ecological advantages, insects may develop an avoidance reaction to the host plant (Karban and De Valpine 2010). Insects may even prefer host plants that are nutritionally suboptimal but not visited by some of their natural enemies, providing an "enemy-free space" and better overall survival rates. Studies have shown significant differences in parasitism levels between forest trees, suggesting that the strong influence of the host plant on the risk of attack by parasitoids is a potentially important selective force in the evolution of herbivore diet breadth. These differences may result from plant morphological differences, such as the presence or absence of trichomes or plant hairs, which may affect the efficacy of insect parasitoids and predators (Lill et al. 2002). Moreover, the modified root structures in legume plants trees also allow root-feeding insects to thrive, roots cuticles, and the beneficial microorganisms surrounding the rhizosphere may provide compounds beneficial for defense (Negi et al. 2021; Dhiman et al. 2022) or induce systemic resistance.

10.2.4 Host Plant as Microclimate

In contrast to conventional meteorological measurements, plants offer their commensals distinct microclimates. Because of transpiration and photosynthetic processes, plant surfaces have boundary layers of relatively still air that differ greatly in temperature and relative humidity from surrounding conditions. Even with closed stomata, emissions through the cuticle can result in high concentrations of monoterpenes in the boundary layer. The thickness of the leaf boundary layer is influenced by wind speed, leaf size, and leaf structure (Callaway 2007).

Depending on the transpiration rate, size, shape, reflectance, and height above the ground, a leaf's upper surface can be either warm or cool. The underside is more humid and colder than the top surface. There can be significant variations in temperature at the leaf surface, and the relative humidity near the leaf surface can be higher than that of the surrounding air (Jones 1985).

The microclimate that small insects, like aphids and the early instars of folivorous insects, live in greatly affects their water balance and temperature. Due to vegetation's ability to block radiation from reaching the ground, the ground surface beneath vegetation is very different from that of bare soil. Additionally, vegetation creates humidity, temperature, and wind gradients (Cahon et al. 2018).

10.2.5 Plants' Response to Herbivore Damage

Plants have remarkable regeneration powers as long as insects do not attack their meristemic tissues or apical tips. Plants are modular organisms consisting of repetitive multicellular units, each with its own meristem, which reduces the adverse effects of herbivory considerably and allows for easy recovery from tissue removal (Kozlowski et al. 2012). This is due to the presence of dormant meristems and the ability to redirect resources to regrowing tissues. Under moderate or good resource conditions, plants can partially or wholly compensate for losses to insect feeding or even overcompensate by producing more biomass than lost. However, demonstrating overcompensation is more difficult than expected, as it requires that insect-damaged plants exhibit a significant increase in fitness compared to undamaged controls. Factors such as anatomical characteristics, reproduction mode, timing of herbivory, stored reserves, and availability of water, nutrients, and light contribute to a plant's capacity to partially compensate for tissue losses (Kant et al. 2015; Sharma et al. 2023b). Therefore, it is premature to make generalizations regarding the ecological or agricultural importance of compensatory responses.

Compensation responses to insects that do not destroy photosynthetic machinery, such as phloem sap-feeding species, are harder to measure (Sharma et al. 2023a). However, changes in photosynthetic performance may offer insight. In holly trees infested by scale insects, photosynthesis increased, compensating for nutrient losses to herbivory. This effect extended beyond the infested leaf and stimulated photosynthesis in scale-free leaves adjacent to infested leaves (Retuerto et al. 2004).

10.2.6 Anti-Herbivore Protection

Physical attributes that serve as barriers or repellents to insects, such as the leaf''s thickness, spines, hard bark, trichomes, wax coating, thorn-modified branches, or the plant's overall design, are the end results of biochemical processes regulated by genes. As a result, morphological and chemical resistance components are seamlessly integrated into a continuous defense spectrum. Physical defenses are often the first line of defense against insect stress. These can include structural features like thorns or spines that deter insects from feeding or tough, fibrous leaves that are difficult for insects to chew through (Doughari 2015).

10.2.7 Cuticular Waxes

Most vascular plants' cuticles are covered with a thin layer of hydrophobic wax, which protects against insect herbivory, pathogen invasion, and desiccation. These layers are variable in thickness and amount, ranging from a fraction to several percent of a plant's dry weight. They are not homogeneous structures but consist of distinct chemical and mechanical layers (Yeats and Rose 2013).

Wax crystals, a type of wax, are found on all aerial surfaces and can cause insects to struggle to feed from leaves covered with large amounts of wax. For example, epicuticular wax in Eucalyptus is slippery and hinders herbivorous psyllids during leaf development. Hence, this causes starvation in insect herbivores compared to adult leaves, which are lower in wax. Dense and easily abraded microscopic wax crystals can indirectly affect herbivore populations by impairing their adhesion, mobility, and effectiveness (Brennan and Weinbaum 2001).

10.2.8 Leaf Thickness and Strength

Plant cell walls are strengthened by macromolecules like cellulose lignin, which constitute most proportion on tissues. Other macromolecule structure includes suberin, callose, and sclerenchymatous fibers, which make plants resistant to mechanical injury and herbivorous insects. Leaf toughness, which is the best predictor of interspecific variation in herbivory rates, acts as an effective factor in reducing herbivory. Leaf toughness is a key functional trait in plant adaptation to herbivore pressures and physically stressful environmental factors, such as strong wind. It varies by plant form and climatic zone, with variations between species. The internal leaf structure and photosynthetic characteristics are related to an increase in leaf toughness as trees grow in height (Alberts et al. 2002; Keegstra 2010).

10.2.9 Trichomes

Trichomes, often known as plant hairs, are uni- or multicellular appendages that emerge from different plant species' epidermal cells. They differ greatly in terms of form, size, location, and purpose. The collective trichome cover of a plant surface is referred to as "pubescence." These trichomes can be distinguished as nonglandular and glandular. Glandular trichomes have the potential to be toxic and disruptive to several kinds of herbivores through the secretion of allelochemicals. Nonglandular trichomes act as a barrier to protect plants from insects, and extensive pubescence can keep small species that feed on plant piercings from penetrating the epidermis (Hülskamp 2004; Shanower 2004).

10.2.10 Plant Galls Formation

Plant galls, which are growths or swellings caused by the enlargement or multiplication of plant cells, are induced by organisms that use them for nourishment and shelter. These galls, which can be induced by various insects, are particularly complex when induced by gall midges or cynipid wasps. Unlike plant tumors, which are generally unorganized, insect-induced galls have a distinct morphological organization and physiological function. Within the gall, the insect regulates not only the growth of the gall structure but also the flow of nutrients and assimilates (Raman 2021). Additionally, galling insects seem to control the way that defense chemicals are made in their victims. The shape of a gall is believed to be actively determined by the galling insect, as evidenced by the fact that even closely related insect species can induce vastly different galls in the same host plant. For example, *Urophora quadrifasciata* induces a primitive achene gall in *Centaurea jacea* flower head, and its relative *Urophora jaceana* forms complex gall formation (Korneyev et al. 2005).

10.2.11 Plant Size and Architecture

The size and architecture of a plant affect the number of insect species that inhabit it. Insect preferences and survival may be influenced by plant features such as canopy spacing, stem, branch angle, complex surface, leaf and bud shapes and dimensions, and branching angles. The diversity of plant species in an area can also impact the diversity of insect species. The more plant species present in grasslands and forests, the more insect species find a habitat there. This is not only true for the number of insect species but also for the number of insect individuals (Barthélémy and Caraglio 2007; Debinski 2023).

Larger and more complex plants tend to have higher insect species richness due to the limited ecological niches they exploit. This is because herbivorous insects exploit restricted parts of their hosts, and the range of ecological niches is correlated with plant complexity. The increasing size and architectural complexity of plants lead to an increase in the diversity of the associated insect fauna. Specific architectural characteristics can also influence herbivores by impacting their natural predators. Some plant species are able to tolerate insects by attracting their natural predatory enemies; based on their architecture the predation rate differs in similar plant species (Stinson and Brown 1983).

10.3 Plant Responses to Insect Stress

Plants lack the ability to physically defend themselves or escape and often have long generation times and low recombination rates and seem to be at a disadvantage compared to the capabilities of herbivores. Furthermore, insects benefit from having wings, allowing them to disperse and invade potential food sources even at significant distances from their birthplace and larval home. However, despite plants' apparent susceptibility to herbivore attacks, the world's flora has evolved into a lush and highly diverse covering. It's clear that plants have an effective resistance system in place, which is based on a combination of physical, chemical, and developmental characteristics.

If physical defenses are breached, plants can resort to chemical warfare. Many plants produce secondary metabolites—compounds not required for the plant's immediate survival but that provide other benefits. These can include toxins that harm or deter insects or compounds that taste bad and discourage further feeding. Plants produce a variety of chemicals that can influence insect behavior. These chemicals can be broadly categorized into two types: direct and indirect defenses (Aljbory and Chen 2018).

Direct Defense Compounds: These compounds, such as glucosinolates or protease inhibitors, directly influence the insect's performance and feeding behavior. For example, some plants produce bitter or toxic substances that deter herbivores from eating them.

Indirect Defense Compounds: Some plants employ indirect defenses against insect stress. For example, when attacked by herbivorous insects, some plants release volatile organic compounds (VOCs) into the air. These VOCs can attract natural enemies of the herbivores, such as predatory insects or parasitoids, effectively "calling for help" from these allies. VOCs are emitted by plants after an herbivore attack. These VOCs can serve as attractants for parasitic wasps, which in turn predate on the attacker. Individual VOCs can be directly toxic or deterrent, serve as signal molecules to attract natural enemies, and/or be perceived by distal plant tissues as a priming signal to prepare for expected herbivory.

10.3.1 Plant Primary Metabolites Interact with Insects

Plants harvest the solar energy via photosynthesis to form sugars, which are used to convert nitrogen into amino acids, the building blocks of proteins. Sugar, in turn, is converted into biomass comprising of organic polymers such as lignocellulose, the most abundant organic polymers. Cellulose, hemicellulose, and lignin constitute most of the dry mass in plants. Other primary metabolites produced are proteins, carbohydrates, and lipids, which are essential nutrients for herbivores and are formed by primary plant metabolites. The variation in these compounds can significantly impact insect preference and performance. Many insects feed on plant tissues, consuming the sugars and other primary metabolites produced by photosynthesis. This feeding can induce biochemical and physiological changes in the host plants, affecting processes such as photosynthesis (Zhou et al. 2015).

10.3.2 Plant Secondary Metabolites Interact with Insects

Plant chemistry and structure offer defense mechanisms against insect herbivores, drawing in parasitoids and herbivore predators as well. Insects with specific functions can also profit from plant secondary metabolites. In their interactions with insect herbivores, VOCs derived from plants are essential. Specific VOCs have several functions, including acting as signaling molecules to draw in natural enemies, directly toxic or deterrent, and/or priming signals to distant plant tissues to get ready for anticipated herbivory. The two-way interactions between insect herbivores and entomopathogens are strongly influenced by bottom-up plant effects. Plant defensive chemicals can affect the persistence and infectivity of entomopathogens in plants (Wink 2018).

Secondary metabolites are classified as allelochemic due to their nonnutritive nature for the plants but affect the other species. Their role in ecology is essential as a single plant species is able to produce a large number of these metabolites (Kong et al. 2019). Secondary metabolites, typically produced from universally present precursors like acetyl-coenzyme A, amino acids, or shikimate, can be classified based on their biosynthetic pathways. Alkaloids are known to be produced in many plants and insects. Alkaloids can be used as a toxin or to deter insects at lower concentrations. Many insects are unable to produce steroids that are naturally found in plants and have to feed on cholesterol or sitosterol for the synthesis of ecdysone (Singh et al. 2022e). Phenolics such as rotenone, phaseolin, and catechin are feeding deterrents and are used as the natural insecticides—compounds such as tannins cause an astringent taste in many plants (Dar et al. 2017). Glucosinolates found in the Brassicaceae family are toxic to insects, butterflies, and oviposition stimulants (Badenes-Pérez 2023). Most of the plants synthesize cyanogenic compounds such as cyanogenic glycoside prunasin. These compounds hydrolyze to form hydrogen cyanide, which is a toxic and potent feeding deterrent to many different insects and is released upon tissue damage.

10.3.3 Induced Resistance

Induced resistance in plants is a defense mechanism that develops after exposure to pathogens, pests, beneficial microorganisms, chemical agents, physical wounding, or herbivory. This process increases the plant's level of basal resistance against future attacks compared to non-stimulated plants. Using a variety of morphological, biochemical, and molecular defense mechanisms, plants withstand herbivore attacks. These diverse, extremely dynamic defense mechanisms against herbivores can be either direct or indirect. The herbivore's growth and development are impacted by direct defense because of antibiosis, which is caused by secondary metabolites that are either constitutively produced or induced upon infestation by insect pests. Enlisting the natural enemies of the insect pests is part of the indirect defense strategy (Rashid and Chung 2017; Singh et al. 2022c).

In response to insect exposure, elicitors activate plant defense mechanisms, resulting in physiological and biochemical processes like the stimulation of the salicylic acid (SA) and jasmonic acid (JA) pathways, calcium ion, reactive oxygen species (ROS) formation, activation of mitogen-activated protein kinase (MAPK), and other reactions (Wang et al. 2023). The increased production of antioxidants provides a better physiological state in plants (Sharma et al. 2022). Inducers of plant resistance, also called elicitors, are substances that enhance defense against pathogen or pest attacks by activating host defense mechanisms. Crop-infecting agents such as viruses, bacteria, fungi, oomycetes, nematodes, and herbivores can all be effectively combated by these products (Siah et al. 2018).

Also, induced direct resistance and induced indirect resistance have been observed in numerous plant species across various families, indicating that they are common defense mechanisms in plants. A variety of insects, including biting, chewing, cell piercing, and phloem feeders, can be effectively controlled by both forms of induced resistance. They provide resistance against stem borers, caterpillars, and aphids.

10.3.4 Induced Direct Resistance

Plant-induced direct resistance is a remarkable defense mechanism employed by plants to safeguard themselves against various pests and pathogens. This intricate system relies on the plant's ability to actively recognize and respond to threats, thus reducing the risk of damage or infection. When a plant senses the presence of potential invaders, it can trigger a cascade of biochemical reactions. One of the key components of this defense strategy is the production of specialized compounds, such as antimicrobial proteins, phytochemicals, and secondary metabolites. These substances act as formidable deterrents, inhibiting the growth and development of pests and pathogens upon contact (Belete 2018; Das et al. 2022; Kashyap et al. 2023).

Plants can activate the same chemical resistance mechanisms through induction when faced with threats. For example, when Colorado potato beetles feed on potato plants, the plants produce more proteinase inhibitors that disrupt the beetles' digestion and affect other herbivorous insects as well. When the plant's response is suppressed, such as by introducing a gene that interferes with the induction process, it significantly benefits the caterpillar. Alternatively, certain potato varieties respond to Colorado potato beetle oviposition by undergoing a hypersensitive response, causing the plant tissue around the eggs to die, leading to egg desiccation or their fall to the ground, potentially exposing them to soil-dwelling predators (Sablon et al. 2012).

10.3.5 Induced Indirect Resistance

Indirect resistance involves the plant triggering responses that indirectly affect the herbivores by attracting natural enemies or altering the plant's chemistry. One common example is the release of VOCs by plants when they are under attack by herbivores. These VOCs serve as chemical signals, attracting predators and parasitoids that feed on the herbivores, thus reducing their numbers and deterring future attacks. Additionally, the plant can alter its chemical composition in response to herbivore damage, making it less palatable or even toxic to the herbivores. Induced indirect resistance highlights the intricate and dynamic interactions between plants, herbivores, and their natural enemies in ecosystems, showcasing how plants can manipulate their environment to enhance their survival and minimize damage from insect pests (Huntly 1991). This ecological strategy underscores the importance of understanding the interplay between plants, herbivores, and their natural enemies in shaping the balance of ecosystems (Poelman 2015).

1. VOCs: When plants are attacked by herbivores, they often release VOCs into the air. These VOCs act as chemical signals, attracting predators and parasitoids that are natural enemies of the herbivores. For example, the scent of damaged plants can attract ladybugs, parasitic wasps, or predatory insects like lacewings. These

beneficial insects prey on or lay eggs in herbivores, reducing their populations and providing a form of biological control.

2. Induced Chemical Changes: Reactions of plants to damage from herbivores can also alter their molecular makeup. Secondary metabolites that can repel or even be toxic to herbivores include phenolic compounds, terpenes, and alkaloids, among others. These chemical alterations may decrease the plant's appeal as a food source and lower the fitness of the herbivore.
3. Allelopathy: Allelopathic compounds are released by certain plants into the soil, where they can impede the growth of neighbouring plants. Inadvertently lessening competition for resources this can promote the growth of the focal plant. Although herbivores are not the target of this tactic, it does indirectly affect herbivore populations by influencing the plant community.

10.4 Conclusion

The green habitat is vast flora surrounded by a diversity of organisms that thrive on them. Such interactions may benefit or counter the flora they surround. As insects utilize the biomass produced by plants to grow and survive, the plants, in turn, have developed physical barriers and biochemical responses to resist the insects from harming them. The plant architecture, size, and characteristics affect insect preferences and survival. They respond to herbivore damage through biochemical responses, increased photosynthetic machinery, production of toxic compounds, or changes in insect behavior. Also, different species of plants inhabit insect-feeding predators. The primary metabolites produced by plants are utilized by insects for nourishment and secondary metabolites for the requirement of peculiar compounds that insects lack. Some beneficial insects provide protection from pests. The physical damage to some plants triggers different organic compounds to repel insects and are toxic to some, providing immunity through induced resistance. Plants have evolved to counter the adverse effects of insects as well as to gain from them; such interaction is important for a dynamic ecosystem.

References

Alberts B, Johnson A, Lewis J, Raff M, Roberts K, Walter P (2002) The plant cell wall. In: Molecular biology of the cell, 4th edn. Garland Science, New York

Aljbory Z, Chen MS (2018) Indirect plant defense against insect herbivores: a review. Insect Sci 25(1):2–3

Andrew NR, Terblanche JS (2013) The response of insects to climate change. Climate of change: living in a warmer world. David Bateman, Auckland, pp 38–50

Badenes-Pérez FR (2023) Plant glucosinolate content and host-plant preference and suitability in the small white butterfly (Lepidoptera: Pieridae) and comparison with another specialist lepidopteran. Plan Theory 12(11):2148

Barthélémy D, Caraglio Y (2007) Plant architecture: a dynamic, multilevel and comprehensive approach to plant form, structure and ontogeny. Ann Bot 99(3):375–407
Belete T (2018) Defense mechanisms of plants to insect pests: from morphological to biochemical approach. Trends Tech Sci Res 2(2):30–38
Belsky AJ (1986) Does herbivory benefit plants? A review of the evidence. Am Nat 127:870–892
Bendix C, Lewis JD (2018) The enemy within: phloem-limited pathogens. Mol Plant Pathol 19(1):238–254
Blossey B, Hunt-Joshi TR (2003) Belowground herbivory by insects: influence on plants and aboveground herbivores. Annu Rev Entomol 48(1):521–547
Brennan EB, Weinbaum SA (2001) Effect of epicuticular wax on adhesion of psyllids to glaucous juvenile and glossy adult leaves of *Eucalyptus globulus* Labillardière. Aust J Entomol 40(3):270–277
Cahon T, Caillon R, Pincebourde S (2018) Do aphids alter leaf surface temperature patterns during early infestation? Insects 9(1):34
Callaway RM (2007) Positive interactions and interdependence in plant communities. Springer Science & Business Media, Berlin
Chapman RF (2009) Foraging and food choice in phytophagous insects. Chem Ecol 1:72–101
Chauhan V, Kaushal D, Dhiman VK, Kanwar SS, Singh D, Dhiman VK, Pandey H (2022) An insight in developing carrier-free immobilized enzymes. Front Bioeng Biotechnol 10:794411
Chauhan D, Singh D, Pandey H, Khan Z, Srivastava R, Dhiman VK, Dhiman VK (2023) Impact of transcription factors in plant abiotic stress: a recent advancement for crop improvement. In: Plant transcription factors, pp 271–286
Clark EL, Karley AJ, Hubbard SF (2010) Insect endosymbionts: manipulators of insect herbivore trophic interactions? Protoplasma 244:25–51
Dar SA, Rather BA, Wani AR, Ganie MA (2017) Resistance against insect pests by plant phenolics and their derivative compounds. Chem Sci Rev Lett 6(22):1073–1081
Das R, Deb P, Pandey H, Shyam P, Singh D (2022) Botanical synthesis of silver nanoparticles (Agnps) and its antifungal effect against *Alternaria porri* causing purple blotch of onion: an in vitro and natural epiphytic study. J Agric Food Res 10:100390
Debinski DM (2023) Insects in grassland ecosystems. In: Rangeland wildlife ecology and conservation, pp 897–929
Del-Claro K, Torezan-Silingardi HM (2021) An evolutionary perspective on plant-animal interactions. In: Plant-animal interactions: source of biodiversity. Springer International, Cham, pp 1–15
Denno RF, Kaplan I (2007) Plant-mediated interactions in herbivorous insects: mechanisms, symmetry, and challenging the paradigms of competition past. In: Ecological communities: plant mediation in indirect interaction webs, vol 4, pp 19–50
Denno RF, McClure MS, Ott JR (1995) Interspecific interactions in phytophagous insects: competition revisited and resurrected. Annu Rev Entomol 40:297–331
Dhiman VK, Rana N, Dhiman VK, Pandey H, Verma P, Singh D (2022) Effect of rhizobial isolates and nitrogen fertilizers on nursery performance, nodulation behavior and nitrogenase activity of Dalbergia sissoo Roxb. seedlings. Plant Stress 4:100080
Doughari J (2015) An overview of plant immunity. J Plant Pathol Microbiol 6(11):10–4172
Getanjaly VL, Sharma P, Kushwaha R (2015) Beneficial insects and their value to agriculture. Res J Agric For Sci 2320:6063
Heinze J, Simons NK, Seibold S, Wacker A, Weithoff G, Gossner MM, Prati D, Bezemer TM, Joshi J (2019) The relative importance of plant-soil feedbacks for plant-species performance increases with decreasing intensity of herbivory. Oecologia 190:651–664
Hülskamp M (2004) Plant trichomes: a model for cell differentiation. Nat Rev Mol Cell Biol 5(6):471–480
Huntly N (1991) Herbivores and the dynamics of communities and ecosystems. Annu Rev Ecol Syst 22(1):477–503

Jones MB (1985) Plant microclimate. In: Techniques in bioproductivity and photosynthesis, pp 26–40
Kant MR, Jonckheere W, Knegt B, Lemos F, Liu J, Schimmel BC, Villarroel CA, Ataide LM, Dermauw W, Glas JJ, Egas M (2015) Mechanisms and ecological consequences of plant defence induction and suppression in herbivore communities. Ann Bot 115(7):1015–1051
Karban R, De Valpine P (2010) Population dynamics of an Arctiid caterpillar–tachinid parasitoid system using state-space models. J Anim Ecol 79(3):650–661
Karthik S, Reddy MS, Yashaswini G (2021) Climate change and its potential impacts on insect-plant interactions. In: The nature, causes, effects and mitigation of climate change on the environment. IntechOpen, London, p 10
Kashyap P, Shirkot P, Das R, Pandey H, Singh D (2023) Biosynthesis and characterization of copper nanoparticles from *Stenotrophomonas maltophilia* and its effect on plant pathogens and pesticide degradation. J Agric Food Res 1:100654
Keegstra K (2010) Plant cell walls. Plant Physiol 154(2):483–486
Kong CH, Xuan TD, Khanh TD, Tran HD, Trung NT (2019) Allelochemicals and signaling chemicals in plants. Molecules 24(15):2737
Korneyev V, Zwölfer H, Seitz A (2005) Phylogenetic relationships, ecology, and ecological genetics of cecidogenous Tephritidae. In: Biology, ecology and evolution of gall-inducing arthropods. Science Publishers, Enfield, pp 321–371
Kozlowski TT, Kramer PJ, Pallardy SG (2012) The physiological ecology of woody plants. Academic press, New York
Kumar J, Murali-Baskaran RK, Jain SK, Sivalingam PN, Mallikarjuna J, Kumar V, Sharma KC, Sridhar J, Mooventhan P, Dixit A, Ghosh PK (2021) Emerging and re-emerging biotic stresses of agricultural crops in India and novel tools for their better management. Curr Sci 121:26–36
Kumari P, Jasrotia P, Kumar D, Kashyap PL, Kumar S, Mishra CN, Kumar S, Singh GP (2022) Biotechnological approaches for host plant resistance to insect pests. Front Genet 13:914029
Lampert E (2012) Influences of plant traits on immune responses of specialist and generalist herbivores. Insects 3(2):573–592
Lill JT, Marquis RJ, Ricklefs RE (2002) Host plants influence parasitism of forest caterpillars. Nature 417(6885):170–173
Manosathiyadevan M, Bhuvaneshwari V, Latha R (2017) Impact of insects and pests in loss of crop production: a review. In: Sustainable agriculture towards food security, pp 57–67
Marwal A, Srivastava AK, Gaur RK (2020) Improved plant tolerance to biotic stress for agronomic management. Agrica 9(2):84–100
Mello MO, Silva-Filho MC (2002) Plant-insect interactions: an evolutionary arms race between two distinct defense mechanisms. Braz J Plant Physiol 14:71–81
Negi S, Singh D, Pandey H et al (2021) Biofortification for high Fe and Zn in various Poaceae crops by using different molecular breeding and biotechnological approaches. Plant Physiol Rep 26:636–646. https://doi.org/10.1007/s40502-021-00625-w
Paganizza V (2017) Insects in agriculture: traditional roles and beyond. In: Agricultural law: current issues from a global perspective, pp 163–179
Pitteloud C, Descombes P, Sànchez-Moreno S, Kergunteuil A, Ibanez S, Rasmann S, Pellissier L (2020) Contrasting responses of above-and below-ground herbivore communities along elevation. Oecologia 194:515–528
Poelman EH (2015) From induced resistance to defence in plant-insect interactions. Entomol Exp Appl 157(1):11–17
Raman A (2021) Gall-inducing insects and plants: the induction conundrum. Curr Sci 120:66–78
Rashid MH, Chung YR (2017) Induction of systemic resistance against insect herbivores in plants by beneficial soil microbes. Front Plant Sci 8:1816
Retuerto R, Fernanez-Lema B, Rodrjguez-Roiloa, Obeso JR (2004) Increased photosynthetic performance in holly trees infested by scale insects. Funct Ecol 18:664–669

Sablon L, Dickens JC, Haubruge É, Verheggen FJ (2012) Chemical ecology of the Colorado potato beetle, *Leptinotarsa decemlineata* (say)(Coleoptera: *Chrysomelidae*), and potential for alternative control methods. Insects 4(1):31–54

Sangma RR, Manpoong C, Sharma A, Devadas VS, Singh D, Pandey H (2022) Performance of black rice (*Oryza sativa*) varieties grown in Namsai district of Arunachal Pradesh, India. Res Crops 1:11–14

Sarwar M, Salman M (2015) Insecticides resistance in insect pests or vectors and development of novel strategies to combat its evolution. Int J Bioinform Biomed Eng 1(3):344–351

Schoonhoven LM, Van Loon JJ, Dicke M (2004) Insect-plant biology. Oxford University Press, Oxford

Shanower T (2004) Trichomes and insects. Encycl Entomol 3:232–235

Sharma G, Malthankar PA, Mathur V (2021) Insect–plant interactions: a multilayered relationship. Ann Entomol Soc Am 114(1):1–6

Sharma SK, Singh D, Pandey H, Jatav RB, Singh V, Pandey D (2022) An overview of roles of enzymatic and non-enzymatic antioxidants in plant. In: Antioxidant defense in plants: molecular basis of regulation, vol 5, pp 1–3

Sharma A, Pandey H, Nampoothiri Devadas VA, Singh D, Kartha BD, Wangsu M (2023a) Regulations and factors affecting of light emitting diodes artificial light in crop production under soilless culture. ACS Agric Sci Technol 3(8):616–630

Sharma A, Pandey H, Manpoong C, Vashishth A, Singh D, Bang NT (2023b) Biometric analysis, biosynthetic pathway and multipurpose uses of buckwheat local varieties in Namsai district of eastern Himalayas in India. Trends Food Sci Technol 8:251. https://doi.org/10.1016/j.tifs.2023.04.020

Shikano I (2017) Evolutionary ecology of multitrophic interactions between plants, insect herbivores and entomopathogens. J Chem Ecol 43:586–598

Siah A, Magnin-Robert M, Randoux B, Choma C, Rivière C, Halama P, Reignault P (2018) Natural agents inducing plant resistance against pests and diseases. In: Natural antimicrobial agents, pp 121–159

Singh D, Pandey H, Dhiman VK, Thakur K, Sharma I, Dhiman VK, Singh V, Pandey D (2022a) Assessment of environmental impacts of metal/metalloid pollution on plants. In: Metals metalloids soil plant water systems, pp 217–232

Singh D, Dhiman VK, Pandey H, Dhiman VK, Pandey D (2022b) Crosstalk between salicylic acid and auxins, cytokinins and gibberellins under biotic stress. In: Auxins, cytokinins and gibberellins signaling in plants, pp 249–262

Singh D, Pandey H, Dhiman VK, Bang NT, Dhiman VK, Singh V, Pandey D (2022c) Secondary metabolite engineering for plant immunity against various pathogens. In: Metabolic engineering in plants, pp 123–143

Singh D, Pandey H, Thakur K, Sharma I, Naraian R (2022d) Aquaporins and their functions in water transportation in different plant species. Environ Sustain 5(4):443–456

Singh D, Pandey H, Singh V (2022e) Natural products that target cancer stem cells. In: Pandurangan A, Anandasadagopan S, Alhumaydhi F (eds) Handbook of research on natural products and their bioactive compounds as cancer therapeutics. IGI Global, pp 169–186. https://doi.org/10.4018/978-1-7998-9258-8.ch008

Stinson CS, Brown VK (1983) Seasonal changes in the architecture of natural plant communities and its relevance to insect herbivores. Oecologia 56:67–69

Stotz HU, Kroymann J, Mitchell-Olds T (1999) Plant-insect interactions. Curr Opin Plant Biol 2(4):268–272

Strauss SY, Zangerl AR (2002) Plant-insect interactions in terrestrial ecosystems. In: Plant-animal interactions: an evolutionary approach, pp 77–106

Sunya H, Sharma A, Manpoong C, Devadas VS, Pandey H, Singh D (2022) Evaluation of improved varieties of rice (*Oryza sativa* L.) under environmental condition of Namsai, Arunachal Pradesh, India. Ecol Environ Conserv:907

Tlak Gajger I, Dar SA (2021) Plant allelochemicals as sources of insecticides. Insects 12(3):189

Tudi M, Daniel Ruan H, Wang L, Lyu J, Sadler R, Connell D, Chu C, Phung DT (2021) Agriculture development, pesticide application and its impact on the environment. Int J Environ Res Public Health 18(3):1112

Tyagi S, Keval R (2020) Understanding and utilizing resistance to insects in plants. In: Advances in agricultural entomology, pp 91–115

Venkateswarlu B, Shanker AK, Shanker C, Maheswari M (2011) Crop stress and its management: perspectives and strategies. Springer Science & Business Media, Berlin

Wang H, Shi S, Hua W (2023) Advances of herbivore-secreted elicitors and effectors in plant-insect interactions. Front Plant Sci 14:1176048

War AR, Buhroo AA, Hussain B, Ahmad T, Nair RM, Sharma HC (2020) Plant defense and insect adaptation with reference to secondary metabolites. In: Co-evolution of secondary metabolites, pp 795–822

Wardhaugh CW (2015) How many species of arthropods visit flowers? Arthropod Plant Interact 9(6):547–565

Whitehill JG, Bohlmann J, Krokene P (2023) Forest insect—plant interactions. In: Forest entomology and pathology. Entomology, vol 1, pp 169–204

Wink M (2018) Plant secondary metabolites modulate insect behavior-steps toward addiction? Front Physiol 9:364

Yadav G, Prajapati P, Singh D, Hora S, Singh S, Vishwakarma K, Ahmad IZ (2023) Emerging trends in plant metabolomics and hormonomics to study abiotic stress tolerance associated with rhizospheric probiotics. In: Plant hormones in crop improvement, vol 1, pp 283–306

Yeats TH, Rose JK (2013) The formation and function of plant cuticles. Plant Physiol 163(1):5–20

Zhou S, Lou YR, Tzin V, Jander G (2015) Alteration of plant primary metabolism in response to insect herbivory. Plant Physiol 169(3):1488–1498

Chapter 11
Phytopathogenic Fungi: Stress and Mitigation Approaches

Pooja Yadav, Rupali Katoch, Indu, Namo Dubey, and Kunal Singh

Abstract Phytopathogenic fungi are major threat to the agriculture worldwide. Use of chemical fungicides are common practice to reduce recurrence of phytopathogens. These fungicides have harmful effects to the environment and human health as their trace amount gets precipitated in plant tissues meant for consumption. Green and alternate strategies like application of microbial antagonists acting as biocontrol agents have been under exploration since the last few years. Such microbes produce antagonistic compounds and inhibit the phytopathogenic fungi. Other than microbes, essential oils extracted from plants and nanoparticles have also been utilised to counter fungal pathogens. Gene pool from wild relatives of crop plants is another extensively applied methodology to incorporate resistance against phytopathogenic fungi. Therefore, major antagonistic approaches towards phytopathogens were discussed with emphasised attention to microbial biological control agents and their mode of action.

Keywords Fungal phytopathogens · Biocontrol agents · Plant immunity · Essential oil

11.1 An Overview of Fungal Phytopathogens

Interaction of fungi with plants may result in both beneficial and harmful effects. Those fungi colonising plants with detrimental effect on their growth, development or reproduction are categorised as phytopathogens. Fungal phytopathogens cause

P. Yadav · N. Dubey · K. Singh (✉)
CSIR-Institute of Himalayan Bioresource Technology, Palampur, India

Academy of Scientific and Innovative Research (AcSIR), Ghaziabad, India
e-mail: kunal@ihbt.res.in

R. Katoch · Indu
CSIR-Institute of Himalayan Bioresource Technology, Palampur, India

P. Singh Chauhan et al. (eds.), *Plant-Microbe Interaction and Stress Management*, Rhizosphere Biology,
https://doi.org/10.1007/978-981-97-4239-4_11

some of the most devasting diseases to crop plants hindering the food production worldwide (Fisher et al. 2012). As per an estimate, 8000 fungal species, belonging to 64 genera of fungi frequently attack the plants (Arraiza et al. 2018). Most of the known phytopathogenic fungi belongs to Ascomycota and Basidiomycota. Prominent pathogens like *Magnaporthe oryzae*, *Botrytis cinerea*, *Fusarium oxysporum* belong to ascomycetes, whereas *Puccinia graminis*, *Ustilago maydis*, *Melampsora lini* are members of basidiomycetes. Severity of fungal pathogens depend upon plant species, virulence of pathogenic strain and environmental conditions (Millan et al. 2021). They affect the plant not only during pre-harvest stage but also at post-harvesting stage. They attack and infiltrate the host through various mechanisms including appressoria formation, production of cell wall-degrading enzymes (CWDEs) and mycotoxins (Salvatore and Andolfi 2021). Disease caused by pathogen could be localised as *Perenosclerospora sorghi* attacks on maize plant leaves only, whereas *Phytophthora infestans* causes late blight in potato, affecting the whole plant. Disease symptoms caused by fungal pathogens include leaf spot, canker, mildew, blight, rust, root rot, damping off, wilt, gall, scab and many more (Jain et al. 2019). Majority of plants are susceptible to multiple fungal pathogens (Ho et al. 2021).

To inhibit these pathogens, chemical fungicides are commonly applied. However, they are not sustainable and have hazardous impact on plants and mankind. As an alternative approach, biological control agents have been explored along with nanoparticles and essential oil (EO) to inhibit the phytopathogenic fungi in an eco-friendly way. Some nanoparticles are also reported, being inhibitory effect to the growth of fungal phytopathogens.

11.2 Fungal Pathogen Based on Nutritional Requirements

In the natural ecosystem, the fungi shows a range of nutritional adaptability during interaction with plants. For instance, many fungi proliferate after killing the host dead and deriving nutrition from the same, whereas others may utilise specialised structures known as haustoria to acquire the nutrition from living cell (Rajarammohan 2021). Therefore, depending on the mode of nutrition that arose by mechanism of infection, the fungal pathogens have been classified into three broad groups, namely biotrophs, necrotrophs and hemibiotrophs (Cal et al. 2022).

Biotrophs group of pathogens derive nutrition from living host tissues for their growth and survival. These fungi infect the host plant without causing their death (Perfect and Jonathan 2001). Therefore, biotrophs are masters of subduing host defence in living cells in a subtle way to induce the release of plant nutrients (Koeck et al. 2011). Biotrophic fungi utilised appressoria to penetrate pre-formed physical barriers of the plant parts such as the waxy cuticle, cell wall and cell membrane. During the course of infection, the cell wall disrupted by fungal hyphae is bordered by invaginated plant plasma membrane leading to formation of haustoria. Thus, haustoria get used for water uptake, nutrient absorption and energy source leading to long-term relation between biotrophic fungi and host plant (Perfect and Jonathan 2001). However, haustoria formation is not an essentiality for biotrophic fungal pathogens. Many biotrophs infiltrate apoplastic spaces utilising protein channels and endosomes for nutritional uptake (Yoshida et al. 2016). The formation of infection structure is controlled by complex regulatory pathways regulated by differential gene expression (Lai et al. 2020). Haustoria were reported in many pathogens like *Bremia lactucae* that cause downy mildew disease in *Lactuca sativa* (Lebeda et al. 2014) *Blumeria graminis* f. sp. *hordei* causing powdery mildew disease in barley and *Puccinia poarum* causing disease in *Tussilago farfara* (Mapuranga et al. 2022).

On the contrary, phytopathogen deriving nutrients from dead and dying cells are termed as necrotrophs. In case of necrotrophic fungi, infection process may be categorised in four major steps, namely spore attachment, germination, cell wall maceration and penetration, lesion expansion and sporulation (van Kan 2006). Necrotrophs release phytotoxins for establishment in host. For example, *B. cinerea* during plant infection releases sesquiterpene phytotoxin—botrydial that promotes chlorosis of *Phaseolus vulgaris* (Colmenares et al. 2002). The pathogen also releases CWDEs including cellulase, glucanase, xyloglucanase, polygalacturonase and pectinase with potential to degrade the structural cell wall components pectin, cellulose and hemicellulose (Kubicek et al. 2014). *B. cinerea* also infects the crops after harvesting and causes huge loss in yield of fruits, vegetables and ornamentals (Doehlemann et al. 2017).

Many other fungi behave as biotrophs initially, followed by necrotrophic phase (Glazebrook 2005). These fungi have been classified as Hemibiotrophic fungi (Rajarammohan 2021). They combine life cycle phases of biotrophic stage to counter the immune system of the host cell followed by necrotrophic phase, where the pathogen secretes toxins inducing host cell death (Koeck et al. 2011). One of the best studied examples of hemibiotrophic fungi is *M. oryzae*, causative agent of rice blast disease. In the initial stage of infection this pathogen obtains nutrients from living host. During this stage, primary hyphae expand through host plant cell after successful penetration with the help of penetration peg. Growing hyphae expand into bulbous protruding haustoria. Then, hyphae of *M. oryzae* conquer neighbouring cells and transformed to necrotrophic stage in which they kill the host. Necrotic lesions form on the surface of leaves ultimately causing cell death. At this stage, pathogens derive nutrition from dead tissue and act purely as necrotroph (Fernandez and Orth 2018).

11.3 Microbial Biocontrol Agents for Control of Fungal Pathogens

Harmful fungicides get accumulated in water bodies, soil and plants due to their non-biodegradable nature (Pirttilä et al. 2021). Thus, there is utmost need for safe and eco-friendly agents for control of fungal diseases to minimise the application of harmful chemicals as fungicide (Millan et al. 2021). The term Biological Control Agents, often shortened as biocontrol agents (BCAs) refers to the application of beneficial microbes which suppresses the other harmful microbial species involved in deterioration of the normal plant growth (Pandit et al. 2022). Various fungal and bacterial species have been reported with potential to use as BCAs to subdue harmful pathogens (Heydari and Pessarakli 2010).

11.4 Bacterial Biocontrol Agents Against Fungal Pathogens

Bacteria with utilisation as biocontrol agents are mostly derived from soil with potential to control several fungal pathogens. The bacterial BCAs act on the pathogenic fungi by mycoparasitism, production of siderophores, biofilm formation, volatile organic compounds (VOCs) production, niche exclusion, antibiosis and competition for resources (Pandit et al. 2022; Roy et al. 2018). Species of *Bacillus, Cryptococcus*, *Rhodosporidium*, *Rhodotorula*, *Streptomyces*, *Paenibacillus*, *Rhodobacter*, *Agrobacterium*, *Weissella*, *Rahnella*, *Lactobacillus*, *Pseudomonas* and so forth have been reported to have ability to antagonise the phytopathogenic fungi (Yu et al. 2020). A few examples of bacterial biocontrol agents and their target pathogen with their mechanism of action have been listed in Table 11.1. Bacterial isolates of genera *Bacillus*, *Streptomyces* and *Pseudomonas* were found as most effective for utilisation as biocontrol agents against the fungal pathogens (Albayrak 2020). Among them, *Bacillus* species were known to exhibit considerable biocontrol activity due to their impeding nature against the fungal pathogens and their ability to resist to the hostile environment. *Bacillus* can antagonise many fungi directly and indirectly (Fira et al. 2018). The direct mechanism of inhibition includes systematic resistance and production of lipopeptide group of antifungal compounds surfactins, iturin, fengycin and CWDEs (Meena and Kanwar 2015; Albayrak 2020). For example, *Bacillus subtilis* antagonises *F. oxysporum* by producing CWDEs (Fig. 11.1), chitinase and glucanases (Khan et al. 2018). *Bacillus amyloliquefaciens* antagonises *Botryosphaeria dothidea* through production of antifungal substances fengycin and surfactin. These substances inhibit the conidial germination and growth of mycelia (Li et al. 2016). Genome analysis identified several genes encoding proteins for pectinase, cellulase and xylanase in *Bacillus simplex*, useful against phytopathogens (Khan et al. 2018).

Another genus *Pseudomonas* is a wide spectrum bacterial biocontrol agent. *Pseudomonas* has been applied against multiple fungal pathogens due to their

Table 11.1 Few recent and prominent examples of bacterial biocontrol agents used against fungal phytopathogen inhibition

S. no.	Bacterial biocontrol agents	Fungal plant disease	Disease caused	Host	Mechanism of inhibition	References
1	*Bacillus amyloliquefaciens*	*Fusarium oxysporum*	Vascular wilt and root rot	Wheat	Discharges antibiotics and enzymes and causes hydrolysis of pathogen	Wang et al. (2018)
2	*Bacillus velezensis*	*F. oxysporum*	Vascular wilt diseases	Tomato	Production of cell wall degrading enzymes and volatile organic compounds (VOCs)	Myo et al. (2019)
3	*Pseudomonas putida*	*Dematophora nectarix*	White rot	Apple	Phosphate solubilisation and siderophore production	Verma et al. (2016)
4	*Bacillus subtilis*	*Alternaria solani*	Early blight disease	Potato	VOCs production	Zhang et al. (2020)
5	*Bacillus vallismortis*	*Fusarium graminearum*	Corn stalk rot	Maize	Inhibit mycelial growth	Li et al. (2019)
6	*Bacillus mojavensis*	*Fusarium verticilliodes*	Fusarium ear rot	Maize	Release secondary metabolites	Blacutt et al. (2016)
7	*Paenibacillus polymyxa*	*Verticillium dahliae*	Verticillium wilt	Cotton	Secretion of antibiotics	Zhang et al. (2018)
8	*Pseudomonas jessenii*	*Rhizoctonia solani*	Bottom rot disease	Lettuce	Exhibit antifungal activity	Windisch et al. (2017)
9	*Streptomyces albidoflavus*	*Magnaporthe oryzae*	Rice blast	Rice	Competition, parasitism	Law et al. (2017)
10	*Pseudomonas protegens*	*Botrytis cinerea*	Botrytis bunch rot	Grapevine	Synthesise antifungal secondary metabolites	Andreolli et al. (2019)
11	*Bacillus cereus*	*A. solani*	Early blight	Tomato	Parasitism, biofilm formation	Karthika et al. (2020)
12	*Bacillus and Pseudomonas* spp.	*Cyrtomium falcatum*	Red rot	Sugarcane	Siderophore	Shastri et al. (2020)
13	*S. albidoflavus*	*Magnaporthe oryzae*	Rice blast	Rice	Competition, parasitism	Gao et al. (2020)

(continued)

Table 11.1 (continued)

S. no.	Bacterial biocontrol agents	Fungal plant disease	Disease caused	Host	Mechanism of inhibition	References
14	*Mesorhizobium* spp.	*F. oxysporum*	Fusarium wilt	Chickpea	Induce systemic resistance	Kumari and Khanna (2020)
15	*Achromobacter xylosoxidans*, *B. subtilis*, *Lysinibacillus fusiformis*	*A. solani*	Early blight	Tomato	Induce systemic resistance	Attia et al. (2020)
16	*B. mycoides*	*Colletotrichum gloeosporioides*	Post-harvest disease	Avocado	Reduced the mycelial growth by production of VOCs	Guevara- Avendaño et al. (2019)
17	*Pseudomonas azotoformans*, *Pseudomonas frederiksbergensis*, *Arthrobacter nitroguajacolicus*	*Fusarium and Alternaria* spp.	Sudden wilt disease	Tobacco	Biofilms formation, production of antifungal compound surfactin	Santhanam et al. (2019)
18	*B. safensis*	*Alternaria*	Leaf spot and blight diseases	Lentil	Production of HCN, production of cell wall degrading enzymes	Roy et al. (2018)
19	*B. cereus*	*Alternaria*	Leaf spot and blight diseases	Lentil	VOCs production	Roy et al. (2018)
20	*Bacillus and Pseudomonas* spp.	*S. rolfsii*	Southern blight	Dendrobium officinale	Antibiotics, Siderophore production	Ali et al. (2022)
21	*B. velezensis*	*B. cinerea*	Gray Mould	Grapes	Antibiotic	Ali et al. (2022)

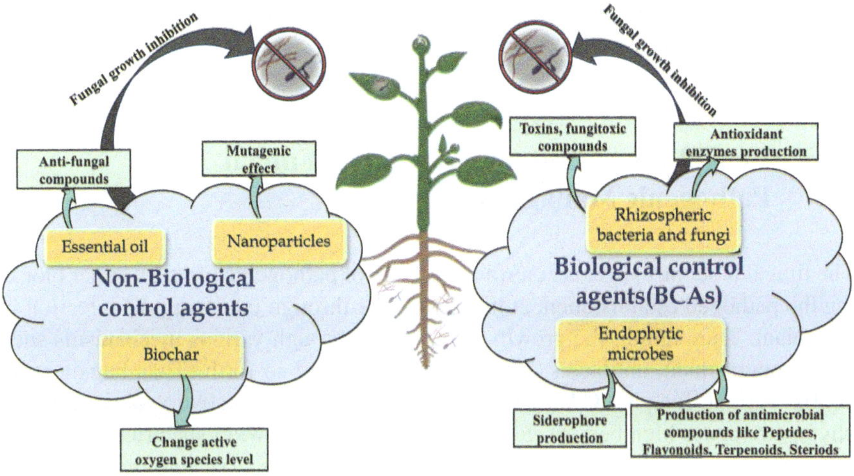

Fig. 11.1 Major biological and non-biological control agents getting utilised against fungal phytopathogens with their modes of action

ability to synthesise antifungal compounds phenazine, pyoluteorin and pyrrolnitrin (Pang et al. 2015; Cimmino et al. 2021). The strains of *Pseudomonas* have also been typified as plant growth promoting bacteria with capability to enhance the plant health and yield, by meliorating the availability and uptake of nutrients, through phosphate solubilisation, production of siderophore compounds, synthesis of auxins (indole-3-acetic acid) and 1-aminocyclopropane-1-carboxylate (Andreolli et al. 2019; Bensidhoum et al. 2016). For example, *Pseudomonas* sp. protects the seedlings of grass from fungal infection caused by *Fusarium* via production of antifungal volatiles (Verma et al. 2018). *Pseudomonas aeruginosa* produces a siderophore, Pyochelin, an antifungal agent against *Phellinus noxius,* causative agent of brown root rot in rubber tree (Ho et al. 2021). *Pseudomonas fluorescens* produces antifungal metabolites such as phenazine, pyoluteorin and pyrrolnitrin against *Pyhtium aphanidernatum*, the causative agent of rhizome rot in turmeric. *Streptomyces* is also been reported as bio-fungicide with potential to inhibit multiple plant pathogenic fungi (El-Tarabily 2006). The antifungal activity arose due to synthesis of toxic compounds and CWDEs, particularly amylases, cellulases, chitinases and glucanases (Oskay 2009). Disfunctioning of mitochondria, inhibition of spore germination and spores bursting and production of VOCs were other reported mode of action (Duan et al. 2020; Meena et al. 2022). As *Rhizoctonia bataticola*, a fungal pathogen that causes mungbean root rot, could be inhibited by *Streptomyces rameus* through production of chitinase (Meena et al. 2022).

Like free living plant growth promoting bacteria (PGPB), endophytic bacteria have also been reported to carry biocontrol potential. Common mechanism employed by them include production of hydrolytic enzymes, antibiotics, nutrient cut-back for the pathogens (Liaqat and Eltem 2016). For instance, many species of *Bacillus*, *Paenibacillus*, *Serattia* and *Pseudomonas* with endophytic property isolated from wheat, showed antagonistic activity towards the *Fusarium graminearum* (Whitaker and Bakker 2019).

11.5 Fungal Biocontrol Agents Against Fungal Pathogenic Management

The fungal biocontrol agents control the growth of pathogenic fungi by either blocking the pathogen establishment at initial stage or through inhibiting the infection in host plant. They inhibit the growth of pathogen through various mechanisms such as mycoparasitism, antibiosis, induced resistance and so forth (Thambugala et al. 2020). Various fungal BCAs have been reported to be effective against multiple phytopathogens. Members of genus *Trichoderma* have wide spectrum capability and commercially available with 25 species. Till date, *Trichoderma* has been reported with biocontrol activity against approximately 100 fungal pathogens (Thambugala et al. 2020). *Trichoderma* can also behave as endophyte with the ability to colonise various plants such as wheat, cotton, maize and soyabean. *Trichoderma* can induce local and systemic resistance in plants (Pascholati et al. 2019; Nawrocka et al. 2019). Under mycoparasitism, it forms a hook-like structure to enter the host fungal cell wall during penetration and grows along the pathogenic fungi hyphae (Bailey et al. 2008). Many members secrete CWDEs and toxic metabolites (Omann et al. 2012). *Trichoderma virens* and *Trichoderma harzianum* were reported to induce the resistance against pathogens *Rhizoctonia solani*, *Colletotrichum lindemuthianum* and *B. cinerea* in cotton and bean, respectively (Harman et al. 2004). *Trichoderma atroviride*, an endophyte of adult red pine has shown antifungal properties against canker pathogen *Diplodia pinea* (Santamaria et al. 2012). A few fungal BCAs with target species were discussed in Table 11.2.

Many endophytic fungi release secondary metabolites (peptides, flavonoids, terpenoids, steroids etc.) with potential to act as antibacterial and antifungal agent (Fadiji and Babalola 2020). *Phomopis cassia*, an endophyte isolated from *Cassia spectabilis* produces sesquiterpenes with antifungal ability against *Cladosporium sphaerospermum* and *Cladosporium cladosporioides* (Silva et al. 2006). The endophytic fungi *Penicillium* sp., *Hypocrea* sp. and *Lasiodiplodia theobromae* being isolated from tomato was found to be effective against wilt pathogen *F. oxysporum* (Abro et al. 2019). *Piriformospora indica*, a widely used endophyte was reported to induce the resistance against *F. oxysporum* f. sp. *cubense* in *Musa* sp. by increasing the antioxidant enzyme activities (Fontana et al. 2021). It also enhanced resistance against *B. graminis* attack in barley and *Thanatephorus cucumeris* attack in rice (Nassimi and Taheri 2017). Interestingly, *F. oxysporum*, a known pathogen with wide spectrum pathogenicity, has been reported as endophyte also in some recent reports. As EF119 strain of *F. oxysporum* was reported as endophytic with the ability to act as BCA against *P. infestans* in tomato (Kim et al. 2007). Similarly, *Metarhizium anisopliae*, an entomopathogenic fungus, was reported to associate as

Table 11.2 Some fungal biocontrol agents applied for fungal pathogens management

S. no	Fungal biocontrol agent	Pathogen	Disease	Host	Reference
1	*Alternaria brassicae*	*Verticillium dahliae*	*Verticillium* wilt of cotton	Cotton	Li et al. (2014)
2	*Alternaria compacta*	*V. dahliae*	*Verticillium* wilt of cotton		Li et al. (2014)
3	*Alternaria atra*	*Botrytis aclada*	Neck rot of onion	Onion	Köhl (2004)
4	*A. atra*	*Botrytis cinerea*	Leaf spots of onion		Köhl (2004)
5	*A. atra*	*Botrytis squamosa*	Leaf blight of onion		Köhl (2004)
6	*A. atra*	*L. theobromae*	Crown rot of banana	Banana	Alvindia and Natsuaki (2009)
7	*Pythium nunn*	*Globisporangium ultimum* (*Pythium ultimum*)	Damping-off disease of cucumber	Cucumber	Paulitz and Baker (1987) and Thambugala et al. (2020)
8	*Pythium oligandrum*	*Aphanomyces cochlioides*	Seedling disease of sugar beet	Sugarbeet	Takenaka et al. (2006) and Thambugala et al. (2020)
9	*P. oligandrum*	*A. cochlioides*	Seedling and taproot diseases of sugar beet	Sugarbeet	
10	*Pythium oligandrum*	*B. cinerea*	Grey mould of grapevine	Grapevine	Mohamed et al. (2007)
11	*Pythium oligandrum*	*B. cinerea*	Grey mould of strawberry	Strawberry	Meszka and Bielenin (2010)
12	*Pythium oligandrum*	*Fusarium graminearum*	*Fusarium* head blight		Takenaka et al. (2003)
13	*Pythium acanthicum*	*Fusarium culmorum* and *Microdochium nivale*	*Fusarium* ear blight of wheat	Wheat	Diamond and Cooke (2003)
14	*Chaetomium globosum*	*Puccinia recondite*	Wheat leaf rust		Park et al. (2005)
15	*Clonostachys rosea*		Tomato grey mould	Tomato	Borges et al. (2015)
16	*Trichoderma asperellum* TaspHu1	*Alternaria alternata*	Leaf spot disease	Tomato	Yu et al. (2021)
17	*Trichoderma* sp. NRCB3 and *T. asperellum* Prr2	*F. oxysporum* f. sp. *cubense*	*Fusarium* wilt	Banana	Thangavelu and Gopi (2015)
18	*Trichoderma atroviride*	*Rhizoctonia solani*	Damping-off and root rot	Cucumber (*Cucumis sativus*)	Nawrocka et al. (2019)

(continued)

Table 11.2 (continued)

S. no	Fungal biocontrol agent	Pathogen	Disease	Host	Reference
19	*T. asperellum*	*Sclerotium cepivorum*	Onion white rot	Onion	Rivera-Mendez et al. (2020)
20	*T. asperellum* T8	*Colletrotrichum gloeosporiodes*	Anthracnose	Mango	de los Santos-Villalobos et al. (2013)
21	*T. asperellum*	*Sclerotinia sclerotiorum*	White Mold	Beans	Geraldine et al. (2013)
22	*Trichoderma hamatum*	*Sclerotinia sclerotiorum*	White Mold	Cabbage	Jones et al. (2014)

endophyte in wheat with potential to be used as biocontrol agent against *F. graminearum* (Hao et al. 2021). Many other pathogenic fungi have also been reported to present as endophyte in different plant species. For example, *Colletotrichum gloeosporioides*, *Botryosphaeria ribis* and *Clonostachys rosea* of *Theobroma cacao* provide protection against major pathogens *Moniliophthora roreri*, *Phytophthora palmivora* and *Moniliophthora perniciosa,* the causative agents of frosty pod rot, black pod rot and witches broom diseases, respectively. Many endophytic fungi produce toxins, antibiotics and fungitoxic compounds that directly antagonise the pathogens. Culture filtrate of common bean endophyte *Purpureocillium lilacinum* contains bioactive compounds with the ability to suppress the growth of *Sclerotinia sclerotiorum* (Elsherbiny et al. 2019). The endophytes conferred systemic resistance and induced systemic resistance to plants. Similarly, endophytic *Fusarium solani* provides induced systemic resistance (ISR) to tomato against the *Septoria lycopersici* (Kavroulakis et al. 2007). The arbuscular mycorrhizal fungi (AMF) from division Zygomycota and order Glomales have also been reported to exert biocontrol activity (Mukerji and Ciancio 2007). *Glomus* species have been reported to control various soil borne pathogens such as *Rhizoctonia*, *Phytophthora*, *Sclerotinium* and *Verticillium* (Harrier and Watson 2004). *Glomus deserticola* and *Glomus clarum* suppress the growth of *R. solani* in cowpea, *Fusarium verticillioides* in maize and *R. solani* in cucumber (Olowe et al. 2018; Aljawasim et al. 2020).

Often, the use of simultaneous application of more than one microbe is recommended to suppress the disease incidence. Thus, many multi-strains biocontrol consortium arevfound to be helpful in inhibiting the growth of soil-borne pathogens (Niu et al. 2020). Such consortia of different microbes carry high efficacy. Consortia of *Trichoderma asperellum* and *B. amyloliquefaciens* display more inhibitory effect on *B. cinerea* than alone (Wu et al. 2018). There was a report showing that combination of AMF, *Trichoderma* and Plant Growth-Promoting Rhizobacteria (PGPRs) is more effective to control the pathogen *Fusarium* (root rot) under in-vitro and in-vivo conditions (El-Sharkawy and Abdelrazik 2022). Emerging field of the synthetic biology has also been utilised to identify the genetic traits of various fungal antagonists to be cross-introduced in other fungi. For example, *T. virens* was transformed with multiple genes encoding lytic enzymes. The resulted strain produced a mixture of glucanases and showed more inhibition against *R. solani*, *Pythium ultimum* and *Rhizopus oryzae* (Djonovic et al. 2007).

11.6 Non-Biological Control Agents Against Fungal Pathogens

Many non-biological entities such as EOs of plants and metal nanoparticles have been explored for their utility against phytopathogens. Similarly, Zinc nanoparticles was used to control the *Erythricium salmonicolor* in coffee (Arciniegas-Grijalba et al. 2017). Silver nanoparticles were applied to inhibit the *Aspergillus* sp. and *Rhizopus* sp. growth (Medda et al. 2014). The silver nanoparticles have been reported as antifungal agents against *Candida*, *Bipolaris* and *Magnaporthe grisea* (Panáček et al. 2009). Nanoparticles inhibitory mechanism involves the release of ions by nanomaterials, which binds with the cell membrane and proteins changing the membrane permeability of target microbe. Many nanoparticles affect the mitochondrial membrane potential in response to ROS production and could have harmful mutagenic effect on the target fungi (Cruz-Luna et al. 2021).

EO produced by medicinal and aromatic plants could show a broad spectrum of antifungal activities against many plant pathogenic fungi. The EOs get produced in very less amount in plant parts like leaves, flowers, fruits, seeds and roots (Shaaban et al. 2012). Plant EOs have been used in food industries, and the antifungal activities of EOs attributed to the compounds α-terpinyl acetate, carvacrol, thymol, cymene, linalool, limonene and pinene (Nazzaro et al. 2017). For example, EO with presence of limonene extracted from *Citrus sinensis* inhibits the *Aspergillus niger* growth (Bora et al. 2020). Many EOs exhibiting the antifungal nature have been often derived from members of family Lamiaceae. The antifungal effect of EO against phytopathogenic fungi varies. One plant EO may inhibit single or multiple fungi. For instance, the EO of black caraway and fennel inhibited the growth of pathogen *B. cinerea* (Aminifard and Mohammadi 2012). Likewise, the EO from clove, lemongrass, oregano and thyme inhibited *Aspergillus* sp., whereas ginger and cinnamon oil had no effect (Božik et al. 2017). *Litsea cubeba*-derived EO shows antifungal activity against *F. solani*, *Alternaria alternata*, *Fusarium moniliforme* and *A. niger* (Fig. 11.1). Thymol, a major component of Thyme EO, showed inhibitory effect against *Aspergillus flavus* through ROS production (Nazzaro et al. 2017; Shen et al. 2016). Thymus sp. oil exhibited antifungal activity against many broad spectrum fungal pathogens, namely *Penicillium*, *Trichoderma*, *Fusarium*, *Alternaria*, *Cladosporium*, *Aspergillus*, *Phomopsis*, *Microsporon and Trichophyton* (Soković et al. 2009; Parikh et al. 2021). Similarly, Bouzenna and Krichen (2013) showed that *Pelargonium graveolens* EO with citronellol (35%) and geraniol (28.8%) was inhibitory to *R. solani*. Many of these EOs inhibit pathogens via cell wall and cell membrane disruption, disrupting the DNA, RNA and protein synthesis. Upon spraying, lemongrass and basil EOs checked the growth of pathogens

Phytophthora capsici, *Phytophthora drechsleri* and *Phytophthora melonis* under controlled conditions (Amini et al. 2016). EO from *Mentha x piperita* showed antagonistic effect against *R. solani* and *Macrophomina phaseolina* (Abdolahi et al. 2011).

Application of biochar is another emerging and successful non-biological methodology to control pathogens and pests. Biochar is carbon-rich ash derived from organic material (preferably plant) (Murtaza et al. 2023). The biochar affects the growth of microbes via modifications in soil properties, root exudates and nutrient availability. They have been reported to induce systemic resistance, adsorbing toxins released by pathogens and increasing the microbial diversity in rhizosphere (Jaiswal et al. 2017). Biochar can bring change in active oxygen species level of plants helping in reduction of foliar fungal pathogens (Bonanomi et al. 2015). Causal organisms of root and crown rot disease of *Asparagus* were reduced by application of hardwood char. Application increased the presence of *Pseudomonas* and AMF (Elmer 2016). The holm oak char application increases the bacterial diversity in rhizosphere resulting in increased resistance against *B. cinerea* (De Tender et al. 2016). Plant waste biochar from pepper was reported effective against the pathogens *Podosphaera aphanis* and *Colletotrichum acutatum*, the causative agents of powdery mildew and anthracnose in strawberry (Poveda et al. 2021).

11.7 Plant Genetic Factors Interact and Fight Against Fungal Pathogens

Pathogens secrete an array of molecules known as effectors. They constitute a major component of pathogenesis that modify plant innate immunity and facilitate infection (Petit-Houdenot and Fudal 2017). Plants evolved resistance (R) genes encoding R proteins that recognise such effectors, directly or indirectly. When a pathogen's effector protein gets recognised by plants, a cascade of immune reactions set off known as effector-triggered immunity (ETI) leading to plant defence mediated by hypersensitive response (HR). Pathogens can become more virulent by evolving their effector encoding gene repertoire under pressure from plant R genes (Jones and Dangl 2006). Such receptors sense pathogen and pathogen-derived molecule-like effector and microbial-associated molecular pattern (MAMP). Plants receptor includes receptor-like proteins (RLP) and receptor-like kinases (RLK). These proteins tethered to cell membrane through a transmembrane domain and may consist of intracellular kinase domain along with NBS-LRR domain (Jose et al. 2020). Presence of NLR receptor have been reported in both plants and animals. Direct interaction of one such protein Pita with pathogenic effector AVR-Pita improved plant immunity in rice against *M. grisea*. RPW8.2 from *Arabidopsis thaliana* recognises fungal effector GcR8IP1 from mildew pathogen *Golovinomyces cichoracearum* and provide resistance to plant. RPW8.2 enhanced localised H_2O_2 accumulation and callose deposition. The NLRs are proteins composed of nucleotide-binding and leucine-rich repeat receptor domains with capability to induce programmed cell death (hypersensitive response, HR) and resistance by detecting effectors and elicitors from corresponding pathogen. The NLR can detect a wide range of pathogens, including viruses, bacteria, nematodes, insects, fungi and oomycetes. For example, direct interactions between Pita and pathogen effector AVR-Pita provide resistance to *M. grisea*.

Tools like RenSeq (resistance gene enrichment sequencing) can be used to find genomic variants in NBS-LRR genes to identify the catalogue of genetic variability among immunity conferring proteins (Jupe et al. 2013). RenSeq, for example, was successfully applied to find out *Rpi-amr3i* gene involved in resistance against *P. infestans* from *Solanum americanum* (Witek et al. 2016). Transgenic potato harbouring *Rpi-amr3i* demonstrated impedance to *P. infestans* in greenhouse conditions (Witek et al. 2016). Likewise, application of RenSeq pipeline to introgressed lines of commercial wheat with wild relative *Aegilops peregrina* identified numerous NBS-LRR encoding resistance orthologs effective against rust diseases. Among them, *LrAp* introgression showed resistance to multiple strains of rust causing fungi including *Puccinia triticina* and *P. graminis* (Narang et al. 2020).

Creation of chimeric protein utilising synthetic biology approaches can emerge as major source of producing new and novel resistance proteins. For example, a functional recombinant receptor was produced by fusing the cytoplasmic domain of the rice *Xa21* receptor with the Arabidopsis *EF-Tu* receptor (EFR). Introduction of chimeric receptor EFR::XA21 with chitin binding domain (CBD) part of protein EF-Tu and receptor part from *Xa21* in rice enhanced immunity against rice blast fungus *M. grisea* (Schwessinger et al. 2015).

11.8 Summary

To effectively control the diseases caused by phytopathogenic fungi, multiple strategies could be adopted. Continuous use of chemical fungicides in controlling the diseases caused by fungal pathogens have resulted in their accumulation in soil, water and food chain leading to negative and harmful effects on plants, humans and environment. Therefore, eco-friendly alternatives to suppress phytopathogens are the need of the hour. Microbes including bacterial and fungal members and plant extracts like EO have emerged as viable alternatives against key pathogens. Among microbial biocontrol agents, *Trichoderma* and *Pseudomonas* have been utilised with broad spectrum effect. Exploration of disease resistance gene pool from wild relatives of major crops and economically viable nanoparticles with strong influence against fungal phytopathogens could be key to a sustainable world with ample food security.

References

Abdolahi A, Hassani A, Ghosta Y, Meshkatalsadat MH, Shabani R (2011) Screening of antifungal properties of essential oils extracted from sweet basil, fennel, summer savory and thyme against postharvest phytopathogenic fungi. J Food Saf 31:350–356

Abro MA, Sun X, Li X, Jatoi GH, Guo LD (2019) Biocontrol potential of fungal endophytes against *Fusarium oxysporum* f. sp. cucumerinum causing wilt in cucumber. Plant Pathol J 35(6):598

Albayrak BC (2020) *Bacillus* species as biocontrol agents for fungal plant pathogens. In: *Bacilli* and *Agrobiotechnology*: phytostimulation and biocontrol, part of the *Bacilli* in climate resilient agriculture and bioprospecting book series. Springer, Berlin, pp 239–265

Ali A, Iftikhar Y, Mubeen M, Ali H, Zeshan MA, Asad Z, Rehman MA, Abbas M, Rafique M, Ghani MU (2022) Antagonistic potential of bacterial species against fungal plant pathogens (FPP) and their role in plant growth promotion (PGP): a review. Phyton-Int. J Exp Bot 91(9):1859–1877. https://doi.org/10.32604/phyton.2022.021734

Aljawasim BD, Khaeim HM, Manshood MA (2020) Assessment of arbuscular mycorrhizal fungi (Glomus spp.) as potential biocontrol agents against damping-off disease Rhizoctonia solani on cucumber. J Crop Protect 9(1):141–147

Alvindia DG, Natsuaki KT (2009) Biocontrol activities of *Bacillus amyloliquefaciens* DGA14 isolated from banana fruit surface against banana crown rot-causing pathogens. Crop Prot 28(3):236–242

Amini J, Farhang V, Javadi T, Nazemi J (2016) Antifungal effect of plant essential oils on controlling *Phytophthora* species. Plant Pathol J 32:16–24

Aminifard MH, Mohammadi S (2012) Essential oils to control *Botrytis cinerea in vitro* and *in vivo* on plum fruits. J Sci Food Agric 93:348–353. https://doi.org/10.1002/jsfa.5765

Andreolli M, Zapparoli G, Angelini E, Lucchetta G, Lampis S, Vallini G (2019) *Pseudomonas protegens* MP12: a plant growth-promoting endophytic bacterium with broad-spectrum antifungal activity against grapevine phytopathogens. Microbiol Res 219:123–131. https://doi.org/10.1016/j.micres.2018.11.003

Arciniegas-Grijalba PA, Patiño-Portela MC, Mosquera-Sánchez LP, Guerrero-Vargas JA, Rodríguez-Páez JE (2017) ZnO nanoparticles (ZnO-NPs) and their antifungal activity against coffee fungus *Erythricium salmonicolor*. Appl Nanosci 7:225–241. https://doi.org/10.1007/s13204-017-0561-3

Arraiza MP, González-Coloma A, Andres MF, Berrocal-Lobo M, Alfonso J, Domínguez-Núñez CJ, Navarro-Rocha J, Calderón-Guerrero C (2018) Antifungal effect of essential oils. In: Potential of essential oils. IntechOpen, London. https://doi.org/10.5772/intechopen.78008

Attia MS, El-Sayyad GS, Abd Elkodous M, El-Batal AI (2020) The effective antagonistic potential of plant growth-promoting rhizobacteria against Alternaria solani causing early blight disease in tomato plant. Sci Hortic 266:109289. https://doi.org/10.1016/j.scienta.2020.109289

Bailey B, Bae H, Strem M, Crozier J, Thomas S, Samuels G, Vinyard B, Holmes K (2008) Antibiosis, mycoparasitism, and colonization success for endophytic *Trichoderma* isolates with biological control potential in *Theobroma cacao*. Biol Control 46:24–35

Bensidhoum L, Nabti ED, Tabli N, Kupferschmied P, Weiss A, Rothballer M, Schmid M, Keel C, Hartmann A (2016) Heavy metal tolerant *pseudomonas protegens* isolates from agricultural well water in north eastern Algeria with plant growth promoting, insecticidal and antifungal activities. Eur J Soil Biol 75:38–46. https://doi.org/10.1016/j.ejsobi.2016.04.006

Blacutt AA, Mitchell TR, Bacon CW, Gold SE (2016) Bacillus mojavensis RRC101 lipopeptides provoke physiological and metabolic changes during antagonism against Fusarium verticillioides. Mol Plant Microbe Interact 29(9):713–723

Bonanomi G, Ippolito F, Scala F (2015) A "black" future for plant pathology? Biochar as a new soil amendment for controlling plant diseases. J Plant Pathol 97(2):223

Bora H, Kamle M, Mahato DK, Kumar TP (2020) Citrus essential oils (CEOs) and their applications in food: an overview. Plan Theory 9(3):357

Borges ÁV, Saraiva RM, Maffia LA (2015) Biocontrol of gray mold in tomato plants by *Clonostachys rosea*. Trop Plant Pathol 40:71–76

Bouzenna H, Krichen L (2013) *Pelargonium graveolens L'Her*. And *Artemisia arborescens L.* essential oils: chemical composition, antifungal activity against *Rhizoctonia solani* and insecticidal activity against *Rhysopertha Dominica*. Nat Prod Res 27(9):841–846

Božik M, Císarová M, Tančinová D, Kouřimská L, Hleba L, Klouček P (2017) Selected essential oil vapours inhibit growth of *Aspergillus* spp. in oats with improved consumer acceptability. Ind Crop Prod 98:146–152

Cal AD, Melgarejo P, Jimenez-Gasco MM (2022) Editorial: necrotrophic fungal plant pathogens. Front Plant Sci 13:839674

Cimmino A, Bahmani Z, Castaldi S, Masi M, Isticato R, Abdollahzadeh J, Amini J, Evidente A (2021) Phenazine-1-carboxylic acid (PCA), produced for the first time as an antifungal metabolite by *Truncatella angustata*, a causal agent of grapevine trunk diseases (GTDs) in Iran. J Agric Food Chem 69:12143–12147

Colmenares AJ, Aleu J, Duran-Patron R, Collado IG, Hernandez-Galan R (2002) The putative role of botrydial and related metabolites in the infection mechanism of *Botrytis cinerea*. J Chem Ecol 28:997–1005

Cruz-Luna AR, Cruz-Martínez H, Vásquez-López A, Medina DI (2021) Metal nanoparticles as novel antifungal agents for sustainable agriculture: current advances and future directions. J Fungi 7:1033. https://doi.org/10.3390/jof7121033

de los Santos-Villalobos S, Guzmán-Ortiz DA, Gómez-Lim MA, Délano-Frier JP, de-Folter S, Sánchez-García P, Peña-Cabriales JJ (2013) Potential use of Trichoderma asperellum (Samuels, Liechfeldt et Nirenberg) T8a as a biological control agent against anthracnose in mango (Mangifera indica L.). Biol Control 64(1):37–44

De Tender CA, Debode J, Vandecasteele B, D'Hose T, Cremelie P, Haegeman A, Ruttink T, Dawyndt P, Maes M (2016) Biological, physicochemical and plant health responses in lettuce and strawberry in soil or peat amended with biochar. Appl Soil Ecol 107:1–12

Diamond H, Cooke BM (2003) Preliminary studies on biological control of the Fusarium ear blight complex of wheat. Crop Prot 22(1):99–107. https://doi.org/10.1016/S0261-2194(02)00117-5

Djonovic S, Vittone G, Mendoza-Herrera A, Kenerley CM (2007) Enhanced biocontrol activity of *Trichoderma virens* transformants constitutively coexpressing beta-1,3- and beta-1,6-glucanase genes. Mol Plant Pathol 8:469–480. https://doi.org/10.1111/j.1364-3703.2007.00407

Doehlemann G, Ökmen B, Zhu W, Sharon A (2017) Plant pathogenic fungi. Microbiol Spectr 5(1). https://doi.org/10.1128/microbiolspec

Duan Y, Chen J, Pang Z, Ye X, Zhang C, Hu H, Xie J (2020) Antifungal mechanism of *Streptomyces ma*. FS-4 on *fusarium* wilt of banana. J Appl Microbiol 130(1):196–207. https://doi.org/10.1111/jam.14784

Elmer WH (2016) Effect of leaf mold mulch, biochar, and earthworms on mycorrhizal colonization and yield of asparagus affected by fusarium crown and root rot. Plant Dis 100(12):2507–2512

El-Sharkawy EES, Abdelrazik E (2022) Biocontrol of *fusarium* root rot in squash using mycorrhizal fungi and antagonistic microorganisms. Egypt J Biol Pest Control 32:13. https://doi.org/10.1186/s41938-022-00513-x

Elsherbiny EA, Taher MA, Elsebai MF (2019) Activity of *Purpureocillium lilacinum* filtrates on biochemical characteristics of *Sclerotinia sclerotiorum* and induction of defense responses in common bean. Eur J Plant Pathol 155:39–52. https://doi.org/10.1007/s10658-019-01748-5

El-Tarabily KA (2006) Rhizosphere-competent isolates of streptomycete and non-streptomycete actinomycetes capable of producing cell-wall-degrading enzymes to control *Pythium aphanidermatum* damping-off disease of cucumber. Can J Bot 84(2):211. https://doi.org/10.1139/b05-153

Fadiji AE, Babalola OO (2020) Elucidating mechanisms of endophytes used in plant protection and other bioactivities with multifunctional prospects. Front Bioeng Biotechnol 8:467. https://doi.org/10.3389/fbioe.2020.00467

Fernandez J, Orth K (2018) Rise of a cereal killer: the biology of *Magnaporthe oryzae* biotrophic growth. Trends Microbiol 26(7):582–597

Fira D, Dimkić I, Berić T, Lozo J, Stanković S (2018) Biological control of plant pathogens by *Bacillus* species. J Biotechnol 285:44–55. https://doi.org/10.1016/j.jbiotec.2018.07.044

Fisher MC, Henk DA, Briggs CJ, Brownstein JS, Madoff LC, McCraw SL, Gurr SJ (2012) Emerging fungal threats to animal, plant and ecosystem health. Nature 484(7393):186–194

Fontana DC, de Paula S, Torres AG, de Souza VHM, Pascholati SF, Schmidt D, Dourado Neto D (2021) Endophytic fungi: biological control and induced resistance to phytopathogens and abiotic stresses. Pathogens 10(5):570

Gao Y, Zeng XD, Ren B, Zeng JR, Xu T, Yang YZ, Hu XC, Zhu ZY, Shi LM, Zhou GY, Zhou Q (2020) Antagonistic activity against rice blast disease and elicitation of host-defence response capability of an endophytic Streptomyces albidoflavus OsiLf-2. Plant Pathol 69(2):259–271. https://doi.org/10.1111/ppa.13118

Geraldine AM, Lopes FAC, Carvalho DDC, Barbosa ET, Rodrigues AR, Brandão RS, Ulhoa CJ, Junior ML (2013) Cell wall-degrading enzymes and parasitism of sclerotia are key factors on field biocontrol of white mold by Trichoderma spp. Bio Control 67(3):308–316

Glazebrook J (2005) Contrasting mechanisms of defense against biotrophic and necrotrophic pathogens. Annu Rev Phytopathol 43:205–227. https://doi.org/10.1146/annurev.phyto.43.040204.135923

Guevara-Avendaño E, Bejarano-Bolívar AA, Kiel-Martínez AL, Ramírez-Vázquez M, Méndez-Bravo A et al (2019) Avocado rhizobacteria emit volatile organic compounds with antifungal activity against Fusarium solani, Fusarium sp. associated with Kuroshio shot hole borer, and Colletotrichum gloeosporioides. Microbiol Res 219:74–83. https://doi.org/10.1016/j.micres.2018.11.009

Hao Q, Albaghdady DMD, Xiao Y, Xiao X, Mo C, Tian T, Wang G (2021) Endophytic *Metarhizium anisopliae* is a potential biocontrol agent against wheat fusarium head blight caused by *Fusarium graminearum*. J Plant Pathol 103:875–885

Harman GE, Howell CR, Viterbo A, Chet I, Lorito M (2004) *Trichoderma* species-opportunistic, avirulent plant symbionts. Nat Rev Microbiol 2(1):43–56. https://doi.org/10.1038/nrmicro797

Harrier LA, Watson CA (2004) The potential role of *arbuscular mycorrhizal* (AM) fungi in the bioprotection of plants against soil-borne pathogens in organic and/or other sustainable farming systems. Pest Manag Sci 60:149–157

Heydari A, Pessarakli M (2010) A review on biological control of fungal plant pathogens using microbial antagonists. J Biol Sci 10(4):273–290. https://doi.org/10.3923/jbs.2010.273.290

Ho YN, Hoo SY, Wang BW, Hsieh CT, Lin CC, Sun CH, Peng CC, Lin C, Yang YL (2021) Specific inactivation of an antifungal bacterial siderophore by a fungal plant pathogen. ISME J 15:1858–1861

Jain A, Sarsaiya S, Wu Q, Lu Y, Shi J (2019) A review of plant leaf fungal diseases and its environment speciation. Bioengineered 10(1):409–424

Jaiswal K, Elad Y, Paudel I, Graber ER, Cytryn E, Frenkel O (2017) Linking the belowground microbial composition, diversity and activity to soilborne disease suppression and growth promotion of tomato amended with biochar. Sci Rep 7(1):44382

Jones JD, Dangl JL (2006) The plant immune system. Nature 444(7117):323–329

Jones EE, Rabeendran N, Stewart A (2014) Biocontrol of Sclerotinia sclerotiorum infection of cabbage by Coniothyrium minitans and Trichoderma spp. Biocontrol Sci Technol 24(12):1363–1382

Jose J, Ghantasala S, Roy Choudhury S (2020) Arabidopsis transmembrane receptor-like kinases (RLKs): a bridge between extracellular signal and intracellular regulatory machinery. Int J Mol Sci 21(11):4000. https://doi.org/10.3390/ijms21114000

Jupe F, Witek K, Verweij W, Śliwka J, Pritchard L, Etherington GJ, Maclean D, Cock PJ, Leggett RM, Bryan GJ, Cardle L, Jones JD (2013) Resistance gene enrichment sequencing (Ren Seq) enables reannotation of the NB-LRR gene family from sequenced plant genomes and rapid mapping of resistance loci in segregating populations. Plant J 76(3):530–544

Karthika S, Midhun SJ, Jisha MS (2020) A potential antifungal and growth-promoting bacterium Bacillus sp. KTMA4 from tomato rhizosphere. Microb Pathog 142:104049. https://doi.org/10.1016/j.micpath.2020.104049

Kavroulakis N, Ntougias S, Zervakis GI, Ehaliotis C, Haralampidis K, Papadopoulou KK (2007) Role of ethylene in the protection of tomato plants against soil-borne fungal pathogens conferred by an endophytic *Fusarium solani* strain. J Exp Bot 58:3853–3864. https://doi.org/10.1093/jxb/erm230

Khan N, Martínez-Hidalgo P, Ice TA, Maymon M, Humm EA, Nejat N, Sanders ER, Kaplan D, Hirsch AM (2018) Antifungal activity of *Bacillus* species against *fusarium* and analysis of the potential mechanisms used in biocontrol. Front Microbiol 9:2363. https://doi.org/10.3389/fmicb.2018.02363

Kim HY, Choi GJ, Lee HB, Lee SW, Lim HK, Jang KS, Son SW, Lee SO, Cho KY, Sung ND (2007) Some fungal endophytes from vegetable crops and their anti-oomycete activities against tomato late blight. Lett Appl Microbiol 44:332–337. https://doi.org/10.1111/j.1472-765X.2006.02093.x

Koeck M, Hardham AR, Dodds PN (2011) The role of effectors of biotrophic and hemibiotrophic fungi in infection. Cell Microbiol 13(12):1849–1857. https://doi.org/10.1111/j.1462-5822.2011.01665

Köhl J (2004) Biological control of botrytis spp. by *Ulocladium atrum* through competitive colonisation of necrotic plant tissues. Plant Research International, Wageningen

Kubicek CP, Starr TL, Glass NL (2014) Plant cell wall-degrading enzymes and their secretion in plant-pathogenic fungi. Annu Rev Phytopathol 52:427–451. https://doi.org/10.1146/annurev-phyto-102313-045831

Kumari S, Khanna V (2020) Induction of systemic resistance in chickpea (Cicer arietinum L.) against Fusarium oxysporum f. sp. ciceris by antagonistic rhizobacteria in assistance with native mesorhizobium. Current Microbiol 77(1):85–98. https://doi.org/10.1007/s00284-019-01805-6

Lai Y, Cao X, Chen J, Wang L, Wei G, Wang S (2020) Coordinated regulation of infection-related morphogenesis by the KMT2-Cre1-Hyd4 regulatory pathway to facilitate fungal infection. Sci Adv 6(13):1659

Law JWF, Ser HL, Khan TM, Chuah LH, Pusparajah P, Chan KG, Goh BH, Lee LH (2017) The potential of Streptomyces as biocontrol agents against the rice blast fungus, Magnaporthe oryzae (Pyricularia oryzae). Front Microbiol 8:3. https://doi.org/10.3389/fmicb.2017.00003

Lebeda A, Křístková E, Kitner M, Mieslerová B, Jemelková M, Pink DA (2014) Wild Lactuca species, their genetic diversity, resistance to diseases and pests, and exploitation in lettuce breeding. Eur J Plant Pathol 138:597–640

Li C, He X, Luo X, Xu L, Liu L, Min L, Jin L, Zhu L, Zhang X (2014) Cotton WRKY1 mediates the plant defense-to-development transition during infection of cotton by Verticillium dahliae by activating JASMONATE ZIMDOMAIN1 expression. Plant Physiol. 166:2179–2194. https://doi.org/10.1104/pp.114.246694

Li X, Zhang Y, Wei Z, Guan CY, Liao X (2016) Antifungal activity of isolated *bacillus amyloliquefaciens* SYBCH47 for the biocontrol of peach gummosis. PLoS One 11(9):e0162125. https://doi.org/10.1371/journal.pone.0162125

Li Y, Wang R, Liu J, Xu L, Ji P, Sun L, Jiang B, Li L (2019) Identification of a biocontrol agent Bacillus vallismortis BV23 and assessment of effects of its metabolites on Fusarium graminearum causing corn stalk rot. Biocontrol Sci Tech 29(3):263–275

Liaqat F, Eltem R (2016) Identification and characterization of endophytic bacteria isolated from *in vitro* cultures of peach and pear rootstocks. 3 Biotech 6:1–8

Mapuranga J, Zhang L, Zhang N, Yang W (2022) The haustorium: the root of biotrophic fungal pathogens. Front Plant Sci 13:963705

Medda S, Hajra A, Dey U, Bose P, Mondal NK (2014) Biosynthesis of silver nanoparticles from Aloe vera leaf extract and antifungal activity against *Rhizopus* sp. and *Aspergillus* sp. Appl Nanosci 5:875–880. https://doi.org/10.1007/s13204-014-0387-1

Meena KR, Kanwar SS (2015) Lipopeptides as the antifungal and antibacterial agents: applications in food safety and therapeutics. Biomed Res Int 2015:473050. https://doi.org/10.1155/2015/473050

Meena LI, Rajeswari E, Devi PA, Kamalakannan A, Kalaiselvi T (2022) Antifungal potential of *Streptomyces rameus* GgS 48 against mungbean root rot [*Rhizoctonia bataticola* (Taub.) Butler]. J Biosci 47:10. https://doi.org/10.1007/s12038-021-00244-5

Meszka B, Bielenin A (2010) Polyversum WP a new biological product against strawberry grey mould. Phytopathol 58:13–19

Millan AF, Larraya L, Farran I, Ancin M, Veramendi J (2021) Successful biocontrol of major postharvest and soil-borne plant pathogenic fungi by antagonistic yeasts. Biol Control 160:104683

Mohamed N, Lherminier J, Farmer MJ, Fromentin J, Béno N, Houot V, Milat ML, Blein JP (2007) Defense responses in grapevine leaves against *Botrytis cinerea* induced by application of a *Pythium oligandrum* strain or its elicitin, oligandrin, to roots. Phytopathology 97(5):611–620

Mukerji KG, Ciancio A (2007) Mycorrhizae in the integrated pest and disease management. In: General concepts in integrated pest and disease management, pp 245–266

Murtaza G, Ahmed Z, Eldin SM, Ali B, Bawazeer S, Usman M, Iqbal R, Neupane D, Ullah A, Khan A, Hassan MU (2023) Biochar-soil-plant interactions: a cross talk for sustainable agriculture under changing climate. Front Environ Sci 11:1059449

Myo EM, Liu B, Ma J, Shi L, Jiang M, Zhang K, Ge B (2019) Evaluation of Bacillus velezensis NKG-2 for biocontrol activities against fungal diseases and potential plant growth promotion. Biol Cont 134:23–31

Narang D, Kaur S, Steuernagel B, Ghosh S, Bansal U, Li J, Zhang P, Bhardwaj S, Uauy C, Wulff BB, Chhuneja P (2020) Discovery and characterisation of a new leaf rust resistance gene introgressed in wheat from wild wheat *Aegilops peregrina*. Sci Rep 10(1):7573

Nassimi Z, Taheri P (2017) Endophytic fungus *Piriformospora indica* induced systemic resistance against rice sheath blight via affecting hydrogen peroxide and antioxidants. Biocontrol Sci Tech 3157:256–267

Nawrocka J, Gromek A, Małolepsza U (2019) Nitric oxide as a beneficial signaling molecule in *Trichoderma atroviride* TRS25-induced systemic defense responses of cucumber plants against *Rhizoctonia solani*. Front Plant Sci 10:421

Nazzaro F, Fratianni F, Coppola R, Feo VD (2017) Essential oils and antifungal activity. Pharmaceuticals 10(4):86. https://doi.org/10.3390/ph10040086

Niu B, Wang W, Yuan Z, Sederoff RR, Sederoff H, Chiang VL, Borriss R (2020) Microbial interactions within multiple-strain biological control agents impact soil-borne plant disease. Front Microbiol 11:585404

Olowe OM, Olawuyi OJ, Sobowale AA, Odebode AC (2018) Role of arbuscular mycorrhizal fungi as biocontrol agents against *Fusarium verticillioides* causing ear rot of *Zea mays L.* (maize). Curr Plant Biol 15:30–37. https://doi.org/10.1016/j.cpb.2018.11.005

Omann MR, Lehner S, Rodríguez CE, Brunner K, Zeilinger S (2012) The seven-transmembrane receptor Gpr1 governs processes relevant for the antagonistic interaction of Trichoderma atroviride with its host. Microbiol 158:107. https://doi.org/10.1099/mic.0.052035-0

Oskay (2009) Antifungal and antibacterial compounds from Streptomyces strains. Afr J Biotechnol 8(13):3007–3017

Panáček A, Kolář M, Večeřová R, Prucek R, Soukupova J, Kryštof V (2009) Antifungal activity of silver nanoparticles against *Candida* spp. Biomaterials 30:6333–6340. https://doi.org/10.1016/j.biomaterials.2009.07.065

Pandit MA, Kumar J, Gulati S, Bhandari N, Mehta P, Katyal R, Rawat CD, Mishra V, Kaur J (2022) Major biological control strategies for plant pathogens. Pathogens 11(2):273. https://doi.org/10.3390/pathogens11020273

Pang AH, Garneau-Tsodikova S, Tsodikov OV (2015) Crystal structure of halogenase PltA from the pyoluteorin biosynthetic pathway. J Struct Biol 192(3):349–357. https://doi.org/10.1016/j.jsb.2015.09.013

Parikh L, Agindotan BO, Burrows ME (2021) Antifungal activity of plant-derived essential oils on pathogens of pulse crops. Plant Dis 105(6):1692–1701

Park JH, Choi GJ, Jang KS, Lim HK, Kim HT, Cho KY, Kim JC (2005) Antifungal activity against plant pathogenic fungi of chaetoviridins isolated from Chaetomium globosum. FEMS Microbiol lett 252(2):309–313

Pascholati SF, Souza VHM, Filho JC (2019) *Trichoderma*: Uso na Agricultura. In: Indução de resistência por *Trichoderma*, vol 1. Embrapa Soja, Brasília, pp 235–254

Paulitz TC, Baker R (1987) Biological control of Pythium damping-off of cucumbers with Pythium nunn: population dynamics and disease suppression. Phytopathol 77(2):335–340. https://doi.org/10.1094/Phyto-77-335

Perfect SE, Jonathan R (2001) Green infection structures of biotrophic and hemibiotrophic fungal plant pathogens. Mol Plant Pathol 2(2):01–108

Petit-Houdenot Y, Fudal I (2017) Complex interactions between fungal avirulence genes and their corresponding plant resistance genes and consequences for disease resistance management. Front Plant Sci 8:1072

Pirttilä AM, Tabas HMP, Baruah N, Koskimäki JJ (2021) Biofertilizers and biocontrol agents for agriculture: how to identify and develop new potent microbial strains and traits. Microorganisms 9:817

Poveda J, Martínez-Gómez Á, Fenoll C, Escobar C (2021) The use of biochar for plant pathogen control. Phytopathol 111(9):1490–1499

Rajarammohan S (2021) Redefining plant-necrotroph interactions: the thin line between hemibiotrophs and necrotrophs. Front Microbiol 12:673518. https://doi.org/10.3389/fmicb.2021.673518

Rivera-Mendez W, Obregon M, Moran-Diez ME, Hermosa R, Monte E (2020) Trichoderma asperellum biocontrol activity and induction of systemic defenses against Sclerotium cepivorum in onion plants under tropical climate conditions. Bio Control 141:104145

Roy T, Bandopadhyaya A, Sonawaneb PJ, Majumdarc S, Mahapatrab NR, Alamd S, Dasa N (2018) Bio-effective disease control and plant growth promotion in lentil by two pesticide degrading strains of *Bacillus* sp. Biol Control 127:55–63

Salvatore MM, Andolfi A (2021) Phytopathogenic fungi and toxicity. Toxins 13(10):689

Santhanam R, Menezes RC, Grabe V, Li D, Baldwin IT et al (2019) A suite of complementary biocontrol traits allows a native consortium of root-associated bacteria to protect their host plant from a fungal sudden-wilt disease. Mol Ecol 28(5):1154–1169. https://doi.org/10.1111/mec.15012

Santamaria O, Smith DR, Stanosz GR (2012) Interaction between *Diplodia pinea* or *Diplodia scrobiculata* and fungal endophytes isolated from pine shoots. Can J For Res 42(10):1819–1826. https://doi.org/10.1139/x2012-132

Schwessinger B, Bahar O, Thomas N, Holton N, Nekrasov V, Ruan D, Ronald PC (2015) Transgenic expression of the dicotyledonous pattern recognition receptor EFR in rice leads to ligand-dependent activation of defense responses. PLoS Pathog 11(4):e1004809. https://doi.org/10.1371/journal.ppat.1004809

Shaaban HAE, El-Ghorab AH, Shibamoto T (2012) Bioactivity of essential oils and their volatile aroma components: review. J Essent Oil Res 24:203–212

Shen Q, Zhou W, Li H, Hu L, Mo H (2016) ROS involves the fungicidal actions of thymol against spores of *Aspergillus flavus* via the induction of nitric oxide. PLoS One 11:0155647. https://doi.org/10.1371/journal.pone.0155647

Shastri B, Kumar R, Lal RJ (2020) Isolation, characterization and identification of indigenous endophytic bacteria exhibiting PGP and antifungal traits from the internal tissue of sugarcane crop. Sugar Tech 22(4):563–573. https://doi.org/10.1007/s12355-020-00824-z

Silva GH, Teles HL, Zanardi LM, Young MCM, Eberlin MN, Hadad R, Araújo ÂR (2006) Cadinane sesquiterpenoids of *Phomopsis cassiae*, an endophytic fungus associated with *Cassia spectabilis* (Leguminosae). Phytochemistry 67(17):1964–1969

Soković MD, Vukojević J, Marin PD, Brkić DD, Vajs V, Van Griensven LJLD (2009) Chemical composition of essential oils of *thymus* and *Mentha* species and their antifungal activities. Molecules 14:238–249. https://doi.org/10.3390/molecules14010238

Takenaka S, Nishio Z, Nakamura Y (2003) Induction of defense reactions in sugar beet and wheat by treatment with cell wall protein fractions from the mycoparasite Pythium oligandrum. Phytopathol 93(10):1228–1232. https://doi.org/10.1094/PHYTO.2003.93.10.1228

Takenaka S, Nakamura Y, Kono T, Sekiguchi H, Masunaka A, Takahashi H (2006) Novel elicitin-like proteins isolated from the cell wall of the biocontrol agent Pythium oligandrum induce defence-related genes in sugar beet. Mol Plant Pathol 7(5):325–339. https://doi.org/10.1111/j.1364-3703.2006.00340.x

Thambugala KM, Daranagama DA, Phillips AJL, Kannangara SD, Promputtha I (2020) Fungi vs. fungi in biocontrol: an overview of fungal antagonists applied against fungal plant pathogens. Front Cell Infect Microbiol 10:604923. https://doi.org/10.3389/fcimb.2020.604923

Thangavelu R, Gopi M (2015) Combined application of native Trichoderma isolates possessing multiple functions for the control of Fusarium wilt disease in banana cv. Grand Naine. Biocont Sci Tech 25(10):1147–1164

Van Kan JA (2006) Licensed to kill: the lifestyle of a necrotrophic plant pathogen. Trends Plant Sci 11(5):247–253

Verma PP, Thakur S, Kaur M (2016) Antagonism of Pseudomonas putida against Dematophora nectarix a major apple plant pathogen and its potential use as a biostimulent. J Pure Appl Microbiol 10(4):2717–2727

Verma SK, Kingsley KL, Bergen MS, Kowalski KP, White JF (2018) Fungal disease prevention in seedlings of Rice (*Oryza sativa*) and other grasses by growth-promoting seed-associated endophytic bacteria from invasive *Phragmites australis*. Microorganisms 6(1):21. https://doi.org/10.3390/microorganisms6010021

Wang X, Wang C, Li Q, Zhang J, Ji C, Sui J, Liu Z, Song X, Liu X (2018) Isolation and characterization of antagonistic bacteria with the potential for biocontrol of soil-borne wheat diseases. J Applied Microbiol 125(6):1868–1880

Windisch S, Bott S, Ohler MA, Mock HP, Lippmann R, Grosch R, Smalla K, Ludewig U, Neumann G (2017) Rhizoctonia solani and bacterial inoculants stimulate root exudation of antifungal compounds in lettuce in a soiltype specific manner. Agronomy 7(2):44

Whitaker BK, Bakker MG (2019) Bacterial endophyte antagonism toward a fungal pathogen in vitro does not predict protection in live plant tissue. FEMS Microbiol Ecol 95(2):237. https://doi.org/10.1093/femsec/fiy237

Witek K, Jupe F, Witek AI, Baker D, Clark MD, Jones JD (2016) Accelerated cloning of a potato late blight–resistance gene using RenSeq and SMRT sequencing. Nat Biotechnol 34(6):656–660

Wu Q, Ni M, Dou K, Tang J, Ren J, Yu C, Chen J (2018) Co-culture of *Bacillus amyloliquefaciens* ACCC11060 and *Trichoderma asperellum* GDFS1009 enhanced pathogen-inhibition and amino acid yield. Microb Cell Fact 17:155. https://doi.org/10.1186/s12934-018-1004-x

Yoshida S, Cui S, Ichihashi Y, Shirasu K (2016) The haustorium, a specialized invasive organ in parasitic plants. Annu Rev Plant Biol 67:643–667

Yu Z, Wang Z, Zhang Y, Wang Y, Liu Z (2021) Biocontrol and growth-promoting effect of Trichoderma asperellum TaspHu1 isolate from Juglans mandshurica rhizosphere soil. Microbiol Res 242:126596

Yu L, Qiao N, Zhao J, Zhang H, Tian F, Zhai Q, Chen W (2020) Postharvest control of Penicillium expansum in fruits: a review. Food Biosci 36:100633

Zhang F, Li XL, Zhu SJ, Ojaghian MR, Zhang JZ (2018) Biocontrol potential of Paenibacillus polymyxa against Verticillium dahliae infecting cotton plants. Bio Cont 127:70–77

Zhang D, Yu S, Yang Y, Zhang J, Zhao D, Pan Y, Fan S, Yang Z, Zhu J (2020) Antifungal effects of volatiles produced by Bacillus subtilis against Alternaria solani in potato. Front Microbiol 11:1196

Chapter 12
Advance Techniques Used for Decoding the Hidden Language Used During Plant–Microbe Interaction

Garima Gupta, Durgesh Singh, Kamna Madheshiya, Updesh Chauhan, and Puneet Singh Chauhan

Abstract A myriad of microbes inhabiting in soil may communicate with host plant in a variety of manner in their entire life cycle that may be categorized as neutral, mutual, commensal, beneficial, or harmful relationships. With the rapidly proliferating population and increased food demand versus supply ratio, the conventional and chemical-based agriculture practices to achieve goals of sustainable agriculture are quite far in economical and environment-friendly terms. For the past few decades, plant growth-promoting rhizobacteria (PGPRs) are being used to enhance the crop production, improve the nutrient content of crops, bioremediation, or biocontrol of pathogens. Researchers found that PGPRs have immense potential as bioinoculants, but there are very few biofertilizers used at commercial scale. It would be extremely important to untangle the complex interaction between plant and microbes for the development of a successful biofertilizer and to ultimately enhance crop production. Earlier cultivable techniques were used mainly focusing on a very small portion of microbes present in rhizosphere, but the last few decades are dedicated for the development of robust, accurate, precise, and high-throughput techniques like recent microscopy (SEM, TEM, CLSM), omics techniques (metagenomics, metaproteomics, metabonomics, etc), next generation sequencing (NGS),

G. Gupta (✉)
Faculty of Bio-Sciences, Institute of Bio-Sciences and Technology, Shri Ramswaroop Memorial University, Barabanki, Uttar Pradesh, India

Microbial Technology Division, CSIR-National Botanical Research Institute, Lucknow, Uttar Pradesh, India
e-mail: garimagupta.ibst@srmu.ac.in

D. Singh · K. Madheshiya · U. Chauhan
Faculty of Bio-Sciences, Institute of Bio-Sciences and Technology, Shri Ramswaroop Memorial University, Barabanki, Uttar Pradesh, India

P. Singh Chauhan
Microbial Technology Division, CSIR-National Botanical Research Institute, Lucknow, Uttar Pradesh, India
e-mail: puneet@nbri.res.in

P. Singh Chauhan et al. (eds.), *Plant-Microbe Interaction and Stress Management*, Rhizosphere Biology,
https://doi.org/10.1007/978-981-97-4239-4_12

and clustered regularly interspaced short palindromic repeats (CRISPR). Here, the present chapter focuses on the advanced techniques and cutting-edge technologies that have opened new avenues in making a clear-cut understanding of interactions taking place among PGPRs and the host plant.

Keywords PGPRs · NGS · Omics · Rhizosphere · Plant–microbe interactions

12.1 Introduction

Rhizosphere constitutes an active region of soil apart from bulk soil in the surrounding of plant roots harboring a wide array of microbes like bacteria and fungi (Hakim et al. 2021). Generally, the plant–microbe interactions taking place in rhizospheric sites can be either beneficial or harmful to the plants; here the major focus of present review is to summarize the techniques available to decode the code language used by host plants and microbes for the interaction and establishment (Joshi et al. 2023). However, the biggest setback of unravelling the mechanism of plant–microbe interaction is that scientists are unable to cultivate the inhabiting microbes in the rhizosphere under laboratory conditions. The study of plant–microbe interaction might have several implications in learning various processes like ecosystem, nutrient cycling, and carbon sequestration (Backer et al. 2018). In the era of molecular techniques and omics technology, advancements in the technology provides excellent opportunities to decipher the mechanisms involved in plant–microbe interactions and combinatorial approach of different techniques provide new insights in the field of plant–microbe interaction (Qi et al. 2023).

12.2 Isolation of Microbes Using Cultivation Techniques

Previously, most of the studies in plant–microbe interactions were cultivation-dependent techniques. These studies impart a brief knowledge about the microflora present in rhizosphere as only 1% of soil microbes can be cultivated in lab and also suffers from great biasness due to inability in cultivating huge percent of microbial community (Backer et al. 2018). Plant growth-promoting rhizobacteria (PGPR) isolation techniques involve sample collection, surface sterilization (using sodium hypochlorite or chloramine-T), and regular serial dilution practice; pure culture techniques are used for the screening and selection of microbes based on different types of media as per the requirement (Gupta et al. 2013). Further, some other cultivation-dependent techniques like BIOLOG and plate count methods are also used. To track PGPR in the host plant, antibiotic resistance markers like rifampicin, kanamycin, and streptomycin based on spontaneous chromosomal antibiotic-resistant mutants or Tn5 mutagenesis approach can also be used (Preston-Mafham et al. 2002; Godino et al. 2016). It can be used to analyze the population count and

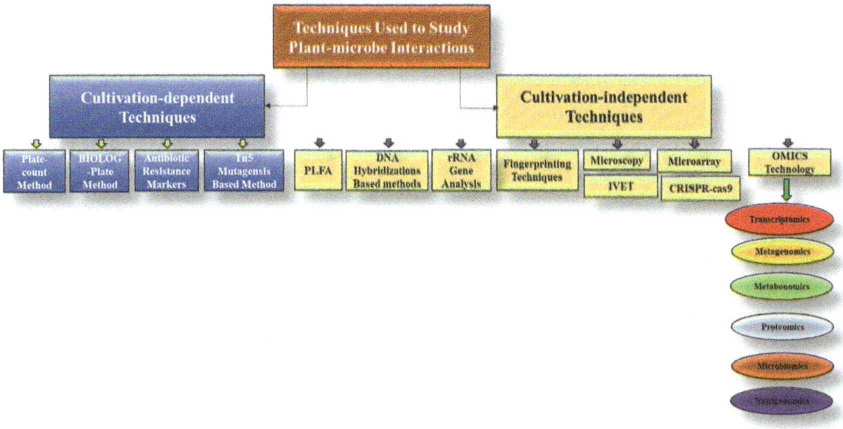

Fig. 12.1 Techniques used to study plant-microbe interaction

its presence in different parts of plants by colony forming unit (CFU) counting on selective media containing antibiotic. This technique can be used for cultivable bacteria only, and it may also pose serious threat by developing multiple antibiotic resistance in the bacteria present in the environment through horizontal gene transfer. Consequently, the deliberate release of PGPB tagged with antibiotic resistance genes is not recommended (Mahdi et al. 2022). Culture-independent analyses, phospholipid fatty acid analysis (PLFA) and DNA hybridizations and re-association kinetics, and its combination with cultivation techniques has less resolution in understanding microbial diversity, so rRNA gene analysis using polymerase chain reaction (PCR), fingerprinting techniques denaturing gradient gel electrophoresis (DGGE), terminal restriction fragment analysis (T-RFLP), amplified rDNA restriction analysis (ARDRA), fluorescence in-situ hybridization (FISH) using probes and fluorescence microscopy, FISH with microautoradiography, microarray, stable isotope probing (SIP-[^{13}C]-labeled), and so forth (Singh et al. 2004) are used for microbial diversity studies. PhyloChips (Ribosomal RNA-targeted oligonucleotide microarrays) gives information which allows detection of microbes for analyzing complex microbial communities in a particular environmental niche (Rincon-Florez et al. 2013) (Fig. 12.1).

12.3 Visualization Techniques Using Microscopy

Most of the plant–microbe interactions occur at cellular level, and these interactions at cell level might be extremely significant for the development of survival strategies required for microbial growth. Although biochemical assays of tissues do not provide clear picture of cellular process occurring during plant–microbe interaction, visual detection techniques at single-cell level can have several implications. Here, the electron microscope became the instrument of choice to examine subcellular interactions between plants and microbes. Previously light microscope has been used for gene expression detection at the cell level by in situ hybridization or by using detectable reporter genes such as GUS (glucuronidase), but the resolution of light microscope is quite less in comparison of electron microscope (Heath 2000).

12.4 Imaging Techniques

There are various imaging techniques used to visualize single cells and its components below:

(A) *Video Microscopy*: Light microscope along with video camera CCD (solid-state charge-coupled device), converting optical image to electrical video signal.
(B) *Confocal Laser Scanning Microscopy (CLSM)*: Fluorescent samples in particular areas of specimen are detected and digitized; some studies use fluorescent probes.
(C) *Laser Trapping*: Laser beams use photo-driven force to "capture" subcellular particles to focus on a particular spot (Gilroy 1997).

12.5 Image Analysis Software

WinRhizo is a software used to study the root traits like morphology, topology, and architecture for avoiding human mistakes. Roots are one of the main entrance or association point of microbes, so all the root properties get affected in the presence of microbes. There are several image analysis software programs available, like WinRHIZO, which is an image acquisition system, with high-quality scanner, lighting system convert image into morphology data including total root length, area, volume, diameter, branching, topology, architecture, and color analyses. Although the system is quite costly, gentle root washing and no overlapping in fine roots should happen. Some other image analysis software freely available are ImageJ (formerly NIHImage) (not user-friendly), RhizoVision Explorer (easily used, fast image processing, and reliable), GIS and RTK-GNSS software (Gamalero et al. 2022).

12.6 In Vivo Expression Technology

It is a gene fusion technology used to study gene expression of bacteria present in the rhizosphere. It is a promoter-trapping technology based on selection of promoters induced by environment by genetic complementation of conditional mutation.

This is done by fusing in vivo-expressed promoter to the sequence complementing the mutation. It has great importance as it can be used to study the entire set of genes and physiological conditions required for survival in vivo present in microbes inhabiting in natural environment and can be used to study genes further even mutation may be lethal. The main disadvantage of this study is that only activated genes can be studied, and downregulated or unexpressed genes cannot be identified by this method. *Pseudomonas fluorescens* SBW25 genes (Type III secretion, oxidative stress, and transport), *P. fluorescens* PF0–1 (nutrient utilization, transport, detoxification, and metabolism) have been studied using IVET technology (Kiely et al. 2006).

12.7 Transcriptomics

Transcriptome is the analysis of all mRNA transcripts present in the cell and helps in analyzing the transcripts produced by genes present in bacteria under a specified environmental condition. Transcriptomic studies are of two types: oligonucleotide microarrays and complementary DNA (cDNA) microarrays. In microarray technique, isolation of mRNA is done from treated and control samples, and then cDNA is formed using reverse transcription approach. Then, cDNA is tagged with fluorescent molecules and hybridized using DNA chips. Different plant hosts such as rice, wheat, sugarcane, tobacco, and tomato and PGPB like *Herbaspirillum seropedicae*, *Azospirillum brasilense*, *Burkholderia anthina*, Paraburkholderia *kururiensis*, *Bacillus subtilis*, and so forth have been studied using transcriptomic analysis, and studies provided the details regarding nutrient transport, defense responses, hormone signaling, and developmental pathways, which can impart better understanding about the plant–PGPB interactions elaborately (Mukherjee 2022).

12.8 PGPB Whole Genomes Sequencing

It includes metagenome shotgun sequencing that involves DNA or RNA extraction from microbial community present in a particular condition, library construction, and then, short-read sequencing of all genome sequences or transcripts are done. This provides the random information about short random DNA/cDNA fragments, which can be used as markers for particular microorganisms and metabolic pathways after assembling. Whole genome sequencing of various PGPRs belonging to a variety of genus belonging to *Azoarcus*, *Bacillus*, *Pseudomonas*, *Klebsiella*, *Glucanoacetobacter*, and so forth has been done previously, and still, every year many PGPR genomes are being sequenced (Gupta et al. 2012).

12.9 Next Generation Sequencing

Another successful "omics" approach could be next generation sequencing (NGS) technique for identifying (a) genes involved in biocontrol and plant growth enhancement in PGPRs, (b) mechanistic insights into various process, (c) estimation of success rate in field trials and stable results, and (d) risk analysis. Thus, NGS technologies can be used to mine and annotate the respective genes responsible for carrying out plant–microbe interactions, whole genome studies, or metagenomics studies of various plant associated bacteria. Even reverse gene engineering approach can be used to establish the metabolic activities and gene regulation mechanisms of associated bacteria. Instruments used for NGS can be classified as second- and third-generation sequencing technologies. After Sanger and Maxam-Gilbert method (First generation sequencing), second (Roche 454, Illumina, and Life Technologies based on sequencing by Oligonucleotide Ligation and Detection [SOLiD] and Ion Torrent sequencers) and third generation sequencing (PacBio RS by Pacific Biosciences) technologies were established (Kumari et al. 2017).

12.10 Shotgun Sequencing and Metagenomics

A few studies using metagenomics approach specifically shotgun sequencing was used to explore plant-associated microorganisms, for example, Roche 454 sequencing technology. Microbial communities associated with *Arabidopsis thaliana*, clover, rice, tomato, and soybean have been reported using metagenomics technology. Earlier sanger sequencing was used for understanding the strategies used by PGPR in nutrient acquisition and promoting plant growth. Previously reported studies revealed the importance of phytic acid in plant growth promotion (Knief 2014). Maize rhizosphere using shotgun metagenomic sequencing revealed that the soil property N-NO_3 immensely affects the diversity and composition of genes in the microbial samples studied and influenced by anthropogenic activities. It will divulge the relation between nutrient cycling, plant health, rhizosphere microbiome, and anthropogenic activities, ultimately affecting the food demand and crop yield to suffice the burgeoning human population (Chukwuneme et al. 2021). Recently, highest abundant bacteria *Pseudomonas* and *Burkholderia* as well as zinc/manganese transport system were more in suppressive than conducive soils, and phosphate transport system was absent in conducive soils. It suggests that PGPR might be involved in Foc growth suppression by phosphate solubilization and releasing secondary metabolites such as siderophores (Nisrina et al. 2021). Metagenomic study revealed the microbial community genes involved in assimilatory sulfate reduction, sulfur reduction, SOX systems, sulfur oxidation, and organic sulfur transformation are responsible for mitigating the harmful effects of salt on grapevine with or without salt stress (Wang et al. 2023).

12.11 Studies Through Amplicon Sequencing

Amplicon sequencing method is used to study plant–microbe interaction and discriminate the rhizosphere and phyllosphere communities. Most researchers used Roche 454 sequencing, whereas some used Ion PGM platform or the Illumina MiSeq. Amplicon size in phyllosphere studies range from ~1000 to 10,000 reads per sample, whereas 454 FLX+ instrument was used for longer reads. It also imparts the mechanism behind plant colonization in phyllosphere, microbes biogeographical distribution, factors affecting plant-microbe interaction, plant growth, nutrition, soil types, microbiota selection, role of treatments in crop cultivation like irrigation (Knief 2014). Zhang et al. (2019) studied the Pepper seedlings inoculated with *Bacillus velezensis* NJAU-Z9, which had stable and higher abundance, and changed rhizospheric microbial community and stimulated host plant to select distinctive microbes (*Sphingomonas*, *Sphingopyxis*, *Bradyrhizobium*, *Chitinophaga*, *Dyadobacter*, *Streptomyces*, *Lysobacter*, *Pseudomonas* and *Rhizomicrobium*, *Cladorrhinum*, *Cladosporium* and *Aspergillus*) in microbial community resulting in enhanced plant growth throughout the growth period induced by pre-colonization led to yield enhancement (Zhang et al. 2019). There are various studies related to 16S sequencing conducted in various plants like *Zea mays* L. (soil fortification) (Josephine and Thomas 2019), Oilseed rape (*Brassica napus* L.) (Liu et al. 2021), rice (Imchen et al. 2019) (changes in microbial community in development stages), wheat (Reid et al. 2021), and so forth.

12.12 Metabolomics/Metabonomics

It is a well-established technique for understanding the plant–microbe interactions involving various types of microbial interactions with host plant like fungi, mycorrhiza, PGPR, and so forth. It has immense importance in exploring the information related to nutrient acquisition, effect on host growth, and receptor-based host-microbe recognition, discriminating pathogenic or symbiotic interaction. Metabolomics may have two main perspectives experimentally: targeted (quantification and detection of pre-defined and characterized metabolites from biological samples and prior of metabolites is mandatory) and untargeted approaches (involve global metabolic profiling and analysis of each detectable metabolite from the sample). It gave reproducible results, such as mass spectrometry (MS), including liquid or gas chromatography (L/GC), nuclear magnetic resonance (NMR) used for recognizing compounds with identical masses, structures, and so forth (Khanijou et al. 2022).

Chamam et al. (2013) studied two *Azospirillum* strains enhancing plant growth of two rice (*Oryza sativa*) cultivars, although metabolome behaved differently. Similarly, metabolism of maize treated with nitrogen-fixing bacteria showed more changes than with non-nitrogen-fixing mutants. *B. velezensis* 5113 also led to metabolic modulation and explored similar kind of metabolites and regulatory mechanisms to increase tolerance capacity of wheat under different types of abiotic stress, and it may different mechanisms like express stress defense proteins and energy release, alter amino acids metabolic pathways, GABA accumulation, enhancing photosynthetic capacity for growth promotion, and so forth (Abd El-Daim et al. 2019). It can also be used for revealing biocontrol mechanisms like *Rhizobium leguminosarum* bv phaseoli RRE6, and *R. leguminosarum* bv trifolii ANU 843 enhanced the amount of phenolic compounds involved in providing resistance against the pathogen *Rhizoctonia solani* (Mishra et al. 2006). Metabolic changes during colonization by *H. seropedicae* SmR1 (fix+) or a fix− mutant strain (SmR54) were studied in model grass species *Setaria viridis* and compared with uninoculated roots. Compounds such as purine, zeatin, and riboflavin pathways were high in inoculated plants, whereas nitrogen, starch, and sucrose get reduced on inoculation with fix− strain or uninoculated, due to N limitation. However, they were high in indole-alkaloid biosynthesis colonized by the fix− strain indicating plant defense response; thus, metabolomics can also reveal the mechanistic insights of plant growth promotion by PGPRs (Agtuca et al. 2020). Apart from that, PGPR volatile organic compounds (VOC) metabolome study can also be conducted involving various plant growth-promotion strategies (Márquez et al. 2022), understanding host plant-pathogenic microbe and beneficial microbe tripartite interaction (Adeniji et al. 2020), abiotic interactions like *Pseudomonas oryzihabitans* AXSa06-inoculated plants enhanced antioxidant enzymes and molecules providing salt tolerance by producing metabolites like amino acids and genes involved in ethylene or abscisic acid pathway (Mellidou et al. 2021). Numerous studies in different plants like in Maize-*Azospirillum lipoferum* CRT1 involve benzoxazinoids (Couillerot et al. 2013); rice-*Azospirillum* involved tryptophan (Chamam et al. 2013); *Medicago truncatula-Sinorhizobium meliloti* involved amino acids, organic acids, sugars, lipids, and flavonoids; *Solanum lycopersicum-B. amyloliquefaciens* involved surfactins; *Neotyphodium lolii-Lolium perenne* involved mannitol; peramine and perloline, and so forth revealed the importance of metabolomics (Gupta et al. 2022).

12.13 Proteomics in Plant-Microbe Interaction

Basically, it is identifying and characterizing a set of proteins expressed in a particular condition in comparison to control, although it's quite challenging to work on huge dynamic ranges of proteins present in abundances as well as in discriminating the plant and microbe proteins produced during plant–microbe interaction.

Peptide/Protein Separation Techniques: Proteomics studies have provided vast information about a wide array of proteins of varying hydrophobic nature, size, and charge by using classical gel-based methods and later on by MS. However, MS technologies are frequently more powerful for large-scale proteomics when combined with gel-free separation techniques, but gel-free separation along with MS are more efficient for large-scale proteomics. Proteomics comprises of three steps mainly: (1) protein extraction, (2) protein fragmentation and peptides identification by MS, and (3) quantifying protein expression and PTM (post-translational modifications). Specifically in plant–PGPR interaction, most of the proteomics studies have been done in legume–Rhizobia interaction, abiotic stress in plants, biotic stress by fungal, and bacterial pathogens (Jayaraman et al. 2012). Most of the proteomics studies for plant–PGPR interaction are done on genome-sequenced plants like rice, maize, wheat, and PGPR models. There are various proteomics techniques: (1) two-dimensional (2D) electrophoresis involves protein separation based on isoelectric points (first dimension) by isoelectric focusing (IEF) and then (second dimension) according to their subunit molecular masses by SDS-PAGE (sodium dodecyl sulfate polyacrylamide gel electrophoresis). (2) MS: has high sensitivity, speed, and accuracy use techniques for desorption like electrospray ionization (ESI) and matrix-assisted laser desorption ionization (MALDI), which transform macromolecules to ions in gas phase for MS analyses. (3) DIGE Difference gel electrophoresis (DIGE) improves reproducibility by comparing two or more samples in the same gel. Some other quantitative or qualitative proteomic techniques like ICAT (isotope-coded affinity tag), iTRAQ (isobaric tag for relative and absolute quantification), SILAC (stable isotope labeling with amino acids in cell culture), LC-MS (label-free comparative), protein phosphorylation identification, and protein microarrays can be used to elucidate the mechanism involved in plant–bacterial interactions (Cheng et al. 2010).

In most of the studies, 2D electrophoresis/MALDI-TOF is the proteomic technique used for protein separation as it can separate proteins and isoforms with great resolution and more reproducibility, and label-free quantitation and PTM analyses can be done. Several proteomics studies have been conducted. For instance, *Saccharum* spp.–*G. diazotrophicus* interaction showed proteins involved in metabolism and signaling pathways. Similarly, *H. seropedicae* SmR1-rice involved protein in nitrogen fixation and assimilation. In maize–*A. brasilense* FP2 malate dehydrogenase interaction, H+-transporting F-type ATPase beta and major outer membrane protein OmaA precursor were found important in root adsorption and cell aggregation. In *Sinorhizobium meliloti* 1021–rice interaction, auxin-induced protein and IAA-amino acid hydrolase indicated the importance of phytohormones in plant–PGPR interaction. Proteomics studies identified the transcription factors differentiating nitrogen-fixing nodules, nitrogen metabolism proteins mainly NifHDK and regulatory PII proteins, and also stress and defense proteins (Alberton et al. 2020).

Other omics technologies include diverse areas. Microbiomics studies the microbial communities in ecosystems, interactions between native and invasive microbial communities, and their effects on ecosystem. Nutrigenomics is the combination of various fields like bioinformatics, genetics, molecular biology to deduce the interactions among plants, soil, and microbes concerning nutrient acquisition and metabolism (Qi et al. 2023).

12.14 Rhizobacteria Specific Growth Chambers

Rhizotrons and mini Rhizotrons is a building for the visualization and measurement of plant's underground parts through a transparent surface. These are subterranean labs or plane containers with transparent glass or plastic walls so that roots growing in soil can be easily visualized. They are equipped with sensors for estimating soil water activity, real time, temperature, oxygen, pH, and so forth. Minirhizotrons means transparent tubes fixed in soil along with fiber-optic borescopes or video cameras. However, this system does not provide three-dimensional (3D) images and is costly, too. 3D images can be obtained by quite costly technologies like magnetic resonance imaging (MRI), X-ray computed tomography (CT), or neutron tomography. Rhizobox are the containers of varying sizes and shapes for monitoring root growth in an experimental time course without disturbing plant. Petri dishes, half-cylinders, flat rectangular plots, and also plastic CD cases filled with semi-solid nutrient media, soil, sand, or other substrates, can be used to form a rhizobox. The rhizoboxes can be used to study root growth dynamics during abiotic or biotic stress, different nutritional conditions, and during interaction with microbes (Gamalero et al. 2022).

12.15 Microfluidic Technology

This technology has great advantages for studying plant–PGPR interactions: Transparent chips and observation machines are used for real-time imaging of plant–PGPR interactions. Huge number of rhizosphere samples can be processed, and accurate manipulation chemical environment in rhizosphere can be assessed to understand the interaction process. The microfluidic chips are made up of a polymer polydimethylsiloxane (PDMS) and combined with glass also (Shi et al. 2023). There are different types of microfluidic platforms used for imaging or growth and metabolism of root-microbiomes by phenotypic analysis such as Plantchip, RootChip, and RootArray. PlantChip is a vertical transparent microfluidic chip used for phenotypic analysis of Arabidopsis plants, changes occurring during growth, development and plant–microbe interactions, and so forth. RootChip is a microfluidic chip containing glucose fluorescence sensors used for monitoring sugar levels in root tips infused with glucose. It can also be used for live cell imaging, metabolomics study of exudates, and providing controlled environment. RootArray is a microfluidic device used for computing real-time imaging of gene expression. It has been used to study Arabidopsis seedlings for identifying position of roots, constructing 3D structure of roots, and real-time gene expression in each cell (Ke et al. 2021) (Table 12.1).

Table 12.1 Advantages and limitations of techniques used to study plant–microbe interactions

Cultivable techniques	Plate Count Method	• Only includes viable cells not dead cells • Easy quantitation • Good sensitivity with a specific media	• Chance of contamination • More time and resources required • Depends on phenotypic and biochemical assessment • Requires skill • Only 1% of microbes present in microbial community can be analyzed
	BIOLOG-Plate Method	• Quick and convenient	• Only for cultivables • Only for microbes growing in high-nutrient conditions
	Antibiotic Resistance Markers	• Good reproducibility with minimum error • Target specific organisms	• Increased chances of multiple antibiotic resistance in environment
	Tn5 Mutagenesis	• Capable to induce single hit mutations • Add selectable markers and recover genes after mutagenesis	• Very low frequency of transposition in organisms • Inaccuracy in most transposition systems
Non-cultivable techniques	IVET	• Give information of entire set of genes and physiological conditions required for survival in vivo for microbial growth	• Only activated genes can be analyzed • Downregulated or unexpressed genes cannot be identified
	Next-generation sequencing (NGS)	• Higher throughput • Cheap and highly sensitive • Flexible	• Cannot determine the activity of bacteria identified • Every species and their interaction with others is not very clear • Predictive data only
	Whole metagenome shotgun sequencing	• Highly sensitive • Provide comprehensive information	• Expensive • Gives false-positive results also • Genes showing expression are not shown
	Transcriptomic	• Identifies active genes and pathways	• Less RNA content in soil • mRNA is less stabile • Enzyme translation cannot be understood
	Metabolomics	• Identifies metabolites and metabolic pathways	• No standard method for metabolite preparation • Not easy to identify metabolite produced by which microbe • Purification of metabolite is not easy
	Proteomic	• Large-scale quantification and identification of proteins • Gives information about the gene function and metabolic pathways	• Less level proteins in the soil • Protein extraction and purification is not easy • Chances of contamination in soil • Lack of protein database

12.16 SRS (Split-Root Systems)

This system cultivates the plant roots growing into separate and isolated compartments. Initially, seed germination and development of primary root take place in a single pot. Then, primary root is cut longitudinally, and then, two separate root sections are grown in different pots, or primary root tip is removed which triggers root branching making lateral roots to grow in separate pots. Therefore, using SRS, two separate treatments can be given on the root while having the same upper portion of plant. It can be used to identify the systemic and local responses, indicating that the effect is only on the treated roots while the entire plant system is unaffected, thus suggesting the possibilities of local responses (Saiz-Fernández et al. 2021).

12.17 Clustered Regularly Interspaced Short Palindromic Repeats (CRISPR)

The CRISPR-Cas system (where Cas [nuclease enzyme]) utilizes CRISPR sequences and identifies the site for cleaving DNA strands complementary to the CRISPR sequence for edit. In general, CRISPR-Cas system of *Streptococcus pyogenes* was used as a genome engineering tool with two RNA molecules: the single guide RNA (sgRNA), which is hybrid of the two components known as crRNA and transactivating crRNA (tracrRNA). Together, these form a crRNA complex that guides the Cas9 endonuclease to the target site. After binding to target sequence, endonuclease enzyme does a double-stranded break in target DNA, and this activates DNA repair either through homologous recombination or by nonhomologous end joining. Here, the cellular repair systems can be utilized to replace, disrupt, or insert desired DNA sequence at the target site (Sander and Joung 2014). CRISPR/Cas system can be used to tag bacteria for understanding their interactions over time in soil and with plant host. This can involve creating PGPB mutants to understand the importance of their activities, to create mutants overexpressing protein encoded by PGPB gene, create variants of endogenous PGPB gene, and introduce other organism's gene in PGPB genome. In *B. subtilis* strain HS3, the CRISPR-cas9 system was used to knock out the sfp gene, which encodes the phosphopantetheinyl transferase enzyme essential for the synthesis of antifungal lipopeptides (fengycin and surfactin). This abolished the antifungal activity against fungal pathogens *Rhizoctonia solani* and *Fusarium colmorum*. Similarly, in the future, CRISPR/Cas system can be used to transfer various important traits from one PGPB to other (Gamalero et al. 2022).

References

Abd El-Daim IA, Bejai S, Meijer J (2019) *Bacillus velezensis* 5113 induced metabolic and molecular reprogramming during abiotic stress tolerance in wheat. Sci Rep 9:16282. https://doi.org/10.1038/s41598-019-52567-x

Adeniji AA, Babalola OO, Loots DT (2020) Metabolomic applications for understanding complex tripartite plant-microbes interactions: strategies and perspectives. Biotechnol Rep 25:e00425. https://doi.org/10.1016/j.btre.2020.e00425

Agtuca BJ, Stopka SA, Tuleski TR, Do Amaral FP, Evans S, Liu Y, Xu D, Monteiro RA, Koppenaal DW, Paša-Tolić L, Anderton CR, Vertes A, Stacey G (2020) In-situ metabolomic analysis of *Setaria viridis* roots colonized by beneficial endophytic bacteria. MPMI 33:272–283. https://doi.org/10.1094/MPMI-06-19-0174-R

Alberton D, Valdameri G, Moure VR, Monteiro RA, Pedrosa FDO, Müller-Santos M, De Souza EM (2020) What did we learn from plant growth-promoting Rhizobacteria (PGPR)-grass associations studies through proteomic and metabolomic approaches? Front Sustain Food Syst 4:607343. https://doi.org/10.3389/fsufs.2020.607343

Backer R, Rokem JS, Ilangumaran G, Lamont J, Praslickova D, Ricci E, Subramanian S, Smith DL (2018) Plant growth-promoting Rhizobacteria: context, mechanisms of action, and roadmap to commercialization of biostimulants for sustainable agriculture. Front Plant Sci 9:1473. https://doi.org/10.3389/fpls.2018.01473

Chamam A, Sanguin H, Bellvert F, Meiffren G, Comte G, Wisniewski-Dyé F, Bertrand C, Prigent-Combaret C (2013) Plant secondary metabolite profiling evidences strain-dependent effect in the *Azospirillum–Oryza sativa* association. Phytochemistry 87:65–77. https://doi.org/10.1016/j.phytochem.2012.11.009

Cheng Z, McConkey BJ, Glick BR (2010) Proteomic studies of plant–bacterial interactions. Soil Biol Biochem 42:1673–1684. https://doi.org/10.1016/j.soilbio.2010.05.033

Chukwuneme C, Ayangbenro A, Babalola O (2021) Metagenomic analyses of plant growth-promoting and carbon-cycling genes in maize rhizosphere soils with distinct land-use and management histories. Genes 12:1431. https://doi.org/10.3390/genes12091431

Couillerot O, Ramírez-Trujillo A, Walker V, Von Felten A, Jansa J, Maurhofer M, Défago G, Prigent-Combaret C, Comte G, Caballero-Mellado J, Moënne-Loccoz Y (2013) Comparison of prominent *Azospirillum* strains in *Azospirillum–pseudomonas–glomus* consortia for promotion of maize growth. Appl Microbiol Biotechnol 97:4639–4649. https://doi.org/10.1007/s00253-012-4249-z

Gamalero E, Bona E, Glick BR (2022) Current techniques to study beneficial plant-microbe interactions. Microorganisms 10:1380. https://doi.org/10.3390/microorganisms10071380

Gilroy S (1997) Fluorescence microscopy of living plant cells. Annu Rev Plant Biol 48:165–190. https://doi.org/10.1146/annurev.arplant.48.1.165

Godino A, Príncipe A, Fischer S (2016) A ptsP deficiency in PGPR *Pseudomonas fluorescens* SF39a affects bacteriocin production and bacterial fitness in the wheat rhizosphere. Res Microbiol 167:178–189. https://doi.org/10.1016/j.resmic.2015.12.003

Gupta G, Panwar J, Akhtar MS, Jha PN (2012) Endophytic nitrogen-fixing bacteria as biofertilizer. In: Lichtfouse E (ed) Sustainable agriculture reviews. Springer, Dordrecht, pp 183–221. https://doi.org/10.1007/978-94-007-5449-2_8

Gupta G, Panwar J, Jha PN (2013) Natural occurrence of *Pseudomonas aeruginosa,* a dominant cultivable diazotrophic endophytic bacterium colonizing *Pennisetum glaucum* (L.) R. Br. Appl Soil Ecol 64:252–261. https://doi.org/10.1016/j.apsoil.2012.12.016

Gupta S, Schillaci M, Roessner U (2022) Metabolomics as an emerging tool to study plant–microbe interactions. Emerg Top Life Sci 6:175–183. https://doi.org/10.1042/ETLS20210262

Hakim S, Naqqash T, Nawaz MS, Laraib I, Siddique MJ, Zia R, Mirza MS, Imran A (2021) Rhizosphere engineering with plant growth-promoting microorganisms for agriculture and ecological sustainability. Front Sustain Food Syst 5:617157. https://doi.org/10.3389/fsufs.2021.617157

Heath MC (2000) Advances in imaging the cell biology of plant-microbe interactions. Annu Rev Phytopathol 38:443–459. https://doi.org/10.1146/annurev.phyto.38.1.443

Imchen M, Kumavath R, Vaz ABM, Góes-Neto A, Barh D, Ghosh P, Kozyrovska N, Podolich O, Azevedo V (2019) 16S rRNA gene amplicon based metagenomic signatures of rhizobiome community in rice field during various growth stages. Front Microbiol 10:2103. https://doi.org/10.3389/fmicb.2019.02103

Jayaraman D, Forshey KL, Grimsrud PA, Ané JM (2012) Leveraging proteomics to understand plant–microbe interactions. Front Plant Sci 3:44. https://doi.org/10.3389/fpls.2012.00044

Josephine R, Thomas J (2019) 16S microbial phylogeny of multifunctional plant-growth-promoting rhizobacteria from the rhizosphere of maize (*Zea mays* L.) for agricultural soil fortification. Biotechnologia 100:143–154. https://doi.org/10.5114/bta.2019.85324

Joshi N, Ruparelia JA, Saraf M, Jha CK (2023) Techniques to study plant–microbe interactions that lead to efficient sustainable agriculture. In: Chhabra S, Prasad R, Maddela NR, Tuteja N (eds) Plant microbiome for plant productivity and sustainable agriculture, microorganisms for sustainability. Springer Nature Singapore, Singapore, pp 401–421. https://doi.org/10.1007/978-981-19-5029-2_17

Ke J, Wang B, Yoshikuni Y (2021) Microbiome engineering: synthetic biology of plant-associated microbiomes in sustainable agriculture. Trends Biotechnol 39:244–261. https://doi.org/10.1016/j.tibtech.2020.07.008

Khanijou JK, Kulyk H, Bergès C, Khoo LW, Ng P, Yeo HC, Helmy M, Bellvert F, Chew W, Selvarajoo K (2022) Metabolomics and modelling approaches for systems metabolic engineering. Metab Eng Commun 15:e00209. https://doi.org/10.1016/j.mec.2022.e00209

Kiely PD, Haynes JM, Higgins CH, Franks A, Mark GL, Morrissey JP, O'Gara F (2006) Exploiting new systems-based strategies to elucidate plant-bacterial interactions in the rhizosphere. Microb Ecol 51:257–266. https://doi.org/10.1007/s00248-006-9019-y

Knief C (2014) Analysis of plant microbe interactions in the era of next generation sequencing technologies. Front Plant Sci 5:216. https://doi.org/10.3389/fpls.2014.00216

Kumari A, Sumer S, Jalan B, Nongbri PL, Laskar MA (2017) Impact of next-generation sequencing technology in plant–microbe interaction study. In: Kalia VC, Kumar P (eds) Microbial applications, vol 1. Springer International, Cham, pp 269–294. https://doi.org/10.1007/978-3-319-52666-9_13

Liu Y, Gao J, Bai Z, Wu S, Li X, Wang N, Du X, Fan H, Zhuang G, Bohu T, Zhuang X (2021) Unraveling mechanisms and impact of microbial recruitment on oilseed rape (*Brassica napus* L.) and the rhizosphere mediated by plant growth-promoting Rhizobacteria. Microorganisms 9:161. https://doi.org/10.3390/microorganisms9010161

Mahdi I, Fahsi N, Hijri M, Sobeh M (2022) Antibiotic resistance in plant growth promoting bacteria: a comprehensive review and future perspectives to mitigate potential gene invasion risks. Front Microbiol 13:999988. https://doi.org/10.3389/fmicb.2022.999988

Márquez DGP, Munaro D, Uarrota VG (2022) Metabolomics as a tool to study volatile organic compounds produced by plant growth-promoting Rhizobacteria. In: Sayyed RZ, Uarrota VG (eds) Secondary metabolites and volatiles of PGPR in plant-growth promotion. Springer International, Cham, pp 21–44. https://doi.org/10.1007/978-3-031-07559-9_2

Mellidou I, Ainalidou A, Papadopoulou A, Leontidou K, Genitsaris S, Karagiannis E, Van De Poel B, Karamanoli K (2021) Comparative transcriptomics and metabolomics reveal an intricate priming mechanism involved in PGPR-mediated salt tolerance in tomato. Front Plant Sci 12:713984. https://doi.org/10.3389/fpls.2021.713984

Mishra RPN, Singh RK, Jaiswal HK, Kumar V, Maurya S (2006) Rhizobium-mediated induction of Phenolics and plant growth promotion in rice (*Oryza sativa* L.). Curr Microbiol 52:383–389. https://doi.org/10.1007/s00284-005-0296-3

Mukherjee A (2022) What do we know from the transcriptomic studies investigating the interactions between plants and plant growth-promoting bacteria? Front Plant Sci 13:997308. https://doi.org/10.3389/fpls.2022.997308

Nisrina L, Effendi Y, Pancoro A (2021) Revealing the role of plant growth promoting Rhizobacteria in suppressive soils against *Fusarium oxysporum* f.sp. cubense based on metagenomic analysis. Heliyon 7:e07636. https://doi.org/10.1016/j.heliyon.2021.e07636

Preston-Mafham J, Boddy L, Randerson PF (2002) Analysis of microbial community functional diversity using sole-carbon-source utilisation profiles—a critique. FEMS Microbiol Ecol 42:1–14. https://doi.org/10.1111/j.1574-6941.2002.tb00990.x

Qi S, Wang J, Zhang Y, Naz M, Afzal MR, Du D, Dai Z (2023) Omics approaches in invasion biology: understanding mechanisms and impacts on ecological health. Plan Theory 12:1860. https://doi.org/10.3390/plants12091860

Reid TE, Kavamura VN, Abadie M, Torres-Ballesteros A, Pawlett M, Clark IM, Harris J, Mauchline TH (2021) Inorganic chemical fertilizer application to wheat reduces the abundance of putative plant growth-promoting Rhizobacteria. Front Microbiol 12:642587. https://doi.org/10.3389/fmicb.2021.642587

Rincon-Florez V, Carvalhais L, Schenk P (2013) Culture-independent molecular tools for soil and rhizosphere microbiology. Diversity 5:581–612. https://doi.org/10.3390/d5030581

Saiz-Fernández I, Černý M, Skalák J, Brzobohatý B (2021) Split-root systems: detailed methodology, alternative applications, and implications at leaf proteome level. Plant Methods 17:7. https://doi.org/10.1186/s13007-020-00706-1

Sander JD, Joung JK (2014) CRISPR-Cas systems for editing, regulating and targeting genomes. Nat Biotechnol 32:347–355. https://doi.org/10.1038/nbt.2842

Shi X, Zhao Y, Xu M, Ma L, Adams JM, Shi Y (2023) Insight into plant-microbe interactions in rhizosphere for sustainable agriculture in the new crops era. New Crops 1:100004. https://doi.org/10.1016/j.ncrops.2023.11.002

Singh BK, Millard P, Whiteley AS, Murrell JC (2004) Unravelling rhizosphere–microbial interactions: opportunities and limitations. Trends Microbiol 12:386–393. https://doi.org/10.1016/j.tim.2004.06.008

Wang B, Wang X, Wang Z, Zhu K, Wu W (2023) Comparative metagenomic analysis reveals rhizosphere microbial community composition and functions help protect grapevines against salt stress. Front Microbiol 14:1102547. https://doi.org/10.3389/fmicb.2023.1102547

Zhang Y, Gao X, Shen Z, Zhu C, Jiao Z, Li R, Shen Q (2019) Pre-colonization of PGPR triggers rhizosphere microbiota succession associated with crop yield enhancement. Plant Soil 439:553–567. https://doi.org/10.1007/s11104-019-04055-4

Chapter 13
Sodic-Tolerant Plant Growth-Promoting Rhizobacteria Mediated Sodic Stress Alleviation in Plants

Gunasekaran Yazhini, Thiyageshwari Subramanium, Ariyan Manikandan, Duraisamy Selvi, and Rangasamy Anandham

Abstract Sodic stress is a salt stress affecting crop production and productivity. Though rice is considered a sodic soil reclamation crop, the growth and yield are impaired. The prevailing climate change has increased the salt-affected area worldwide. The occurrence of higher pH-specific ion toxicity and nutrient imbalance in sodic soil adversely affects plant growth. The utilization of plant growth-promoting bacteria could enhance plant growth directly and indirectly and is considered one of the proven biological methods to alleviate the sodic stress in crops. Plant Growth-Promoting Rhizobacteria (PGPR) aids plants to withstand sodic stress through the production of organic acid, Indole Acetic Acid (IAA), siderophore, osmolyte accumulation, biofilm formation, and Exopolysaccharide (EPS) production. The sodium ions can bind to biofilm which reduces the sodium uptake by plants. This chapter highlighted the holistic nature of sodic soil impacts on soil properties, plant growth, and the mechanisms of sodic-tolerant plant growth-promoting bacteria that help plants survive under sodic conditions. Understanding the sodic environment and mechanisms of sodic tolerant plant growth-promoting bacteria leads to achieving optimum crop production and yield under sodic stress conditions.

Keywords Sodic soil · Plant defense mechanism · Sodic-tolerant bacteria · Plant growth-promoting traits

G. Yazhini · T. Subramanium (✉) · D. Selvi
Department of Soil Science and Agricultural Chemistry, Tamil Nadu Agricultural University, Coimbatore, Tamil Nadu, India

A. Manikandan
The Institute of Ecology and Earth Sciences, University of Tartu, Tartu, Estonia

Department of Agricultural Microbiology, Tamil Nadu Agricultural University, Coimbatore, Tamil Nadu, India

R. Anandham
Department of Agricultural Microbiology, Tamil Nadu Agricultural University, Coimbatore, Tamil Nadu, India

P. Singh Chauhan et al. (eds.), *Plant-Microbe Interaction and Stress Management*, Rhizosphere Biology,
https://doi.org/10.1007/978-981-97-4239-4_13

13.1 Introduction

The presence of sodic soil poses a major threat to global food security. In the realm of salt-accumulated land, the regions of South and Southeast Asia collectively accounted for 48 million hectares, constituting a significant portion of the global total of 190 million hectares. Sodicity affects an estimated land area of around 3.77 million hectares in India. These types of soils are predominantly located in the peninsular, plain, desert, and semiarid regions of India. According to Ghosh et al. (2012), the Indo-Gangetic Plain region is home to around 47% of the sodic soil. The primary cause for the formation of salt in the land is attributed to pedological, geological, and hydrological processes. The utilization of low-quality irrigation water in agricultural fields over a prolonged period constitutes an additional contributing factor to the development of sodic soil in irrigated land regions. Nevertheless, the exacerbation of climate change in recent decades has led to the expansion of salt-affected areas globally, surpassing the boundaries of existing salt-affected regions. Salinization predominantly impacts irrigated land. According to Rengasamy (2010), the expansion of salt-affected areas has been exacerbated by climate change, resulting in adverse effects on agricultural development and food production. The increase in global temperatures resulting from global warming has led to elevated levels of evaporation and reduced or inconsistent precipitation. As a result, the leaching of soluble salts has been diminished, leading to the accumulation of salts on the soil surface. This accumulation of salts in the root zone is ensured, as noted by Kang et al. (2009). Under conditions of elevated pH, the nutrients are present in an inaccessible form. The phenomenon induces adverse effects such as ion toxicity, nutritional shortage, and osmotic stress, hence impacting the processes of aeration, water, and nutrient transportation in plants. The disruption of nutrient availability induces immediate physiological stress in plants by perturbing the ion homeostasis of plant cells, resulting in significant constraints on plant development and agricultural output (Amini et al. 2016; Qadir et al. 2007).

Various management approaches, such as cultural practices, contribute to the recovery of sodic soil by improving infiltration and reducing soluble salts. Various chemical substances, such as gypsum, lime, calcium chloride, sulfuric acid, and sulfur, can be utilized for the reclamation of sodic soil. Plant growth-promoting rhizobacteria (PGPR) are commonly employed to improve the systemic tolerance of plants to plant stress (Yang et al. 2009) and enhance plant development in sodic soil conditions. PGPR bacteria are a group of microorganisms that can colonize the rhizosphere or the internal tissues of plant roots. Based on previous research findings, various bacterial genera, including *Microbacterium*, *Pantoea*, *Achromobacter*, *Rhizobium*, *Pseudomonas*, *Bacillus*, *Paenibacillus*, *Enterobacter*, *Burkholderia*, *Methylobacterium*, *Azospirillum*, and *Variovorax*, have been identified as contributors to the enhancement of host-plant tolerance under abiotic stress conditions (Akram et al. 2016; Shahid et al. 2018a). The utilization of these microbes in agricultural settings has proven to be beneficial in mitigating various abiotic stresses, as evidenced by studies conducted by Banaei-Asl et al. (2015).

PGPR play a crucial role in enhancing plant growth and development by facilitating many physiological processes such as nutrient uptake, nitrogen fixation, mineralization of organic compounds, solubilization of mineral nutrients, production of siderophores, and synthesis of phytohormones (Shahid et al. 2018a, b). Furthermore, the synthesis of exopolysaccharides resulted in a decrease in the absorption of sodium ions due to the binding of these ions to the exopolysaccharide matrix (Tewari and Arora 2014; Timmusk et al. 2014). This paper elucidates the fundamental characteristics of sodic soil and its impact on plant growth, as well as the techniques employed by sodic tolerant bacteria to mitigate sodic stress in plants.

13.2 Sodic Soil

A soil is defined as sodic when it possesses the following characteristics: electrical conductivity (EC) of less than 4 dS m^{-1}, pH greater than 8.5, exchangeable sodium percentage (ESP) above 15, and sodium absorption ratio (SAR) surpassing 13. The predominant salts found in sodic soil are sodium carbonate and sodium bicarbonate salts. The presence of an excessive amount of sodium ions (Na^+) has notable implications for the physical properties of soil, which in turn affect several chemical and biological components of the soil. The inclusion of sodium carbonate and sodium bicarbonate salts in the soil results in the dispersion of soil particles and the formation of a less cohesive soil structure. Consequently, the dispersion of soil particles in sodic soil leads to the formation of an inadequate aggregate structure, resulting in diminished aeration, hydraulic conductivity, soil permeability, and infiltration rate. This, in turn, has an impact on various soil properties such as pH, redox potential, and concentration of dissolved organic carbon, ultimately affecting microbial activity (Rietz and Haynes 2003) as well as causing deficiencies in micronutrients within the soil. Furthermore, the study conducted by McNeal and Coleman (1966) revealed that an increase in exchangeable sodium percentage (ESP) resulted in a decrease in hydraulic conductivity and permeability. One further significant issue seen in sodic soil is to the elevated soil pH levels. The availability of plant nutrients, including calcium, magnesium, and certain micronutrients, is diminished under elevated pH conditions. This is due to the reaction of these nutrients with carbonate and bicarbonate ions found in sodic soil, resulting in the formation of insoluble compounds that are not readily accessible to plants (Bernstein 1975).

13.3 Effects of Sodicity on Plant Growth

Salt stress often encompasses osmotic stress and ion damage, as described by Munns (2002) and Munns and Tester (2008). On the other hand, sodic stress induces similar stress components, but in addition to that, it also imposes high-pH stress. The presence of high pH levels and the particular ion action of sodium (Na^+) in sodic environments have a significant influence on the accessibility of both macro- and micronutrients. This, in turn, leads to physiological and biological alterations that ultimately hinder the growth of plants (Bayuelo-Jiménez et al. 2003). According to Alam (1999), the presence of sodium in plant leaves at a concentration range of 0.25–0.5% on a dry weight basis has been seen to result in necrosis and leaf edge burning, which are indicative of toxicity symptoms. In addition to the immediate impacts of sodicity on nutritional imbalances, the compromised aggregate structure diminishes porosity, impairs water movement, and restricts aeration in the proximity of plant roots (Machado and Serralheiro 2017). Sodium ions induce targeted ion toxicity in plants. The accumulation of Na^+ is more likely to occur in leaves than to roots. The transpiration process within the xylem facilitates the transportation of Na^+ ions from the root to the shoot. Concurrently, the roots sustain the Na^+ content over a prolonged duration. The accumulation of Na^+ in leaf tissues leads to necrosis mostly at the tips and margins of older leaves, eventually affecting entire leaves more than younger leaves. The extent of Na^+ toxicity is determined by the level of Na^+ accumulation and the efficient compartmentalization of Na^+ inside the leaf. The accumulation of sodium ions (Na^+) disrupts the distribution of potassium ions (K^+) and sodium ions (Na^+) within cells, leading to a disruption of ion homeostasis and the overall balance of ions. Under normal physiological conditions, plants maintain a high K^+/Na^+ ratio in their cytoplasm by exhibiting elevated levels of K^+ (100–200 mM) and reduced quantities of Na^+ (1–10 mM) (Blumwald 2000). The elevated concentration of Na^+ ions in the extracellular region resulted in a notable alteration in the electrochemical potential at the plasma membrane (−140 mV), which promoted the passive absorption of Na^+ from the surrounding environment into the cytosol. Furthermore, the transport protein exhibited challenges in discriminating between Na^+ and K^+ ions due to their comparable hydrated ionic radii. The aforementioned factors constituted the primary causes of sodium (Na^+) toxicity in plants. According to Bhandal and Malik (1988), the activation of over 50 enzymes is specifically facilitated by the presence of K^+ ions, which cannot be replaced by Na^+ ions. According to Wyn Jones and Pollard (1983), protein synthesis is hindered when greater concentrations of Na^+ (>100 mM) are present due to competition with K^+ for binding sites. The optimal K^+ content for protein synthesis was shown to be within the range of 100–150 mM. According to Zhang et al. (2013), the growth and development of plants can be influenced by nutritional imbalance and ionic stress, which can lead to changes in plant physiological and biochemical systems.

13.4 Plant Effect Defense Mechanism Against Sodicity

The rice seedling stage is adversely impacted by high pH conditions, leading to injury and death of root cells. Consequently, rice seedlings experience withering of the entire plant and eventual mortality (Wei et al. 2015). According to Munns (2002), in conditions of saline stress, there is a competitive interaction between Na^+ and K^+ for uptake into the roots. In a study conducted by Yang and Guo (2018), it was observed that the application of sodic stress by the use of 120 mM of Na^+ did not result in any significant changes in the absorption of K^+ in *Chloris virgata*. Furthermore, the response of *C. virgata* to sodic stress was shown to be similar to its response to saline stress. Nevertheless, when the concentration of Na^+ exceeded 120 mM, it became evident that the combined influence of elevated Na^+ levels and pH had a more pronounced impact on plant stress compared to alone experiencing salt stress. In a separate investigation, the stress experienced was found to induce significant harm to the root cells by the amplification of membrane injury; accumulation of malondialdehyde (MDA); upregulation of the cell death-associated genes OsKOD1, OsHsr203j, OsCP1, and OsNAC4; and the downregulation of the cell death-suppressor gene, OsBI1. Crops employ many abiotic stress-related survival mechanisms to acclimate to their environment, including ion transport and compartmentalization, synthesis and accumulation of osmotic solutes, and synthesis and accumulation of compatible solutes (Munns and Termaat 1986). A variety of suitable solutes encompass N-containing molecules, including amino acids and amides. According to Lauchli and Luttge (2002), nitrogen metabolism is of significant importance in situations characterized by stress. In their study, Wang et al. (2012b) observed a decrease in potassium (K^+) uptake and ammonium ($NH4^+$) assimilation in rice plants due to the accumulation of sodium (Na^+) and chloride (Cl^-) ions in both the leaves and roots. However, they found that this ion accumulation led to an increase in the expression of compatible solutes, specifically OsNADH-GOGAT (glutamate synthase), OAS (asparagine synthetase), OsGS1 (glutamine synthetase), and OsGDH (glutamate dehydrogenase). In a separate investigation, it was shown that the presence of excessive Na^+ ions resulted in a decrease in the expression of OsNiR in mature leaves while promoting its expression in juvenile leaves. The diminished assimilation of NH_4^+ in mature leaves has a consequential impact on the OAS glutamine synthase pathway, resulting in a reduction in the synthesis of osmolytes (Wang et al. 2012a). Nevertheless, the overall concentration of amino acids was increased in response to stressful conditions through the upregulation of proline synthesis within cellular systems (Thu Hoai et al. 2003).

The high levels of salt induce a state of stress in plants, leading to the generation of increased amounts of reactive oxygen species (ROS). These ROS molecules have detrimental effects on the cellular and metabolic processes of plants (Sahin et al. 2018). In a study conducted by Zhang et al. (2017a), it was shown that the application of 15 mM sodium carbonate resulted in a significant reduction in seedling viability, shoot development, root vitality, and antioxidant enzyme activity, specifically superoxide dismutase (SOD), peroxidase (PO), catalase (CAT), and ascorbate peroxidase (APX). Plants have developed many defense mechanisms to address these problems. One such system involves the maintenance of ion homeostasis through the appropriate compartmentalization of sodium ions (Na^+) within the vacuole. According to Rahneshan et al. (2018), the maintenance of membrane potential, osmotic and turgor pressures, enzyme activation, and tropisms in plants are influenced by the increased K^+/Na^+ ratio. According to Rahneshan et al. (2018), osmolytes such as proline and soluble carbohydrates play a protective role in plant cells under conditions of salt stress, by facilitating the activation of antioxidants by enzymatic and nonenzymatic processes. The enzymes SOD, PO, CAT, and APX have been identified in previous studies (Rahneshan et al. 2018; Sarker and Oba 2020; Shafeiee and Ehsanzadeh 2019; Shaheen et al. 2013). Likewise, in response to the stress induced by sodium carbonate, rice plants activate their antioxidant defense mechanisms as a means of mitigating cellular harm. Nevertheless, the extended generation of ROS during periods of stress has the potential to surpass the protective capabilities of the antioxidant defense system, resulting in detrimental effects on rice seedlings. In addition, it has been observed that the tolerant variety exhibits a lower accumulation of antioxidant defense enzymes compared to the sensitive cultivars, as reported by Zhang et al. (2017b).

13.5 Role of Exopolysaccharide (EPS)-Producing Sodic-Tolerant PGPR on Plant Growth Promotion and Soil Health Maintenance

The name "Plant Growth-Promoting Rhizobacteria" encompasses a diverse assemblage of bacterial species that engage in direct or indirect interactions with plants, resulting in beneficial effects on plant nutrition and growth. The beneficial interaction between PGPR and plants is facilitated through multiple mechanisms, which encompass the release of phytohormones like auxin (Asghar et al. 2002), abscisic acid (Belimov et al. 2014), and the utilization of 1-aminocyclopropane-1-carboxylate (ACC) deaminase to mitigate the levels of endogenous ethylene (Brunetti et al. 2021). The initial examination of these organisms focused on their potential to enhance disease resistance (Farag et al. 2013). Subsequent research has revealed their notable capacity to promote plant development through the augmentation of root quantity and leaf surface area (Ruzzi and Aroca 2015) (Fig. 13.1).

13.5.1 The Direct Mechanism of EPS-STB-PGPR in Plant Growth Promotion

Based on research investigations, certain microorganisms are shown to have a significant impact on the process of soil mineral dissolution, hence influencing the cycling of ions and the overall fertility of the soil (Calvaruso et al. 2006; Ehrlich

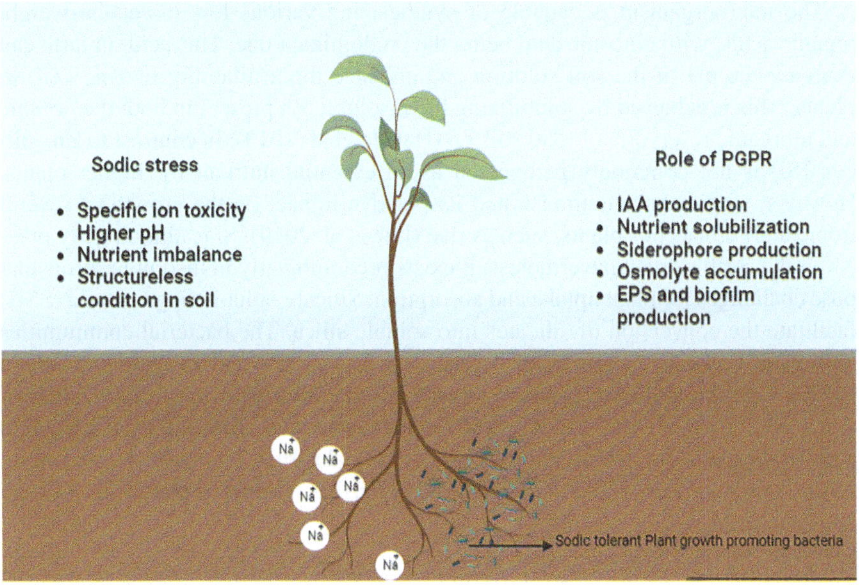

Fig. 13.1 Role of PGPR in sodic stress management in plants

1996; Uroz et al. 2007). Phosphorus is an essential element for several biochemical activities in plants, encompassing energy generation, storage, and transmission (Khan et al. 2009). Nevertheless, it is worth noting that a mere 0.1% of the overall phosphorus content in soil is in a soluble form and may be readily accessed by plants. This is mostly attributed to the extensive fixation of phosphorus in soil and its restricted solubility, as highlighted by Pereira and Castro (2014). The promotion of plant development by plant-associated microbes involves the solubilization of inorganic phosphate. Phosphate solubilizing microorganisms possess the ability to convert inorganic phosphorus into a soluble form through the production of organic acids, extracellular enzymes, chelation, exchange reactions, and the transformation of organic soil phosphorus into the bioavailable forms of $H_2PO_4^-$ and HPO_4^{2-}, which can be taken up by plant roots (Shulse et al. 2019). In a study conducted by Sahay and Patra (2013), it was observed that *Pseudomonas* and *Xanthomonas* spp. obtained from sodic soil were able to effectively decrease the pH of the growth medium from 8 to 5 after 6 days of incubation. This reduction in pH resulted in a significant solubilization of phosphate, with solubilization indices of 2.9 and 3.1 for *Pseudomonas* and *Xanthomonas* spp., respectively. In the present study, a group of microorganisms including *Pseudomonas* sp. strain C-1, *Agromyces tropicus* strain C-5, *Lysinibacillus fusiformis* strain A-11, *Bacillus flexus* C-7, *Lysinibacillus* sp. A-16, *Bacillus tequilensis* C-9, *Bacillus licheniformis* A-18, *Pseudomonas mendocina* C-10, and *B. licheniformis* strain M-16 were isolated from a pH 9.3 environment. These microorganisms were found to possess the ability to solubilize phosphate, as observed in the in vitro study.

The microorganism is capable of synthesizing various low molecular-weight organic acids, with gluconic acid being the predominant one. This acid, in turn, can decrease the pH of the soil solution and enhance the availability of zinc (Zn) to plants. This is achieved by solubilizing the insoluble Zn present in both the organic and inorganic reservoirs of total soil Zn (Hafeez et al. 2013). In contrast to Zn, silicon (Si) is not commonly recognized as an essential nutrient for higher plants. However, studies have demonstrated its positive impact on the growth of several tropical graminaceous plants, such as rice (Lee et al. 2019). Si is ubiquitously present in the earth's crust; nevertheless, it exists predominantly in insoluble forms that pose challenges for plant uptake and absorption. Silicate-solubilizing bacteria (SSB) facilitate the conversion of silicates into soluble silica. The bacterial communities mentioned in the study conducted by Naureen et al. (2015) and Vasanthi and Selvan (2020) have the ability to produce an enzyme that facilitates the solubilization of silicates. This enzymatic activity results in the conversion of insoluble silicates into a more soluble form, which may then be readily absorbed by plants. Furthermore, this process has been found to increase the defensive systems of plants.

13.6 Indole Acetic Acid Production

In addition to nutrition supplementation, PGPR synthesize phytohormones, including auxins, cytokinins, gibberellins, and ethylene. These phytohormones modulate cell proliferation within the root structure by inducing excessive formation of lateral roots and root hairs, thereby enhancing the plant's capacity to absorb nutrients and water. Furthermore, it has been demonstrated that this particular stimulus effectively enhances the shoot system, resulting in improved photosynthetic efficiency and overall plant growth (Ansari et al. 2019). According to Damam et al. (2016), phytohormones created by microbes exhibit a sustained and gradual release, rendering them more efficacious compared to chemically synthesized hormones. According to Miransari and Smith (2014), indole acetic acid (IAA) is the predominant naturally occurring auxin, and it plays a significant role in promoting root growth. Root exudates encompass the presence of the amino acid tryptophan, which holds significant value as a precursor for the formation of indole-3-acetic acid (IAA) in bacteria. According to Vessey (2003), around 80% of rhizobacteria possess the ability to synthesize IAA. The presence of IAA, in conjunction with the plant's endogenous IAA, facilitates cell proliferation and enhances nutrient absorption. Upadhyay et al. (2011) identified 11 bacteria, including *Bacillus* sp., *Penibacillus*, *Enterobacter*, and *Burkholderia*, which were isolated from the wheat rhizosphere cultivated in sodic soil. These bacteria were found to possess the ability to create IAA, a plant hormone that can elicit a physiological response in the host plant.

13.7 Siderophore Production

Iron is present in the soil mostly as Fe^{3+} ions, which exhibit low-solubility characteristics. Siderophores are stable complexes with low molecular weight that are synthesized in environments with limited iron availability. They have been categorized into three distinct types: hydroxamate, phenol catecholate, and carboxylate (Kundan et al. 2015). According to Schalk et al. (2011), siderophores exhibit a strong attraction for iron and Zn, hence increasing their bioavailability within the biological cell. The bacteria *Bacillus aryabhattai* NBRI YE4, *Bacillus cereus* NBRI YE1.1, *Bacillus flexus* NBRI NE3.2, *Bacillus paralicheniformis* NBRI YN4.5, *B. aryabhattai* NBRI YN4.4, B. *aryabhattai* NBRI YN2.5, *Bacillus subtilis* NBRI NN3.2, *Bacillus altitudinis* NBRI YB4.5, and *B. subtilis* NBRI NB3.2 were found to produce siderophores and assist in mitigating high pH stress in maize, as reported by Dixit et al. (2020).

13.8 Osmolyte Accumulation

Compatible solutes, such as proline, sucrose, trehalose, and quaternary ammonium compounds (QACs), are organic molecules with low molecular weight, high solubility, and nontoxic properties. These solutes serve to safeguard plants against stress by modulating cellular osmotic balance, detoxifying ROS, preserving membrane integrity, and stabilizing enzymes and proteins. When exposed to osmotic stress, various types of bacteria exhibit a broad spectrum of responses, including the accumulation of proline at high intracellular concentrations. Bano and Fatima (2009) observed that the simultaneous introduction of *Rhizobium* and *Pseudomonas* into *Zea mays* plants led to enhanced salt tolerance. This was attributed to an elevation in proline synthesis, a reduction in electrolyte leakage, the preservation of leaf relative water content, and the selective absorption of potassium ions. According to Gagné-Bourque et al. (2016), the introduction of *B. subtilis* B26 resulted in a decrease in proline accumulation in timothy (*Phleum pratense* L.) plants under stress conditions, as compared to plants that were not inoculated. Decreased proline levels are indicative of reduced cellular tissue damage. Glycine betaine functions as a crucial osmoregulatory compound that provides stability against the detrimental impacts of salinity, dehydration, low temperatures, and high temperatures (Carillo et al. 2008, 2011). The halophilic salt stress condition induces the accumulation of glycine betaine at concentrations exceeding 1.6 M in *Ectothiorhodospira halochloris*, together with significant quantities of trehalose and a newly discovered amino acid called "ectoine" (Imhoff 1986). Therefore, the utilization of sodic-tolerant PGPR has the potential to mitigate abiotic stress through the regulation of hormones, enzymes, photosynthesis, organic acid secretion, and secondary metabolites (Bisht et al. 2019; Dixit et al. 2020). The study reported on the proline production of *Enterobacter* sp. SKU8, which was isolated from sodic soil. It was observed that the proline production increased as the NaCl content increased. Nevertheless, the synthesis of overall soluble sugar and reducing sugar exhibited a decline when the content of NaCl, a salt, increased. *Bacillus inosolitus* SKU12, which was obtained from sodic soils, exhibited elevated levels of total soluble sugar and reduced sugar production.

13.9 Indirect Mechanism of Plant Growth-Promoting Bacteria in Plant Growth Promotion

13.9.1 EPS Production and Biofilm Formation

EPS are intricate macromolecules mostly consisting of polysaccharides, proteins, nucleic acids, and humic acids (Upadhyay et al. 2011). Extracellular polymeric substances (EPSs) serve as a crucial structural element within biofilms, fulfilling various functions in bacterial defense mechanisms. These functions include facilitating cell adhesion, promoting cell aggregation, facilitating biofilm formation, preventing desiccation, and providing nutrient sources (Mathur and Roy 2021; Saha et al. 2020). The intricate layer produced around microorganisms shields them from unfavorable conditions and facilitates the interaction between microorganisms and the adhesion of biofilms to surfaces (Flemming and Wingender 2001). The production of root exudates from the root attracts the bacteria that are connected with the root, leading to the formation of a biofilm through electrostatic interactions (Prigent-Combaret 2013). The production of biofilm surrounding the root is regulated by metabolic signaling quorum sensing (Drogue 2013). Furthermore, the polymers in question served to stabilize the soil aggregates through several interactions, encompassing dispersion forces, electrostatic interactions, and hydrogen bonds, as demonstrated by Wolfaardt et al. (1999). Consequently, the utilization of plant growth-promoting bacteria that produce EPS has the potential to enhance the volume of macropores and the aggregation of rhizosphere soil, therefore facilitating increased water and nutrient accessibility for the crop (Flemming et al. 2000). According to Dixit et al. (2020), *B. aryabhattai* NBRI YN4.4 and *Staphylococcus cohnii* NBRI YB1.6 exhibit increased production of EPS and alginate when exposed to alkaline environments. This heightened production serves as a protective barrier, hence augmenting their ability to withstand stress and increasing their survivability and tolerance capacity. The EPSs discovered in *Rhodopseudomonas* palustris, which has a molecular weight of 18 kDa, consists of galacturonic acid. This EPS can bind with Na^+ cations present in the surrounding aqueous environment, hence contributing to the salt tolerance of the organisms (Nunkaew et al. 2015). Qurashi and Sabri (2012) have revealed that various bacterial taxa, including *Pseudomonas*, *Bacillus*, *Burkholderia*, *Enterobacter*, *Microbacterium*, *Planococcus*, *Halomonas*, and *Azotobacter*, have been seen to create EPSs under conditions of salt stress. In a study conducted by Upadhyay et al. (2012), various bacterial species including *Bacillus* sp., *Paenibacillus* sp., *Microbacterium* sp., *Burkholderia cepacian*, and

Enterobacter sp. were identified from the rhizosphere of wheat plants grown in sodic soil. These bacteria were found to reduce sodium intake by establishing a biofilm in the vicinity of the plant roots. Gunasekaran et al. (2022) identified several bacterial genera, including Bacillus, Pseudomonas, Stenotrophomonas, *Burkholderia, Rhodococcus, Lysinibacillus, Cytobacillus, Ralstonia, Arthrobacter, Niallia, Achromobacter, Staphylococcus, Kocuria, Franconibacter, Planococcus,* and *Acinetobacter*, as proficient producers of EPSs in sodic soil. Subsequent examination has indicated that EPSs with elevated levels of glucose have the potential to form chelates with sodium ions found in saline soil, hence impeding their accessibility to plants. According to Shultana et al. (2020), the introduction of *B. tequilensis* and *B. aryabhattai* into rice plants increased crop output when subjected to saline stress conditions. This improvement can be attributed to the enhancement of photosynthesis, transpiration, and stomatal conductance. The process of biofilm formation and the adhesive properties of bacterial EPSs play a crucial role in facilitating the cohesion of soil particles and their attachment to plant roots, hence enhancing plant growth in saline environments (Khan and Bano 2019; Yadav et al. 2013). The adhesive properties of EPS are contingent upon its composition of sugars, proteins, and lipids (Deng et al. 2015). The isolates *Bacillus paralicheniformis* L1C5L, *Franconibacter helviticus* L2C1L2, *Pseudomonas* sp. L5C14T, *Arthrobacter* sp. L5C8L and *Ralstonia pickettii* L4C6L exhibited a notable capacity for biofilm formation under sodic stress conditions. This ability was primarily attributed to the increased aggregation of cells. Moreover, the adaptation of the cells to severe environments was supported by the scanning electron microscopy (SEM) image, which revealed structural modifications and enhanced aggregation of the cells under stressful conditions. Furthermore, the regulation of swarming, adherence, and aggregation in *B. paralicheniformis* is governed by the regulator genes YmcA, Ylbf, and SinR, as reported by Gunasekaran et al. (2022).

13.9.2 Plant Growth Promotion

The mitigation of salt stress in plants can be achieved through the synthesis of low-molecular-weight organic solutes, such as proline, as well as the activation of enzymes like SOD and PO. Nevertheless, the majority of plants are unable to produce an enough amount of antioxidants to fulfill their growth requirements when subjected to salt-induced stress. To enhance the defense system of plants, the process of bio-inoculation is employed to increase their tolerance to salt. This is achieved by augmenting the activity of antioxidant enzymes, including SOD, PO, and catalase, as well as organic solutes such as proline (Ashraf and Foolad 2007). The strains *B. licheniformis* B642 and *Pseudomonas fluorescens* FAP2 are classified as PGPR due to their ability to produce EPSs. These strains exhibit plant growth-promoting characteristics, including the synthesis of siderophores and IAA. According to Ansari et al. (2019), the introduction of these strains to plants resulted in notable improvements in various physiological parameters, including net photosynthetic rate, chlorophyll content, internal CO_2 concentration, stomatal conductance, leaf water potential, and transpiration rate, as compared to plants that were not inoculated. The bacterial isolates obtained from the rhizosphere of rice plants grown in sodic soil were analyzed to assess their potential for stimulating plant growth. It was shown that *Rhodococcus pyridinivorans* L3C9N and *Stenotrophomonas maltophilia* L3C7T exhibited the highest production of IAA under sodic stress conditions. The study conducted by Gunasekaran et al. (2022).

Inoculation with *B. aryabhattai* NBRI YN4.4 resulted in a significant rise of 47.03% in Chla, 40.91% in Chlb, and 568.67% in carotenoid content under alkaline conditions. Under alkaline stress conditions, maize plants that were infected with *B. aryabhattai* NBRI YN4.4, *B. aryabhattai* NBRI NB2.5, and *B. subtilis* NBRI YE1.3 exhibited proline accumulation of 59.12%, 32.00%, and 11.27%, respectively, in comparison to the uninoculated control group. Furthermore, plants inoculated with PGPR exhibited a significant elevation in the overall concentration of soluble sugars in comparison to control plants that were not inoculated and were cultivated under alkaline conditions. The introduction of *B. aryabhattai* NBRI YN4.4 into maize plants subjected to alkaline stress led to a significant elevation in the concentration of total soluble sugar by 43.59%. This was followed by the inoculation of *B. aryabhattai* NBRI NB2.5, which resulted in a 26.86% increase, and *B.* subtilis NBRI YE1.3, which caused a 25.27% rise in total soluble sugar levels. According to Dixit et al. (2020), the determination of the threshold level for alkaline stress screening in maize was based on observing a 50% drop in seed germination and slower growth at a pH of 9. The use of PGPR in stressful environments resulted in a significant improvement in seed germination rates, with an increase of 50–100% compared to the control group that did not receive PGPR inoculation. Additionally, the seedling biomass showed a notable enhancement of around 20.18–51.37% in comparison to the uninoculated control. In a study conducted by Damodaran et al. (2019), it was demonstrated that the utilization of various salt-tolerant bacteria, including *Pseudomonas* sp. *A. tropicus*, *L. fusiformis*, *B. flexus*, *Lysinibacillus* sp., *B. tequilensis*, *P. mendocina*, and *B. licheniformis*, which were isolated from saline/sodic environments, had a significant positive impact on the growth of wheat seedlings in salt-affected soil (specifically, saline/sodic stress). This enhancement in growth and vigor index was attributed to both direct and indirect mechanisms. The researchers also concluded that the increased root biomass (both fresh and dry weights) observed in plants treated with *Lysinibacillus* fusiformis can be attributed to the formation of IAA by the bio-inoculants. The aforementioned isolates also exhibited phosphate solubilization and a positive siderophore response. It is widely recognized that siderophores function as iron-chelating agents, thereby impeding the growth of pathogens through the establishment of competition for iron resources. Additionally, siderophores facilitate iron acquisition, so promoting plant growth in conditions of stress. According to Gunasekaran et al. (2022), the introduction of *Franconibacter helveticus* L2C1L2 during periods of stress resulted in enhanced development of rice in terms of root, shoot, and dry matter. Additionally, this inoculation exhibited a significant reduction in the absorption of sodium ions by the plant under sodic stress conditions. Table 13.1 presents a comprehensive compilation of

Table 13.1 Response of microbes in plant growth promotion under abiotic stress

Crop	Stress type	Microbe	Response of microbe to plant	Reference
Wheat	Alkaline stress	*Bacillus simplex*	Reduction of pH around rhizosphere, increasing the plant growth and root phosphorus concentration	Hansen et al. (2020)
Wheat	Alkaline stress	*Bacillus clausii*	Shoot and root growth and grain yield improvement	Torbaghan et al. (2017)
	Sodic stress	*Lysinibacillus* sp.	Enhanced seed germination and vegetation	Damodaran et al. (2019)
		Enterobacter sp.	Enhanced photosynthetic rate	Damodaran et al. (2019)
Maize	Sodic stress	*Bacillus* sp.	Increased photosynthetic pigment and sugar content and reduced proline content	Dixit et al. (2020)

the effects of several PGPR on the stimulation of plant development in the presence of abiotic stressors.

13.10 Soil Properties Improvement by PGPR

EPSs produced by microorganisms enhance both the quality and fertility of the soil (Costa et al. 2018). The EPS contains high-molecular-weight polymers, which facilitate the attachment of bacteria to the exterior of soil particles and stabilize the aggregated particles. The ability of the EPSs to bind to the soil surfaces is facilitated by hydrogen bonding, cation bonds, anion adsorption, and van der Waal's forces. This EPS improves soil aggregate stability, aeration, permeation, and root penetration (Martin et al. 1971). EPS helps in soil micro-aggregation formation by forming an organo-mineral sheath that enhances the structure and stability of the soil under stress conditions (Costa et al. 2018). The role of EPS in soil health maintenance is given in Table 13.2. Dixit et al. (2020) reported that the inoculation of *B. aryabhattai* NBRI NB2.5, *B. subtilis* NBRI YE1.3, and *B. aryabhattai* NBRI YN4.4 enhanced the enzymatic activities viz., soil dehydrogenase, alkaline phosphatase, and betaglucosidase in alkaline condition when compared with uninoculated control.

13.11 Conclusion and Future Research Perspectives

In the context of sodic environments, the soil microbiota assumes a vital function in mitigating the adverse effects of sodic stress on plants. This is achieved through the activation of mechanisms that enhance the plants' ability to tolerate salt and decrease the absorption of free sodium ions that exist in the surrounding surroundings. The sodic-tolerant PGPR have a role in supporting plant physiology by producing IAA,

Table 13.2 Role of EPS in soil health maintenance

Microorganism	EPS role in soil health	References
Bacillus sp.	Soil aggregation	Ashraf et al. (2006)
Rhizobium sp.	Increasing the soil aggregation and water-holding capacity of rhizosphere soil	Kaci et al. (2005)
Pseudomonas putida GAP-p45	Soil aggregation, biofilm formation, and increased root adherence stability of sunflower seedling	Sandhya et al. (2009)
Pseudomonas mendocina	Enhanced soil fertility and stabilization	Kohler et al. (2006)
Microbacterium arborescens	Soil particles cementation	Aureen and Saroj (2013)

siderophores, and EPSs and accumulating osmolytes. One potential approach to mitigate sodic stress in plants, while simultaneously promoting environmental conservation and ensuring agricultural sustainability, involves the implementation of an efficient and economically viable strategy. The elucidation of plant–microbial interactions' roles in regulating stress-responsive genes in plants is crucial for enhancing plant defense mechanisms and production.

Acknowledgments This work was supported by Tamil Nadu Agricultural University.

Ethical Approval and Consent to Participate Not applicable.

Conflict of Interests The authors declare that there are no conflicts of interest related to this article.

References

Akram MS, Shahid M, Tariq M, Azeem M, Javed MT, Saleem S, Riaz S (2016) Deciphering *Staphylococcus sciuri* SAT-17 mediated anti-oxidative defense mechanisms and growth modulations in salt stressed maize (*Zea mays* L.). Front Microbiol 7:867

Alam SM (1999) Nutrient uptake by plants under stress conditions. In: Handbook of plant and crop stress, vol 2, pp 285–313

Amini S, Ghadiri H, Chen C, Marschner P (2016) Salt-affected soils, reclamation, carbon dynamics, and biochar: a review. J Soil Sediment 16:939–953

Ansari FA, Ahmad I, Pichtel J (2019) Growth stimulation and alleviation of salinity stress to wheat by the biofilm forming *Bacillus pumilus* strain FAB10. Appl Soil Ecol 143:45–54

Asghar H, Zahir Z, Arshad M, Khaliq A (2002) Relationship between in vitro production of auxins by rhizobacteria and their growth-promoting activities in *Brassica juncea* L. Biol Fertil Soils 35:231–237

Ashraf M, Foolad MR (2007) Roles of glycine betaine and proline in improving plant abiotic stress resistance. Environ Exp Bot 59:206–216

Ashraf M, Hasnain S, Berge O (2006) Effect of exo-polysaccharides producing bacterial inoculation on growth of roots of wheat (*Triticum aestivum* L.) plants grown in a salt-affected soil. Int J Environ Sci Technol 3:43–51

Aureen LG, Saroj B (2013) *Microbacterium arborescens* AGSB sp. nov., isolated from the rhizosphere of sand dune plant, *Ipomoea pes caprae*. Afr J Microbiol Res 7:5154–5158

Banaei-Asl F, Bandehagh A, Uliaei ED, Farajzadeh D, Sakata K, Mustafa G, Komatsu S (2015) Proteomic analysis of canola root inoculated with bacteria under salt stress. J Proteomics 124:88–111

Bano A, Fatima M (2009) Salt tolerance in *Zea mays* (L). Following inoculation with *rhizobium* and *pseudomonas*. Biol Fertil Soils 45:405–413

Bayuelo-Jiménez JS, Debouck DG, Lynch JP (2003) Growth, gas exchange, water relations, and ion composition of *Phaseolus* species grown under saline conditions. Field Crop Res 80:207–222

Belimov AA, Dodd IC, Safronova VI, Dumova VA, Shaposhnikov AI, Ladatko AG, Davies WJ (2014) Abscisic acid metabolizing rhizobacteria decreases ABA concentrations in planta and alter plant growth. Plant Physiol Biochem 74:84–91

Bernstein L (1975) Effects of salinity and sodicity on plant growth. Annu Rev Phytopathol 13:295–312

Bhandal IS, Malik CP (1988) Potassium estimation, uptake, and its role in the physiology and metabolism of flowering plants. Int Rev Cytol 110:205–254

Bisht N, Tiwari S, Singh PC, Niranjan A, Chauhan PS (2019) A multifaceted rhizobacterium *Paenibacillus lentimorbus* alleviates nutrient deficiency-induced stress in *Cicer arietinum* L. Microbiol Res 223:110–119

Blumwald E (2000) Sodium transport and salt tolerance in plants. Curr Opin Cell Biol 12:431–434

Brunetti C, Saleem AR, Della Rocca G, Emiliani G, De Carlo A, Balestrini R, Khalid A, Mahmood T, Centritto M (2021) Effects of plant growth-promoting rhizobacteria strains producing ACC deaminase on photosynthesis, isoprene emission, ethylene formation and growth of *Mucuna pruriens* (L.) DC. in response to water deficit. J Biotechnol 331:53–62

Calvaruso C, Turpault MP, Frey-Klett P (2006) Root-associated bacteria contribute to mineral weathering and to mineral nutrition in trees: a budgeting analysis. Appl Environ Microbiol 72:1258–1266

Carillo P, Mastrolonardo G, Nacca F, Parisi D, Verlotta A, Fuggi A (2008) Nitrogen metabolism in durum wheat under salinity: accumulation of proline and glycine betaine. Funct Plant Biol 35:412–426

Carillo P, Parisi D, Woodrow P, Pontecorvo G, Massaro G, Annunziata MG, Fuggi A, Sulpice R (2011) Salt-induced accumulation of glycine betaine is inhibited by high light in durum wheat. Funct Plant Biol 38:139–150

Costa OYA, Raaijmakers JM, Kuramae EE (2018) Microbial extracellular polymeric substances: ecological function and impact on soil aggregation. Front Microbiol 9:1636

Damam M, Kaloori K, Gaddam B, Kausar R (2016) Plant growth promoting substances (phytohormones) produced by rhizobacterial strains isolated from the rhizosphere of medicinal plants. Int J Pharm Sci Rev Res 37:130–136

Damodaran T, Mishra V, Jha S, Pankaj U, Gupta G, Gopal R (2019) Identification of rhizosphere bacterial diversity with promising salt tolerance, PGP traits and their exploitation for seed germination enhancement in sodic soil. Agric Res 8:36–43

Deng J, Orner EP, Chau JF, Anderson EM, Kadilak AL, Rubinstein RL, Bouchillon GM, Goodwin RA, Gage DJ, Shor LM (2015) Synergistic effects of soil microstructure and bacterial EPS on drying rate in emulated soil micromodels. Soil Biol Biochem 83:116–124

Dixit VK, Misra S, Mishra SK, Tewari SK, Joshi N, Chauhan PS (2020) Characterization of plant growth-promoting alkalotolerant Alcaligenes and *Bacillus* strains for mitigating the alkaline stress in Zea mays. Antonie Van Leeuwenhoek 113:889–905

Drogue B (2013) Host specificity in the cooperation between the PGPR *Azospirillum* and cereals. Claude Bernard University - Lyon

Ehrlich HL (1996) How microbes influence mineral growth and dissolution. Chem Geol 13:5–9

Farag MA, Zhang H, Ryu CM (2013) Dynamic chemical communication between plants and bacteria through airborne signals: induced resistance by bacterial volatiles. J Chem Ecol 39:1007–1018
Flemming HC, Wingender J (2001) Relevance of microbial extracellular polymeric substances (EPSs)-part I: structural and ecological aspects. Water Sci Technol 43:1–8
Flemming HC, Wingender J, Mayer C, Korstgens V, Borchard W (2000) Cohesiveness in biofilm matrix. In: Paper presented at the symposium of the Society for General Microbiology. Cambridge University Press, Cambridge
Gagné-Bourque F, Bertrand A, Claessens A, Aliferis KA, Jabaji S (2016) Alleviation of drought stress and metabolic changes in timothy (*Phleum pratense* L.) colonized with *Bacillus subtilis* B26. Front Plant Sci 7:584
Ghosh S, Wilson B, Ghoshal S, Senapati N, Mandal B (2012) Organic amendments influence soil quality and carbon sequestration in the Indo-Gangetic plains of India. Agric Ecosyst Environ Microbiol 156:134–141
Gunasekaran Y, Thiyageshwari S, Ariyan M, Roy Choudhury A, Park JH, Selvi D, Chithra L, Anandham R (2022) Alleviation of sodic stress in rice by exploring the exopolysaccharide-producing sodic-tolerant bacteria. Agriculture 12:1451
Hafeez B, Khanif YM, Saleem M (2013) Role of zinc in plant nutrition—a review. Am J Exp Agric 3:374
Hansen V, Bonnichsen L, Nunes I, Sexlinger K, Lopez SR, van der Bom FJ, Nybroe O, Nicolaisen MH, Jensen LS (2020) Seed inoculation with *Penicillium bilaiae* and *Bacillus simplex* affects the nutrient status of winter wheat. Biol Fertil Soils 56:97–109
Imhoff JF (1986) Osmoregulation and compatible solutes in eubacteria. FEMS Microbiol Rev 2:57–66
Kaci Y, Heyraud A, Barakat M, Heulin T (2005) Isolation and identification of an EPS-producing *rhizobium* strain from arid soil (Algeria): characterization of its EPS and the effect of inoculation on wheat rhizosphere soil structure. Res Microbiol 156:522–531
Kang Y, Khan S, Ma X (2009) Climate change impacts on crop yield, crop water productivity and food security—a review. Prog Nat Sci 19:1665–1674
Khan N, Bano A (2019) Exopolysaccharide producing rhizobacteria and their impact on growth and drought tolerance of wheat grown under rainfed conditions. PLoS One 14:e0222302
Khan MS, Zaidi A, Wani PA (2009) Role of phosphate solubilizing microorganisms in sustainable agriculture—a review. Sustain Agric:551–570
Kohler J, Caravaca F, Carrasco L, Roldan A (2006) Contribution of *Pseudomonas mendocina* and *Glomus intraradices* to aggregate stabilization and promotion of biological fertility in rhizosphere soil of lettuce plants under field conditions. Soil Use Manage 22:298–304
Kundan R, Pant G, Jadon N, Agrawal PK (2015) Plant growth promoting rhizobacteria: mechanism and current prospective. J Fertil Pestic 6:9
Lauchli A, Luttge U (2002) Salinity: environment-plants-molecules. Kluwer Academic, Dordrecht, pp 229–248
Lee HS, Singh JK, Ismail MA, Bhattacharya C, Seikh AH, Alharthi N, Hussain RR (2019) Corrosion mechanism and kinetics of Al-Zn coating deposited by arc thermal spraying process in saline solution at prolong exposure periods. Sci Rep 9:1–17
Machado RMA, Serralheiro RP (2017) Soil salinity: effect on vegetable crop growth. Management practices to prevent and mitigate soil salinization. Horticulturae 3:30
Martin RS, Peters G, Wilkinson JH (1971) Iterative refinement of the solution of a positive definite system of equations. In: Linear Algebra, pp 31–44
Mathur P, Roy S (2021) Insights into the plant responses to drought and decoding the potential of root associated microbiome for inducing drought tolerance. Physiol Plant 172:1016–1029
McNeal B, Coleman N (1966) Effect of solution composition on soil hydraulic conductivity. Soil Sci Soc Am J 30:308–312
Miransari M, Smith DL (2014) Plant hormones and seed germination. Environ Exp Bot 99:110–121
Munns R (2002) Comparative physiology of salt and water stress. Plant Cell Environ 25:239–250

Munns R, Termaat A (1986) Whole-plant responses to salinity. Funct Plant Biol 13:143–160
Munns R, Tester M (2008) Mechanisms of salinity tolerance. Annu Rev Plant Biol 59:651
Naureen Z, Aqeel M, Hassan MN, Gilani SA, Bouqellah N, Mabood F, Hussain J, Hafeez FY (2015) Isolation and screening of silicate bacteria from various habitats for biological control of phytopathogenic fungi. Am J Plant Sci 6:2850
Nunkaew T, Kantachote D, Nitoda T, Kanzaki H, Ritchie R (2015) Characterization of exopolymeric substances from selected *Rhodopseudomonas palustris* strains and their ability to adsorb sodium ions. Carbohydr Polym 115:334–341
Pereira SIA, Castro PML (2014) Phosphate-solubilizing rhizobacteria enhance *Zea mays* growth in agricultural P-deficient soils. Ecol Eng 73:526–535
Prigent-Combaret C (2013) *Azospirillum*-plant interaction. In: Beneficial plant-microbial interactions, pp 237–269
Qadir M, Oster J, Schubert S, Noble A, Sahrawat K (2007) Phytoremediation of sodic and saline-sodic soils. Adv Agron 96:197–247
Qurashi AW, Sabri AN (2012) Bacterial exopolysaccharide and biofilm formation stimulate chickpea growth and soil aggregation under salt stress. Braz J Microbiol 43:1183–1191
Rahneshan Z, Nasibi F, Moghadam AA (2018) Effects of salinity stress on some growth, physiological, biochemical parameters and nutrients in two pistachio (*Pistacia vera* L.) rootstocks. J Plant Interact 13:73–82
Rengasamy P (2010) Soil processes affecting crop production in salt-affected soils. Funct Plant Biol 37:613–620
Rietz D, Haynes R (2003) Effects of irrigation-induced salinity and sodicity on soil microbial activity. Soil Biol Biochem 35:845–854
Ruzzi M, Aroca R (2015) Plant growth-promoting rhizobacteria act as biostimulants in horticulture. Sci Hortic 196:124–134
Saha I, Datta S, Biswas D (2020) Exploring the role of bacterial extracellular polymeric substances for sustainable development in agriculture. Curr Microbiol 77:3224–3239
Sahay R, Patra DD (2013) Identification and performance of stress-tolerant phosphate-solubilizing bacterial isolates on Tagetes minuta grown in sodic soil. Soil Use Manage 29:494–500
Sahin U, Ekinci M, Ors S, Turan M, Yildiz S, Yildirim E (2018) Effects of individual and combined effects of salinity and drought on physiological, nutritional and biochemical properties of cabbage (*Brassica oleracea* var. *capitata*). Sci Hortic 240:196–204
Sandhya V, SK Z, Grover M, Reddy G, Venkateswarlu B (2009) Alleviation of drought stress effects in sunflower seedlings by the exopolysaccharides producing *Pseudomonas putida* strain GAP-P45. Biol Fertil Soils 46:17–26
Sarker U, Oba S (2020) Phenolic profiles and antioxidant activities in selected drought-tolerant leafy vegetable amaranth. Sci Rep 10:1–11
Schalk IJ, Hannauer M, Braud A (2011) New roles for bacterial siderophores in metal transport and tolerance. Environ Microbiol 13:2844–2854
Shafeiee M, Ehsanzadeh P (2019) Physiological and biochemical mechanisms of salinity tolerance in several fennel genotypes: existence of clearly-expressed genotypic variations. Ind Crop Prod 132:311–318
Shaheen S, Naseer S, Ashraf M, Akram NA (2013) Salt stress affects water relations, photosynthesis, and oxidative defense mechanisms in *Solanum melongena* L. J Plant Interact 8:85–96
Shahid M, Akram MS, Khan MA, Zubair M, Shah SM, Ismail M, Shabir G, Basheer S, Aslam K, Tariq M (2018a) A phytobeneficial strain *Planomicrobium* sp. MSSA-10 triggered oxidative stress responsive mechanisms and regulated the growth of pea plants under induced saline environment. J Appl Microbiol 124:1566–1579
Shahid M, Mahmood F, Hussain S, Shahzad T, Haider MZ, Noman M, Mushtaq A, Fatima Q, Ahmed T, Mustafa G (2018b) Enzymatic detoxification of azo dyes by a multifarious *Bacillus* sp strain MR-1/2-bearing plant growth-promoting characteristics. 3 Biotech 8:1–12

Shulse CN, Cole BJ, Ciobanu D, Lin J, Yoshinaga Y, Gouran M, Turco GM, Zhu Y, O'Malley RC, Brady SM, Dickel DE (2019) High-throughput single-cell transcriptome profiling of plant cell types. Cell Rep 27:2241–2247

Shultana R, Kee Zuan AT, Yusop MR, Saud HM (2020) Characterization of salt-tolerant plant growth-promoting rhizobacteria and the effect on growth and yield of saline-affected rice. PLoS One 15:e0238537

Tewari S, Arora K (2014) Talc based exopolysaccharides formulation enhancing growth and production of *Hellianthus annuus* under saline conditions. Cell Mol Biol 60:73–81

Thu Hoai NT, Shim IS, Kobayashi K, Kenji U (2003) Accumulation of some nitrogen compounds in response to salt stress and their relationships with salt tolerance in rice (*Oryza sativa* L.) seedlings. Plant Growth Regul 41:159–164

Timmusk S, Abd El-Daim IA, Copolovici L, Tanilas T, Kännaste A, Behers L, Nevo E, Seisenbaeva G, Stenström E, Niinemets Ü (2014) Drought-tolerance of wheat improved by rhizosphere bacteria from harsh environments: enhanced biomass production and reduced emissions of stress volatiles. PLoS One 9:e96086

Torbaghan ME, Lakzian A, Astaraei AR, Fotovat A, Besharati H (2017) Salt and alkali stresses reduction in wheat by plant growth promoting haloalkaliphilic bacteria. J Soil Sci Plant Nutr 17:1058–1087

Upadhyay S, Singh J, Singh D (2011) Exopolysaccharide-producing plant growth-promoting rhizobacteria under salinity condition. Pedosphere 21:214–222

Upadhyay SK, Singh JS, Saxena AK, Singh DP (2012) Impact of PGPR inoculation on growth and antioxidant status of wheat under saline conditions. Plant Biol 14:605–611

Uroz S, Calvaruso C, Turpault MP, Pierrat JC, Mustin C, Frey-Klett P (2007) Effect of the mycorrhizosphere on the genotypic and metabolic diversity of the bacterial communities involved in mineral weathering in a forest soil. Appl Environ Microbiol 73:3019–3027

Vasanthi P, Selvan SS (2020) Optimization of mixing parameters in nanosilica toughened cement mortar using Taguchi-grey relational analysis. Silicon 14:127–133

Vessey JK (2003) Plant growth promoting rhizobacteria as biofertilizers. Plant and Soil 25:571–586

Wang J, Mao H, Zhao H, Huang D, Wang Z (2012a) Different increases in maize and wheat grain zinc concentrations caused by soil and foliar applications of zinc in Loess Plateau, China. Field Crop Res 135:89–96

Wang H, Wu Z, Zhou Y, Han J, Shi D (2012b) Effects of salt stress on ion balance and nitrogen metabolism in rice. Plant Soil Environ 58:62–67

Wei K, Lv C, Chen M, Zhou X, Dai Z, Shen D (2015) Development and performance evaluation of a new thermal insulation material from rice straw using high frequency hot-pressing. Energ Buildings 87:116–122

Wolfaardt G, Lawrence J, Korber D (1999) Function of EPS. In: Microbial extracellular polymeric substances. Springer, Berlin

Wyn Jones R, Pollard A (1983) Proteins, enzymes and inorganic ions. In: Encyclopedia of plant physiology. Springer, Berlin

Yadav KK, Mandal AK, Chakraborty R (2013) Copper susceptibility in *Acinetobacter junii* BB1A is related to the production of extracellular polymeric substances. Anton Leeuwenhoek 104:261–269

Yang Y, Guo Y (2018) Unraveling salt stress signaling in plants. J Integr Plant Biol 60:796–804

Yang J, Kloepper JW, Ryu CM (2009) Rhizosphere bacteria help plants tolerate abiotic stress. Trends Plant Sci 14:1–4

Zhang T, Kang Y, Wan S (2013) Shallow sand-filled niches beneath drip emitters made reclamation of an impermeable saline-sodic soil possible while cropping with *Lycium barbarum* L. Agric Water Manag 119:54–64

Zhang H, Liu XL, Zhang RX, Yuan HY, Wang MM, Yang HY, Ma HY, Liu D, Jiang CJ, Liang ZW (2017a) Root damage under alkaline stress is associated with reactive oxygen species accumulation in rice (*Oryza sativa* L.). Front Plant Sci 8:1580

Zhang W, Xie Z, Lang D, Cui J, Zhang X (2017b) Beneficial effects of silicon on abiotic stress tolerance in legumes. J Plant Nutr 40:2224–2236

Chapter 14
Endophytic Microbiome: An Insight into the Hidden World of Microorganisms Within Plants

Padinjarakavil Soumya, K. Jayachandran, and E. K. Radhakrishnan

Abstract Endophytes constitute the concealed microbial community within plants, and this chapter delves into the complex bond between plants and these hidden microbial allies, signifying the importance of the "endophytic microbiome." Emphasizing the crucial role of endophytic microorganisms in fortifying the plant health and resilience, the discussion underscores their potential applications in sustainable agriculture and ecological restoration. The chapter concisely outlines advanced technologies and tools for decoding the genetic makeup and the ecological functions of endophytic microbes, underscoring their importance in understanding plant-microbe interactions. By deepening our comprehension of the network of the endophytic microbiome, this chapter reveals a scope of possibilities to harness its transformative potential.

Keywords Endophytes · Plant–microbe interaction · Stress mitigation · Probiotics · Sustainable agriculture

14.1 Introduction

The endophytic microbiome, a term denoting the complex world of microorganisms that reside within the internal tissues of plants, has emerged as a subject of profound importance in recent years. These hidden partners, primarily comprising bacteria and fungi, maintain an intimate and often mutually beneficial relationship with their plant hosts (Rosenblueth and Martínez-Romero 2006). Unlike pathogens that harm plants, these endophytes quietly coexist with their hosts, sometimes for the entire life cycle of the plant. Their significance reverberates across the areas of agriculture, ecology, and biotechnology. In agriculture, the endophytic microbiome has offered

P. Soumya · K. Jayachandran · E. K. Radhakrishnan (✉)
School of Biosciences, Mahatma Gandhi University, Kottayam, Kerala, India
e-mail: radhakrishnanek@mgu.ac.in

P. Singh Chauhan et al. (eds.), *Plant-Microbe Interaction and Stress Management*, Rhizosphere Biology,
https://doi.org/10.1007/978-981-97-4239-4_14

the novel solutions to age-old challenges. These microorganisms play a significant role in enhancing crop health and productivity. Some endophytes aid in nutrient uptake by the host plants, providing them with essential elements, which can be particularly vital in nutrient-poor soils. Others protect plants from diseases by producing antimicrobial compounds, potentially reducing the need for chemical pesticides. Moreover, they contribute to stress tolerance, enabling plants to withstand harsh environmental conditions, ultimately leading to increased crop yields and more resilient agriculture (Khare et al. 2018). In the broader context of ecosystems, the endophytic microbiome plays a pivotal role in maintaining ecological equilibrium. These microorganisms can influence the health and growth of dominant plant species, thereby shaping the structure of entire ecosystems. Healthy plant communities, with the help of endophytes, support biodiversity and keep the ecosystems in balance (Lata et al. 2018). Beyond agriculture and ecology, the endophytic microbiome holds potential for biotechnological applications as well. Some endophytes have yielded bioactive compounds, including antibiotics and other pharmaceutical agents. They have also been explored for enzymes and biofuel-related products. Additionally, certain endophytes have demonstrated the remarkable ability to bioremediate contaminated soils by degrading or detoxifying the environmental pollutants (Schulz and Boyle 2006).

This chapter aims to shed light on the definition and vital roles of these microbial allies within plants and their far-reaching implications for the health of environment and the well-being of society. As our understanding of the endophytic microbiome deepens, it becomes increasingly clear that these hidden partners are not just passive passengers within plants; they are active participants in the biological processes of life, offering both promise and solutions to some of the most pressing challenges.

14.2 Diversity of Endophytic Microbiomes

The diversity of endophytic microbiomes within plants is truly complex and diversified. Bacteria, fungi, and various other microorganisms create a dynamic community inside plants, each with unique roles. Recent studies have highlighted this diversity, providing valuable insights into the functions of these microbial residents.

Bacteria play a significant part in these inside-the-plant communities. Plants like *Arabidopsis thaliana* host a wide range of bacterial species (Bulgarelli et al. 2015). Some of these bacteria are plant-friendly mutualistic microbes, helping with nutrient uptake and protecting plants from harmful diseases. For instance, bacteria from the genera *Pseudomonas* and *Bacillus* have been found to enhance plant growth and act against pathogens. Different plant species house distinct bacterial endophytes, highlighting the intricate relationships between bacteria and their plant hosts (Edwards et al. 2018). Fungi also play a crucial role in these plant-inhabiting communities. Researchers have delved into the fungal endophytes living in grasses and have demonstrated their importance (Knapp et al. 2021). These fungi can promote plant health and help plants to endure tough situations. Moreover, they can produce bioactive compounds with potential ecological and biotechnological uses. With the help of advanced DNA sequencing techniques, researchers have been able to discover new fungal species residing within plants, broadening our understanding on fungal diversity within these communities. Apart from bacteria and fungi, other microorganisms like archaea and viruses also find their place within endophytic microbiomes. Although less studied compared to bacteria and fungi, these microorganisms are becoming increasingly recognized for their roles. A study by Zhang et al. (2019) have already been highlighted for the presence of archaea in rice plants, suggesting their involvement in nutrient cycling and plant health.

Bacteria and fungi are essential components of the endophytic microbiomes that thrive within plants. Recent research reveals that these microscopic inhabitants play diverse and critical roles in the health and resilience of their host plants. Bacteria, such as those in the genera *Pseudomonas* and *Bacillus*, enhance nutrient uptake, safeguard plants against diseases, and promote growth (Hashem et al. 2019). Fungi within plant tissues also contribute to plant well-being, aiding nutrient absorption and stress tolerance (Baron and Rigobelo 2021). Across the globe, the scientific community has made efforts to identify certain endophytes from various plants species (Table 14.1). Their capacity to produce bioactive compounds with potential applications in pharmaceuticals and biotechnology further underscores their significance. The rich diversity of these microorganisms in endophytic communities adds depth to our understanding on plant–microbe interactions, offering promise for both agriculture and ecosystem management.

14.3 Functions and Interactions of Endophytic Microorganisms in Plant Growth

The mutualistic relationships between plants and their associated microorganisms continue to evolve, and researchers at a global level have uncovered the diverse functions and interactions that endophytes exhibit within the plant host. These microorganisms, residing within the plant tissues without causing harm, have been identified as crucial contributors to enhanced nutrient acquisition, stress tolerance, and overall plant health (Schlaeppi and Bulgarelli 2015).

14.3.1 Significance of Endophytic Microorganisms in Regulating Plant Growth

Within the complex network of the plant–microbe interactions, the role of endophytic microorganisms emerges as a vital determinant of plant growth, development, and resilience. The internal region of plant tissues, endophytes, including

Table 14.1 List of selected endophytes identified from various plant species

Endophytes isolated	Plant species	Reference
Bacillus megaterium	*Medicago satavia*	Stajkovic et al. (2009)
Penicillium sp.	*Huperzia serrata*	Wang and Dai (2011)
Colletotrichum sp. *Colletotrichum gloeosporioides*	*Pasania edulis*	He et al. (2012)
Aspergillus sp.	*Datura stramonium* *Moringa olifera* *Prosopis chilensis*	Mahdi et al. (2014)
Bacillus thuringiensis, *Bacillus subtilis* subsp. *subtilis*	*Musa* sp.	Souza et al. (2014)
Stemphylium globuliferum	*Avicennia marina*	Moussa et al. (2016)
Bacillus velezensis	*Citrus* sp.	Rabbee et al. (2019)
Arthrobacter agilis UMCV2	*Fragaria ananassa*	Hernández-Soberano et al. (2020)
Stenotrophomonas maltophilia strain UN1512	*Fragaria ananassa*	Alijani et al. (2020)
B. velezensis 83	*Mangifera indica*	Balderas-Ruíz et al. (2020)
Staphylococcus seiuri	*Psidium guajava*	Riskuwa-Shehu et al. (2020)
Bacillus sp. *strain* ESA 402	*Sorghum bicolor*	Santana et al. (2020)
Pseudomonas sp. M45	*Camellia sinensis*	Hazarika et al. (2021)
Burkholderia gladioli GI-6	*Triticum* spp	Shah et al. (2021)
Microbacterium laevaniforman	*Oryza sativa*	Zhang et al. (2021)
Sphingomona sp.	*Oryza sativa*	Krishnamoorthy et al. (2021)
Erwinia persicina EU-A3SK3	*Foxtail millet*	Devi et al. (2022)
Proteobacteria sp.	*Oryza sativa and Camellia sinensis*	Jia et al. (2022)
B. subtilis B9	*Saccharum officinarum*	Di et al. (2022)
Bacillus, *Stenotrophomonas*, and *Pseudomonas*	*Calendula officinalis rhizosphere*	Tsalgatidou et al. (2023)
B. velezensis	*Zea mays*	Li et al. (2023b)

bacteria and fungi, engage in a myriad of functions that profoundly influence the overall health and well-being of their host plants. These microscopic partners contribute significantly to the nutrient acquisition by solubilizing the phosphates, fixing atmospheric nitrogen, and mobilizing minerals, thereby enhancing the nutritional profile of plants (Maheshwari et al. 2022). Beyond the nutrient provision, endophytes act as a shield in the defense against environmental stresses, including drought, salinity, and temperature extremes, by producing compounds that mitigate the stress effects and fortify the plant's adaptive responses. Their action extends to the point of plant protection, as endophytes serve as biocontrol agents, producing antimicrobial compounds that oppose the pathogenic invasions and paves way to induce systemic resistance for a broad spectrum of diseases (Yu et al. 2022). The combined effect of chemical signaling, quorum sensing, and host specificity characterizes the multifaceted relationships between endophytes and their plant hosts, shaping the physiological landscape of both partners.

14.3.2 Role of Endophytic Microbiome in Nutrient Uptake of Plants

The microorganisms, residing harmoniously within plant tissues, act as indispensable partners in enhancing nutrient availability through an array of strategies. Endophytes exhibit the capacity for nutrient mobilization, proficiently solubilizing essential elements such as phosphorus, iron, and potassium in the soil (Anuroopa et al. 2022). For instance, certain fungal endophytes like *Piriformospora indica* have demonstrated to have remarkable abilities to enhance the phosphorus uptake in plants, contributing to improved overall nutrient status (Li et al. 2023a). Moreover, the phenomenon of nitrogen fixation is supported by specific endophytic bacteria, exemplified by species like *Azospirillum* and *Burkholderia*, which convert atmospheric nitrogen into a plant-usable form, fortifying the plant's nitrogen nutrition (Daniel et al. 2022). *Pseudomonas fluorescens*, a versatile endophyte, has been identified for its role in nutrient exchange with plants, exhibiting a mutualistic relationship through the atmospheric nitrogen fixation, providing a vital nutrient source (Wu et al. 2022). Additionally, the endophytes enhance the mineral absorption by the plant roots, deploying mechanisms that include heightened root surface area and the release of plant growth-promoting substances. Fungal endophytes like arbuscular mycorrhizal fungi are reported to form symbiotic relationships with plants, facilitating the nutrient absorption in exchange for the carbohydrates (Wahab et al. 2023). Through these symbiotic relationships, the endophytic microbiome emerges as a cornerstone in optimizing plant nutrition by mutualistic exchanges within the intricate web of plant–microbe interactions.

14.3.3 Role of Endophytic Microbiome in Pathogen Resistance

Delving into the mechanism of plant defense, it is found that endophytic microorganisms play a pivotal role in fortifying plants against the pathogens. These microorganisms, residing within the plant tissues, emerge as inevitable allies in the defense mechanisms. The most significant phenomenon is the Induced Systemic Resistance (ISR), a strategic response triggered by endophytes that enable the plants to confront a diverse array of pathogens. Notably, endophytes trigger the synthesis of plant defense compounds and helps in immunizing the cell wall structures thus heightening the plant's resilience. Bacterial endophytes like *Bacillus* and fungal endophytes such as *Trichoderma* have been demonstrated to induce ISR and fortify the defense of their host plants (Choudhary et al. 2023). Several research have unfolded the exploration of antagonistic interactions, where certain endophytes serve as biocontrol agents by producing an arsenal of antimicrobial compounds, including antibiotics, lytic enzymes, and secondary metabolites, acting as formidable deterrents against pathogenic invaders (Sharma et al. 2022). Studying deeper into the mechanism of plant defense, the endophytic microbiome emerges as an architect of cell wall reinforcement, inducing the production of pathogenesis-related proteins that fortify the plant's structural integrity, rendering it less susceptible to penetration by pathogens (Guerreiro and Marhavý 2023). Endophytes also help in the phenomenon of indirect defense signaling, where they serve as silent initiators of certain signaling pathways that elicit the production of volatile organic compounds (VOCs) and other signaling molecules, to combat the pathogenic threats (Jha et al. 2023). Therefore, endophytic microorganisms are the silent guardians, fortifying plants with a robust arsenal against the challenges posed by pathogens.

14.3.4 Role of Endophytic Microbiome in Stress Tolerance

The plant–endophyte associations play a crucial role in strengthening the plant stress tolerance which is an essential trait for survival amidst diverse environmental challenges. In the face of water scarcity, certain endophytes emerge as frontline guardians of drought tolerance by secreting osmoprotectants and synthesizing enzymes with pivotal function in water retention. Bacterial endophytes like *Enterobacter* spp., *Stenotrophomonas* spp., and the fungal endophyte *Serendipita indica* are reported to enhance the water stress resilience in various plant species (Juby et al. 2023; Chaudhary et al. 2022). Endophytes become allies in saline landscapes as well, where they actively engage in ion homeostasis, alleviate oxidative stress, and foster the accumulation of compatible solutes, collectively promoting plant survival in saline conditions. Salt-tolerant endophyte *Bacillus amyloliquefaciens* is reported to enhance the salinity tolerance of host plants through multifaceted mechanisms (Egamberdieva et al. 2019). Further exploration extends to the realm of temperature stress, wherein several endophytes exhibit the ability to produce heat shock proteins and other protective molecules, acting as molecular shields to safeguard plants against the detrimental effects of high temperatures. The remarkable contributions of endophytes are further underscored in their role in metal tolerance, a critical facet in the context of phytoremediation. Metal-tolerant bacterium *Pseudomonas putida* has been reported to actively participate in the accumulation of heavy metals to facilitate the plant's detoxification mechanisms, paving the way for enhanced phytoremediation of contaminated soils (Yan et al. 2020). By exploring these intricate adaptations, it is understood that nature itself glorifies the endophytic microbiomes as unavoidable guardians in empowering plants to deal with the formidable challenges posed by diverse environmental stresses.

14.4 Interactions Between Endophytic Microbiomes and Their Host Plants

The symbiotic interactive network woven between endophytic microbiomes and their host plants describes a higher level of mutualistic relationships that underscore the resilience and vitality of both partners. In these mutualistic interactions, endophytes find a nurturing abode within the plant's internal tissues, establishing a symbiotic equilibrium where the plant in return provides a habitat and essential nutrients for the endophyte's sustenance. Simultaneously, the endophyte respond by contributing to the host plant's health and overall fitness (Kumari et al. 2023). This two-sided mutualistic collaboration, benefits the plants in enhancing nutrient uptake, strengthens the defense mechanism, and promoting optimal plant growth. At the molecular level, signaling pathways act as the initiators of interactions, by directing the exchange of recognition signals and chemical messengers between the endophyte and its host. This communication fosters the activation of specific plant response pathways, tailoring the plant's physiological and biochemical pathways to accommodate the presence of its microbial counterpart. Furthermore, the concept of host specificity adds a layer of complexity to these associations, as endophytes demonstrate preferences for specific plant species, establishing more robust and specialized relationships. It was reported that a nitrogen-fixing endophyte, *Bradyrhizobium japonicum,* exhibited specificity for leguminous plants, exemplifying the selection and tailored nature of these interactions (Sugawara et al. 2011).

14.4.1 Endophyte-Mediated Plant Health Via Microbial Production of Metabolites and Biomolecules

Endophytic microorganisms emerge as engineers of plant health, by supplementing the beneficial effects through the production of metabolites and biomolecules. These microbial engineers contribute to the well-being of their host plants by producing a repertoire of plant growth-promoting substances. Phytohormones, including the auxins, cytokinins, and gibberellins, produced from these endophytes, provide a stimulating effect to plant growth and development (Egamberdieva et al. 2017). Beyond nurturing plant growth, certain endophytes transform into vigilant biocontrol agents by producing certain antimicrobial compounds that act as a potent weapon against the growth of plant pathogens. Secondary metabolites produced by endophytes also showcase their versatility by synthesizing bioactive compounds that contribute not only to plant health but also to the defense mechanisms. This includes the production of antibiotics, alkaloids, and other bioactive molecules (Tsipinana et al. 2023). Indole-3-acetic acid (IAA), a ubiquitous compound in microbial-plant signaling, influences plant growth and development, exemplifying the direct impact of endophytic activity (Spaepen et al. 2007). The 1-aminocyclopr opane-1-carboxylate (ACC) deaminase, antifungal compounds like pyrrolnitrin and

DAPG, and hydrolytic enzymes such as chitinases and proteases collectively showcase the diverse array of biomolecules synthesized by endophytes (Etesami et al. 2014). The diverse compounds expressed by endophytes play a pivotal role in enhancing plant health; and the reported compounds with their activity are listed in Table 14.2. These compounds contribute to the disease suppression, modulate the abiotic stress tolerance, and fortify the overall plant resilience. Siderophores another bioactive molecule is produced by endophytes to enhance the nutrient availability by culminating the loop for direct interactions. Furthermore, endophytes indirectly stimulate systemic resistance, modify plant signaling pathways, and foster competition with pathogens, providing a comprehensive view of endophyte-mediated plant health.

14.5 Methods for Studying Endophytic Microbiome

Exploring the world of endophytic microbiomes involves use of advanced methods to understand how plants and their internal microbes interact. Traditional methods involve culturing and studying the individual microorganisms, but newer techniques use high-tech DNA sequencing to analyze all the microbes in plant tissues without culturing them. This approach, called metagenomics, gives a complete picture of the microbial community. Advanced methods also include functional metagenomics and transcriptomics, helping researchers to see what these microbes do and how they contribute to plant health. By combining these modern techniques with smart computer tools, researchers can gain a better, more straightforward understanding of the hidden world of endophytic microbiomes and their impact on plants.

14.5.1 DNA Sequencing

DNA sequencing technologies have become indispensable tools in exploring the complexity of the endophytic microbiomes, facilitating a more comprehensive understanding on the genetic diversity and functional potential of these microbial communities. Metagenomics, a pivotal method in this domain, involves the shotgun sequencing of environmental DNA, enabling the exploration of the entire genetic repertoire present in each sample (Handelsman et al. 1998). This approach has been significant in discovering the diversity of endophytic microorganisms within plant tissues, overcoming the limitations of traditional cultivation-based techniques. Amplicon sequencing, a targeted DNA sequencing approach, has been widely utilized to dissect the taxonomic composition of endophytic microbiomes. By amplifying and sequencing specific genetic markers, such as the 16S rRNA gene for bacteria and the ITS region for fungi, researchers gain valuable insights into the community structure and diversity of these microbial assemblages (Caporaso et al.

Table 14.2 Bioactive compounds from endophytes with beneficial applications in agriculture

S. no	Compound	Origin	Bioactivity/application	Reference
1	Azadirachtin A and B	*Eupenicillium parvum* from *Azadirachta indica*	Insecticidal activity	Kusari et al. (2012)
2	(1R,4R)-1,4-(2,3)-indolmethane-1-methyl-2,4-dihydro-1H-pyrazino-[2,1-b]-quinazoline-3,6-dione	*Penicillium vinaceum from Crocus sativus*	Antifungal activity	Zheng et al. (2012)
3	b-sitosterol	*Phoma* sp. from *Arisaema erubescens*	Antifungal activity	Wang et al. (2012)
4	Extracellular enzymes and auxins (IAA)	*Penicillum citrinum, Preussia* sp., *Aureobasidium* from *Boswellia sacra*	Plant growth promoting activity	Khan et al. (2016)
5	Lipopeptides	*Bacillus amyloliquefaciens* from *ornamental plants*	Antifungal activity	Zouari et al. (2016)
6	Resveratrol	*Aspergillus niger Wine grape Carbernet sauvignon*	Anticancer lifespan increasing activity	Liu et al. (2016)
7	Vindoline	*Curvularia* sp., from *Choanephora infundibulifera Catharanthus roseus*	Antimitotic activity	Pandey et al. (2016b)
8	Massariphenone ergosterol peroxide	*Verticillium* sp.	Biocontrol activity	Rani et al. (2017)
9	Trichodecenins, Trichorovins, Trichocellins, Trichorzianins A and B, Trichorzins, HA and MA, Tricholongins BI and BII Longibrachinsa	*Trichoderma* sp.	Biocontrol activity against *Rhizoctonia solani, Botrytis cinerea, Phytophthora cinnamomi, Pythium irregulare, Pythium middletonii, Sclerotinia sclerotiorum, Fusarium oxysporum, Bipolaris sorokiniana*	Sood et al. (2020)
10	Fusicoccane diterpenes	*Periconia* sp.	Biocontrol activity against *Salmonella typhimurium, Staphylococcus aureus, Klebsiella pneumoniae*, and *Bacillus subtilis*	Azhari and Supratman (2021)

(continued)

Table 14.2 (continued)

S. no	Compound	Origin	Bioactivity/application	Reference
11	Penicitroamide	*Penicillium* sp.	Biocontrol activity	Deshmukh et al. (2022)
12	3-O-Methylalaternin, Altersolanol A	*Ampelomyces* sp.	Biocontrol activity	Wen et al. (2023)
13	Brefeldin	*Cladosporium* sp.	Antifungal activity	Akram et al. (2023)

2011; Tiwari and Bae 2023). Functional metagenomics extends the scope of DNA sequencing applications by providing insights into the functional potential of the endophytic microbiome. This method involves the sequencing of DNA fragments directly extracted from environmental samples, allowing for the identification of genes involved in key metabolic activities, such as nitrogen fixation, phosphate solubilization, and plant growth promotion (Mathieu et al. 2023). Recent advancements in third-generation sequencing technologies, such as PacBio and Oxford Nanopore, offer longer read lengths and enhanced sequencing accuracy, further improving the resolution and quality of endophytic microbiome analyses (Jain et al. 2018). The integration of diverse DNA sequencing techniques, including metagenomics, amplicon sequencing, and functional metagenomics, has greatly expanded our knowledge of endophytic microbiomes, defining their taxonomic diversity and shedding light on the functional roles of these microbial communities in plant–microbe interactions.

14.5.2 Metagenomics

Metagenomics technology has revolutionized the study of endophytic microbiomes by enabling a holistic analysis of the genetic makeup of microorganisms residing within plant tissues. Whole-genome shotgun sequencing, a prominent metagenomic approach, facilitates the unbiased exploration of the entire microbial community's genomic content. This method has been applied effectively in identifying diverse endophytic microorganisms, including bacteria, fungi, and viruses, without the constraints of traditional cultivation methods (Shakya et al. 2019). In-depth analysis of metagenomic data provides insights into the functional potential encoded in the collective genetic material of endophytic microbiomes. This information can be interpreted for understanding the metabolic activities and interactions between endophytic microorganisms and their plant hosts. The integration of advanced bioinformatics tools with metagenomics data further enhances the accuracy and interpretability of these analyses (Navgire et al. 2022). Recent advanced technologies for obtaining long-read sequences have improved the resolution of metagenomic studies. Longer DNA reads offered by the platforms such as Nanopore aid in better assembly of microbial genomes and provide a more comprehensive understanding of the complex microbial communities residing in plant tissues (Overholt et al. 2020).

14.5.3 Metatranscriptomics

Metatranscriptomics has also emerged as a valuable tool for detailing the functional dynamics of endophytic microbiomes within plant tissues by studying the gene expression profiles of the microbial community. This approach involves the sequencing and analysis of the total RNA extracted from an environmental sample, providing insights into the active functions and metabolic activities of endophytic microorganisms. In the context of endophytic microbiomes, metatranscriptomics allows researchers to identify specific genes being expressed by the microbial community within plant tissues, shedding light on their roles in plant–microbe interactions (Gamalero et al. 2022). By employing metatranscriptomics, researchers have distinguished between the dormant and active members of the endophytic microbiome, providing a dynamic snapshot of the microbial community's functional response to environmental stimuli and plant-host interactions (Guo et al. 2021). This method has been crucial in exploring and determining the molecular mechanisms employed by endophytic microorganisms to promote plant growth, enhance nutrient availability, and contribute to host defense responses (Hardoim et al. 2015). Recent studies have highlighted the utility of metatranscriptomics in understanding the temporal and spatial dynamics of endophytic microbiomes (Hewitt et al. 2023). By capturing the gene expression patterns of microorganisms within different plant tissues and across various growth stages, researchers have gained a clearer understanding of the functional roles of endophytic microorganisms in planta. Thus, metatranscriptomics provides a powerful approach to identify and characterize the functional attributes of endophytic microbiomes, offering a dynamic perspective on the active microbial community within plant tissues.

14.6 Potential Applications and Benefits of Endophytic Microbiomes

Endophytic microbiomes, comprising bacteria and fungi residing within plant tissues without causing harm to the host, offer a multitude of benefits that extend beyond the specific contexts of agriculture, disease management, environmental sustainability, and stress mitigation. These microbial communities form integral relationships with their host plants, influencing various physiological and biochemical processes. The symbiotic interactions between plants and endophytes contribute to enhanced nutrient uptake, improved stress tolerance, and overall plant health.

Furthermore, endophytes play a crucial role in shaping the structure and function of plant ecosystems, exemplifying their significance in maintaining ecological balance.

14.6.1 Agricultural Practices

In the agricultural domain, the pivotal role of endophytic microbiomes in sustainable practices is underscored by their significant contributions in promoting plant growth and increasing crop yield. These microscopic allies, residing within plant tissues without causing harm to the host, actively engage in nutrient acquisition processes that have far-reaching effects on soil fertility and overall agricultural productivity. Berendsen et al. (2012) elaborated on the essential role of endophytic microbiomes in nutrient acquisition, emphasizing their capacity for nitrogen fixation and phosphate solubilization. Nitrogen fixation is a critical process where certain endophytes convert atmospheric nitrogen into a form usable by plants, thereby enhancing nitrogen availability in the soil. Similarly, phosphate solubilization involves the transformation of insoluble phosphates into soluble forms, promoting increased accessibility of phosphorus for plant uptake. This dual action contributes significantly to improved soil fertility, creating an environment conducive to robust plant growth. Another study by Etesami et al. (2020) further substantiates the positive impact of endophytes on diverse crops, including staple grains like rice, maize, and wheat, highlighting the broad applicability of their beneficial effects in optimizing agricultural productivity; also, they provide insights into the multifaceted benefits of endophytic microbiomes, emphasizing their role in enhancing nutrient uptake, modulating plant hormones, and mitigating the effects of environmental stressors on crops. Another work by Parnell et al. (2016) adds onto this broad aspect of endophyte applications by showcasing the positive correlations between the presence of endophytes and increased crop yield in maize. The concept of harnessing endophytic microbiomes as bioinoculants aligns with a natural and eco-friendly approach to agriculture, offering a sustainable alternative to traditional practices which are reliant on chemical fertilizers and pesticides. By promoting the growth of beneficial microorganisms within the plant, bioinoculants contribute to a balanced and resilient ecosystem. This approach not only enhances soil health but also leads to increased crop yield through improved nutrient availability and uptake, ultimately supporting the overarching goal of sustainable and environmentally conscious agriculture. Throughout the world, the agricultural scientists have collectively underscored the significant role of endophytic microbiomes in sustainable agriculture. Their capacity to promote plant growth, enhance nutrient acquisition, and increase crop yield positions these microorganisms as valuable contributors to the ongoing efforts for optimizing agricultural productivity in an environmentally responsible manner. The exploration of endophytes as bioinoculants opens promising avenues for sustainable farming practices that prioritize soil health and long-term agricultural resilience.

14.6.2 Plant Disease Management

Endophytic microbiome emerge as natural defenders against plant diseases, contributing to biocontrol strategies in agriculture. Certain endophytes exhibit antagonistic activities against pathogenic fungi and bacteria by producing antimicrobial compounds (Compant et al. 2019). Moreover, endophytes induce systemic resistance in plants, priming them for a more robust defense response. This inherent ability to enhance plant immunity makes endophytic microbiomes valuable allies in the management of crop diseases, reducing the reliance on chemical pesticides and fostering sustainable agricultural practices (Miliute et al. 2020). The ability of certain endophytes to fine-tune the plant's defense mechanisms, inducing systemic resistance, has been a focal point of recent research. This widespread conditioning not only strengthens the plant's defense against pathogens but also highlights the prospect of creating bio-based elicitors for sustainable crop protection. The genomic characterization of such endophytic strains has allowed researchers to identify the molecular mechanisms behind their biocontrol activities, paving the way for targeted applications in disease management.

14.6.3 Environmental Sustainability

The application of endophytic microbiomes aligns with the principles of environmental sustainability by promoting agroecosystem health. Endophytes contribute to reduced dependency on synthetic fertilizers and pesticides, mitigating environmental pollution and preserving biodiversity (Lugtenberg and Kamilova 2009). Additionally, certain endophytes play a pivotal role in phytoremediation, aiding in the cleanup of contaminated soils by facilitating the degradation or sequestration of pollutants (Weyens et al. 2019). This underscores the potential of endophytic microbiomes to support sustainable agricultural practices that are both economically viable and environmentally responsible. The incorporation of endophytic microbiomes in agricultural ecosystems extends beyond the reduction of chemical inputs. Certain endophytes, through mechanisms such as nitrogen fixation or solubilization of phosphates, contribute to improved nutrient availability, promoting soil fertility and reduce the need for external fertilizers (Chaudhary et al. 2022). This dual benefit of minimizing synthetic inputs and optimizing nutrient utilization aligns with the broader goals of sustainable agriculture. Endophytic microbiomes are also known to confer drought tolerance, enhance resistance to extreme temperatures, and mitigate the impact of soil-borne diseases. By enhancing the plant's natural defenses and adaptive capabilities, endophytes contribute to crop stability in the face of changing climatic conditions, a critical aspect in the context of climate change (Anand et al. 2023).

In addition to their role in agroecosystem health, certain endophytes exhibit unique traits that make them potential candidates for the development of novel bio-products. Bioactive compounds produced by specific endophytic strains have demonstrated pharmaceutical and industrial applications, opening new avenues for biotechnological innovations. The application of endophytic microbiomes transcends conventional pest and disease management. It represents a holistic approach to sustainable agriculture by addressing not only immediate challenges but also contributing to long-term soil health, climate resilience, and the development of valuable bio-products.

14.6.4 Stress Mitigation

Endophytic microbiomes act as key players in enhancing plant resilience to various environmental stresses, contributing to stress mitigation strategies in agriculture. Through the modulation of plant physiological processes, endophytes aid in the alleviation of abiotic stresses such as drought, salinity, and extreme temperatures (Marasco et al. 2012). The underlying mechanisms through which endophytes alleviate these stresses are explored from determining the molecular and biochemical processes by which the endophytes sustain within the host plants. Recent research highlights the potential of endophytic microorganisms to serve as bioinoculants for developing stress-tolerant crops, a critical aspect in adapting agriculture to the challenges posed by climate change (Pandey et al. 2016a; Trivedi et al. 2020). This revelation opens avenues for sustainable agricultural practices, aligning with the urgent need to adapt to the challenges posed by climate change. In the broader context, the integration of endophytes into agricultural strategies not only highlights their significance in stress mitigation but also offers a promising solution for ensuring food security in the face of a dynamically changing climate. The versatile benefits of endophytic microbiomes underscore their prominent role in shaping sustainable and resilient agricultural systems, contributing to global efforts for environmentally conscious and productive food production.

14.7 Future Directions and Emerging Research in Endophytic Microbiome as Probiotics for Plant Health

The exploration of endophytic microbiomes as probiotics for plant health represents a flourishing field of research with significant implications for sustainable agriculture. As concerns about environmental impact and food security rise, understanding the intricate relationships between plants and their associated microorganisms has gained prominence. It is the need of the hour to discuss the future directions and emerging research avenues in harnessing endophytic microbiomes as plant probiotics by highlighting their potential to surpass conventional chemical interventions for sustainable plant health.

14.7.1 Significance of Microbiome Over Chemical Substances

The shift toward utilizing endophytic microbiomes as plant probiotics stems from the recognition of their multidirectional benefits, which is proven superior to the efficacy shown by the chemical substances. Unlike synthetic fertilizers and pesticides that can have detrimental effects on the environment and nontarget organisms, endophytic microbiomes establish mutualistic relationships with plants, promoting growth, enhancing nutrient uptake, and mitigating stress (Berendsen et al. 2012). The microbiome's ability to induce systemic resistance in plants contributes to a more sustainable and eco-friendly approach to crop protection compared to chemical treatments (Lugtenberg and Kamilova 2009). Systemic resistance induced by endophytes not only protects the host plant from various pathogens but also minimizes the reliance on external chemical inputs, aligning with the principles of sustainable agriculture. Several ongoing research is uncovering additional dimensions of endophytic microbiome functionality. Recent studies have explored the role of these microbiomes in improving soil health, nutrient cycling, and even their potential in bioremediation processes (Suman et al. 2022). By considering these emerging aspects, the utilization of endophytic microbiomes as plant probiotics presents a holistic and forward-thinking approach to agriculture, promoting both plant health and environmental sustainability. With the increasing understanding of these intricate relationships, the incorporation of endophytic microbiomes into mainstream agricultural practices holds the promise of revolutionizing crop management for a more resilient and eco-conscious future.

14.7.2 Future Directions for Revolutionizing Endophytic Microbiome-Based Formulations for Agricultural Practices

The future of research in endophytic microbiomes as probiotics for plant health involves delving into the molecular mechanisms with an emphasis on discovering and exploring these complex interactions. Understanding how specific endophytes contribute to plant growth promotion, nutrient acquisition, and stress tolerance at the genetic and biochemical levels can pave way to develop new formulations as per the need for each crop. The advent of advanced omics technologies, including metagenomics, metatranscriptomics, and metabolomics, allows the researchers to unfold the complex mechanisms of plant–microbe interactions in unprecedented detail (Trivedi et al. 2020). Emerging research in this field should focus on expanding the catalogue of beneficial endophytes and elucidating their functional roles in different plant species and environmental conditions. Exploring the diversity of endophytic microbiomes across various ecosystems and their adaptation strategies to different niches will contribute to the development of tailored probiotic formulations for specific crops and environments (Compant et al. 2019). Integrating artificial intelligence and machine learning approaches with microbiome data will enhance our ability to predict and manipulate plant–microbe interactions for optimal agricultural outcomes. Synthetic biology offers another scope of exploration with the exciting possibilities for engineering endophytic microbiomes to enhance their beneficial traits further. Synthetic biology tools can be employed to modify or introduce specific genes into endophytes, aiming to optimize their performance in promoting plant health and resilience (Parnell et al. 2016). Tailoring endophytic microbiomes through synthetic biology approaches holds promise for developing custom-made probiotics that address specific challenges faced by modern agriculture.

14.7.3 Challenges and Opportunities

Several challenges exist while addressing the potential applications of vast spread endophytic microbiomes as plant probiotics. These include understanding the stability of introduced microbial strains in diverse environments, ensuring compatibility with different plant hosts, and addressing potential unintended consequences. However, overcoming these challenges presents opportunities for innovation and the development of sustainable agricultural practices. The future of research in endophytic microbiomes as plant probiotics is promising, with ample opportunities for advancing the existing understanding of plant–microbe interactions and leveraging this knowledge for sustainable agriculture. The shift toward microbiome-based solutions emphasizes a significant paradigm transition from the dependency on chemical substances, and offers a holistic and environmentally conscious approach for enhancing plant health.

14.7.4 Integration of Multi-Omics Data (Genomics, Proteomics, Metabolomics) for a Comprehensive Understanding of Plant Microbial Communities and Their Functions

In the ever-evolving landscape of plant–microbe interactions, the integration of multi-omics data has emerged as a transformative approach, offering unparalleled insights into the intricate dynamics of microbial communities within the plant ecosystem. Genomics, proteomics, and metabolomics collectively provide a comprehensive toolkit for determining the unexplored genomic repertoire, protein expression patterns, and metabolic activities of microbial consortia associated with plants. By leveraging high-throughput sequencing technologies, researchers can explore the taxonomic diversity, functional capacities, and genetic adaptations of these microbes, laying the groundwork for a holistic understanding of their roles in plant health and development. By combining the genomics and proteomics steps into the realm of functional characterization, the actual protein expression patterns can be decoded within the microbial community (Géron et al. 2019). Detailed study on the proteome can provide valuable insights into the molecular machinery at work, offering a direct link between the genomic blueprint and the active functional players within the community. This refined method aids in discerning how microbial populations respond to environmental cues, interact with host plants, and contribute to essential physiological processes. Similarly, metabolomics, the final layer in this integrative approach, can define the chemical fingerprints of microbial activities. By scrutinizing the small molecules synthesized or modified by plant-associated microorganisms, metabolomics allows researchers to decipher the metabolic pathways at play (Gaudêncio et al. 2023). The integration of genomics, proteomics, and metabolomics into a cohesive framework facilitates a better clear understanding of plant microbial communities and their functions. This multi-omics approach transcends the limitations of individual data layers, offering a synergistic platform for deciphering the complexities of microbial interactions within the plant rhizosphere, endosphere, and phyllosphere. To solve the mystery of these relationships, the integration of multi-omics data stands as a powerful tool urging us toward adopting sustainable agricultural practices, biocontrol strategies, and a deeper comprehension of the delicate balance that governs the plant–microbe alliance.

14.8 Conclusion

The exploration of the endophytic microbiome has triggered a complex and dynamic aspect that significantly influences the plant health and ecosystem functioning. The diverse array of microorganisms, including bacteria, fungi, and other entities, forms an enriched community within plant tissues, contributing to the resilience and vitality of host plants. Factors influencing the composition of endophytic microbiomes range from environmental conditions to plant genotype, underscoring the intricate nature of these microbial communities. The functions and interactions of endophytic microorganisms in plant growth are multifaceted. They play a pivotal role and serve as a key mediator in contributing to the overall health of the plant and the surrounding environment. Leveraging the inherent capabilities of endophytes can revolutionize the agricultural approaches, moving away from chemical-intensive methods to eco-friendly and sustainable practices. Looking toward the future, the concept of microbial endophytes as probiotics for plant health stands as a promising avenue for the research and application. Integrating multi-omics data, including genomics, proteomics, and metabolomics, offers a holistic understanding of plant microbial communities, paving the way for precision microbiome engineering and tailored probiotic formulations. This integrative approach holds the key for unlocking the full potential of endophytic microbiomes in promoting sustainable agriculture, enhancing plant health, and contributing to the overall resilience of ecosystems. As we delve deeper into these frontiers, the synergy between scientific advancements and practical applications promises to reshape the landscape of plant–microbe interactions for the benefit of both agriculture and the environment.

Acknowledgement: The authors acknowledge the funding and facilities availed under the scheme RUSA 2.0.

References

Akram S, Ahmed A, He P, He P, Liu Y, Wu Y, Munir S, He Y (2023) Uniting the role of endophytic fungi against plant pathogens and their interaction. J Fungi 9:72. https://doi.org/10.3390/jof901007

Alijani Z, Jahanshir A, Morahem A, Bahramnejad B (2020) Volatile compounds mediated effects of *Stenotrophomonas maltophilia* strain UN1512 in plant growth promotion and its potential for the biocontrol of *Colletotrichum nymphaeae*. Physiol Mol Plant Pathol 112:101555

Anand U, Pal T, Yadav N, Singh VK, Tripathi V, Choudhary KK, Shukla AK, Sunita K, Kumar A, Bontempi E, Ma Y (2023) Current scenario and future prospects of endophytic microbes: promising candidates for abiotic and biotic stress management for agricultural and environmental sustainability. Microb Ecol 86:1455–1486. https://doi.org/10.1007/s00248-023-02190-1

Anuroopa N, Anshu BR, Ranadev P, Ashwin R, Bagyaraj DJ (2022) *Pseudomonas* species in soil as a natural resource for plant growth promotion and biocontrol characteristics—an overview. Madras Agric J 1:10–12

Azhari A, Supratman U (2021) The chemistry and pharmacology of fungal genus Periconia: a review. Sci Pharm 89:34. https://doi.org/10.3390/scipharm89030034

Balderas-Ruíz KA, Patricia B, Rosa IS, Víctor G, Sergio ACF, Salvador BO, Miriam MV, Sergio AO, Ángel AGG, Enrique G, Serrano-Carreón L (2020) *Bacillus velezensis* 83 a bacterial strain from mango phyllosphere, useful for biological control and plant growth promotion. AMB Express 1:1–19

Baron NC, Rigobelo EC (2021) Endophytic fungi: a tool for plant growth promotion and sustainable agriculture. Mycology 13:39–55. https://doi.org/10.1080/21501203.2021.1945699

Berendsen RL, Pieterse CM, Bakker PA (2012) The rhizosphere microbiome and plant health. Trends Plant Sci 17:478–486

Bulgarelli D, Garrido-Oter R, Münch PC, Weiman A, Dröge J, Pan Y, McHardy AC, Schulze-Lefert P (2015) Structure and function of the bacterial root microbiota in wild and domesticated barley. Cell Host Microbe 17(3):392–403. https://doi.org/10.1016/j.chom.2015.01.011

Caporaso JG, Lauber CL, Walters WA, Berg-Lyons D, Huntley J, Fierer N, Knight R (2011) Global patterns of 16S rRNA diversity at a depth of millions of sequences per sample. Proc Natl Acad Sci U S A 108:4516–4522

Chaudhary P, Agri U, Chaudhary A, Kumar A, Kumar G (2022) Endophytes and their potential in biotic stress management and crop production. Front Microbiol 13:933017. https://doi.org/10.3389/fmicb.2022.933017

Choudhary N, Dhingra N, Gacem A, Yadav VK, Verma RK, Choudhary M, Bhardwaj U, Chundawat RS, Alqahtani MS, Gaur RK, Eltayeb LB, Al Abdulmonem W, Jeon BH (2023)

Towards further understanding the applications of endophytes: enriched source of bioactive compounds and bio factories for nanoparticles. Front Plant Sci 14:1193573. https://doi.org/10.3389/fpls.2023.1193573

Compant S, Mitter B, Colli-Mull JG, Gangl H, Sessitsch A, Mülle AG (2019) Endophytes of grapevine flowers, berries, and seeds: identification of cultivable bacteria, comparison with other plant parts, and visualization of niches of colonization. Microb Ecol 77:495–507

Daniel AI, Fadaka AO, Gokul A, Bakare OO, Aina O, Fisher S, Burt AF, Mavumengwana V, Keyster M, Klein A (2022) Biofertilizer: the future of food security and food safety. Microorganisms 10:1220. https://doi.org/10.3390/microorganisms10061220

Deshmukh SK, Dufossé L, Chhipa H, Saxena S, Mahajan GB, Gupta MK (2022) Fungal endophytes: a potential source of antibacterial compounds. J Fungi 8:164. https://doi.org/10.3390/jof8020164

Devi R, Tanvir K, Divjot K, Ajar NY, Suman A (2022) Potential applications of mineral solubilizing rhizospheric and nitrogen fixing endophytic bacteria as microbial consortium for the growth promotion of chilli (*Capsicum annum* L.). Biologia 77:1–11

Di Y, Ling K, Pratiksha S, Lu-feng L, Lin-yan X, Lilian H, Li F (2022) Identification and characterization of *Bacillus subtilis* B9: a diazotrophic plant growth-promoting endophytic bacterium isolated from sugarcane root. J Plant Growth Regul 42:1720–1737

Edwards JA, Santos-Medellín CM, Liechty ZS, Nguyen B, Lurie E, Eason S, Phillips G, Sundaresan V (2018) Compositional shifts in root-associated bacterial and archaeal microbiota track the plant life cycle in field-grown rice. PLoS Biol 16:e2003862. https://doi.org/10.5061/dryad.7q7k1

Egamberdieva D, Wirth SJ, Alqarawi AA, Abd-Allah EF, Hashem A (2017) Phytohormones and beneficial microbes: essential components for plants to balance stress and fitness. Front Microbiol 8:2104. https://doi.org/10.3389/fmicb.2017.02104

Egamberdieva D, Wirth S, Bellingrath-Kimura SD, Mishra J, Arora NK (2019) Salt-tolerant plant growth promoting Rhizobacteria for enhancing crop productivity of saline soils. Front Microbiol 10:2791. https://doi.org/10.3389/fmicb.2019.02791

Etesami H, Mirseyed Hosseini H, Alikhani HA (2014) Bacterial biosynthesis of 1-aminocyclopropane-1-caboxylate (ACC) deaminase, a useful trait to elongation and endophytic colonization of the roots of rice under constant flooded conditions. Physiol Mol Biol Plants 20:425–434. https://doi.org/10.1007/s12298-014-0251-5

Etesami H, Adl SM, Maleki H, Chehregani A, Rastin N (2020) Plant growth-promoting endophytic bacteria enhance salinity stress tolerance in *Oryza sativa*. Environ Exp Bot 178:104124

Gamalero E, Bona E, Glick BR (2022) Current techniques to study beneficial plant-microbe interactions. Microorganisms 10:1380. https://doi.org/10.3390/microorganisms10071380

Gaudêncio SP, Bayram E, Lukić Bilela L, Cueto M, Díaz-Marrero AR, Haznedaroglu BZ, Jimenez C, Mandalakis M, Pereira F, Reyes F, Tasdemir D (2023) Advanced methods for natural products discovery: bioactivity screening, dereplication, metabolomics profiling, genomic sequencing, databases and informatic tools, and structure elucidation. Mar Drugs 21:308. https://doi.org/10.3390/md21050308

Géron A, Werner J, Wattiez R, Lebaron P, Matallana-Surget S (2019) Deciphering the functioning of microbial communities: shedding light on the critical steps in metaproteomics. Front Microbiol 10:2395. https://doi.org/10.3389/fmicb.2019.02395

Guerreiro J, Marhavý P (2023) Unveiling the intricate mechanisms of plant defense. Front Plant Physiol 1:1285373. https://doi.org/10.3389/fphgy.2023.1285373

Guo L, Yu H, Wang B, Vescio K, DeIulio GA, Yang H, Berg A, Zhang L, Edel-Hermann V, Steinberg C, Kistler HC, Ma LJ (2021) Metatranscriptomic comparison of endophytic and pathogenic Fusarium-Arabidopsis interactions reveals plant transcriptional plasticity. Mol Plant Microbe Interact 34:1071–1083. https://doi.org/10.1094/MPMI-03-21-0063-R

Handelsman J, Rondon MR, Brady SF, Clardy J, Goodman RM (1998) Molecular biological access to the chemistry of unknown soil microbes: a new frontier for natural products. Chem Biol 5:R245–R249

Hardoim PR, van Overbeek LS, van Elsas JD (2015) Properties of bacterial endophytes and their proposed role in plant growth. Trends Microbiol 23:329–337

Hashem A, Tabassum B, Fathi Abd Allah E (2019) *Bacillus subtilis*: a plant-growth promoting rhizobacterium that also impacts biotic stress. Saudi J Biol Sci 26:1291–1297. https://doi.org/10.1016/j.sjbs.2019.05.004

Hazarika SN, Kangkon S, Atlanta B, Thakur D (2021) Prospecting endophytic bacteria endowed with plant growth promoting potential isolated from *Camellia sinensis*. Front Microbiol 12:738058

He X, Han G, Lin Y et al (2012) Diversity and decomposition potential of endophytes in leaves of a *Cinnamomum camphora* plantation in China. Ecol Res 27:273–284

Hernández-Soberano C, León FR, ValenciaCantero E (2020) Endophytic bacteria *Arthrobacter agilis* UMCV2 and *bacillus methylotrophicus* M4-96 stimulate achene germination, in vitro growth, and greenhouse yield of strawberry (Fragaria × ananassa). Sci Hortic 261:109005

Hewitt S, Ghogare R, Troxel W, Tenic E, Isaac D, Dhingra A (2023) Metatranscriptomic analysis of tomato rhizospheres reveals insight into plant-microbiome molecular response to biochar-amended organic soil. Front Anal Sci 3:1205583. https://doi.org/10.3389/frans.2023.1205583

Jain M, Olsen HE, Paten B, Akeson M (2018) The Oxford Nanopore MinION: delivery of nanopore sequencing to the genomics community. Genome Biol 19:239

Jha P, Kaur T, Chhabra I, Panja A, Paul S, Kumar V, Malik T (2023) Endophytic fungi: hidden treasure chest of antimicrobial metabolites interrelationship of endophytes and metabolites. Front Microbiol 14:1227830. https://doi.org/10.3389/fmicb.2023.1227830

Jia H, Zuguo X, Jinqing M, Yifan L, Changhong H, Mengqian L, Zheng-Zhu Z, Deng W (2022) Endophytic bacteria from the leaves of two types of albino tea plants, indicating the plant growth promoting properties. J Plant Growth Regul 2:331–343

Juby S, Krishnankutty RE, Kochupurakkal J (2023) Drought-alleviating effects of endophytic bacteria isolated from xerophytic plants on *Capsicum annuum* L. seedlings. Curr Microbiol 80:403. https://doi.org/10.1007/s00284-023-03494-8

Khan AL, Al-Harrasi A, Al-Rawahi A, Al-Farsi Z, Al-Mamari A, Waqas M, Asaf S, Elyassi A, Mabood F, Shin JH, Lee IJ (2016) Endophytic fungi from frankincense tree improves host growth and produces extracellular enzymes and indole acetic acid. PLoS One 11:e0158207. https://doi.org/10.1371/journal.pone.0158207

Khare E, Mishra J, Arora NK (2018) Multifaceted interactions between endophytes and plant: developments and prospects. Front Microbiol 9:2732. https://doi.org/10.3389/fmicb.2018.02732

Knapp DG, Lázár A, Molnár A, Vajna B, Karácsony Z, Váczy KZ, Kovács GM (2021) Above-ground parts of white grapevine *Vitis vinifera* cv. Furmint share core members of the fungal microbiome. Environ Microbiol Rep 13:509–520. https://doi.org/10.1111/1758-2229.12950

Krishnamoorthy A, Abhishek G, Pinaki S, Maiti MK (2021) Metagenomics of two gnotobiotically grown aromatic rice cultivars reveals genotype-dependent and tissue-specific colonization of endophytic bacterial communities attributing multiple plant growth promoting traits. World J Microbiol Biotechnol 4:1–16

Kumari P, Deepa N, Trivedi PK, Singh BK, Srivastava V, Singh A (2023) Plants and endophytes interaction: a "secret wedlock" for sustainable biosynthesis of pharmaceutically important secondary metabolites. Microb Cell Fact 22:226. https://doi.org/10.1186/s12934-023-02234-8

Kusari S, Verma VC, Lamshoeft M, Spiteller M (2012) An endophytic fungus from Azadirachta indica a. Juss. that produces azadirachtin. World J Microbiol Biotechnol 28:1287–1294

Lata R, Chowdhury S, Gond SK, White JF Jr (2018) Induction of abiotic stress tolerance in plants by endophytic microbes. Lett Appl Microbiol 66:268–276. https://doi.org/10.1111/lam.12855

Li L, Feng Y, Qi F, Hao R (2023a) Research progress of *Piriformospora indica* in improving plant growth and stress resistance to plant. J Fungi 9:965. https://doi.org/10.3390/jof9100965

Li X, He P, He P, Li Y, Wu Y, Mu C, Munir S, He Y (2023b) Native endophytes from maize as potential biocontrol agents against bacterial top rot caused by cross-kingdom pathogen *Klebsiella pneumoniae*. Biol Control 178:105131

Liu Y, Nan L, Yan H, Zhang D, Han X (2016) Isolation and identification of resveratrol producing endophytes from wine grape cabernet sauvignon. Springerplus 5:1029. https://doi.org/10.1186/s40064-016-2571-0

Lugtenberg B, Kamilova F (2009) Plant-growth-promoting rhizobacteria. Annu Rev Microbiol 63:541–556

Mahdi T, Mohamed L, Yagi S (2014) Endophytic fungal communities associated with ethno-medicinal plants from Sudan and their antimicrobial and antioxidant prospective. J Forest Prod Ind 3:248–256

Maheshwari R, Pradeep K, Namita B, Suneja P (2022) Exploration of plant growth-promoting endophytic bacteria from *Pisum sativum* and *Cicer arietinum* from South–West Haryana. J Basic Microbiol 7:857–874

Marasco R, Rolli E, Ettoumi B, Vigani G, Mapelli F, Borin S, Hadid A, Ayman F, Behairy EI, Usama A, Claudia S, Ameur C, Graziano Z, Daffonchio D (2012) A drought resistance-promoting microbiome is selected by root system under desert farming. PLoS One 7:e48479. https://doi.org/10.1371/journal.pone.0048479

Mathieu E, Léjard V, Ezzine C, Govindin P, Morat A, Giat M, Lapaque N, Doré J, Blottière HM (2023) An insight into functional metagenomics: a high-throughput approach to decipher food-microbiota-host interactions in the human gut. Int J Mol Sci 24:17630. https://doi.org/10.3390/ijms242417630

Miliute I, Buzaite O, Baniulis D, Stanys V, Borriss R (2020) Next-generation sequencing reveals different endophytic bacteria in oilseed rape and winter wheat. Sci Rep 10:1–11

Moussa M, Ebrahim W, El-Neketi M, Mándi A, Kurtán T, Hartmann R, Lin W, Liu Z, Proksch P (2016) Tetrahydroanthraquinone derivatives from the mangrove-derived endophytic fungus *Stemphylium globuliferum*. Tetrahedron Lett 57:4074–4078

Navgire GS, Goel N, Sawhney G, Sharma M, Kaushik P, Mohanta YK, Mohanta TK, Al-Harrasi A (2022) Analysis and interpretation of metagenomics data: an approach. Biol Proced 24:1–22. https://doi.org/10.1186/s12575-022-00179-7

Overholt WA, Hölzer M, Geesink P, Diezel C, Marz M, Küsel K (2020) Inclusion of Oxford Nanopore long reads improves all microbial and viral metagenome-assembled genomes from a complex aquifer system. Environ Microbiol 22:4000–4013. https://doi.org/10.1111/1462-2920.15186

Pandey P, Kang SC, Maheshwari DK (2016a) Isolation of endophytic plant growth-promoting *Burkholderia* sp. MSSP from root nodules of *Mimosa pudica*. Curr Sci 90:512–518

Pandey SS, Singh S, Babu CS, Shanker K, Srivastava NK, Shukla AK, Kalra A (2016b) Fungal endophytes of *Catharanthus roseus* enhance vindoline content by modulating structural and regulatory genes related to terpenoidindole alkaloid biosynthesis. Sci Rep 25:265–283

Parnell JJ, Berka R, Young HA, Sturino JM, Kang Y, Barnhart DM, DiLeo MV (2016) From the lab to the farm: an industrial perspective of plant beneficial microorganisms. Front Plant Sci 7:1110

Rabbee MF, Ali MS, Baek KH (2019) Endophyte *Bacillus velezensis* isolated from *Citrus* spp. controls streptomycin-resistant *Xanthomonas citri* subsp. citri that causes citrus bacterial canker. Agronomy 8:470. https://doi.org/10.3390/agronomy9080470

Rani R, Sharma D, Chaturvedi M, Yadav JP (2017) Antibacterial activity of twenty different endophytic fungi isolated from *Calotropis procera* and time kill assay. Clin Microbiol 6:1000280. https://doi.org/10.4172/2327-5073.1000280

Riskuwa-Shehu ML, Ismail HY, Ijah UJ (2020) Heavy metal resistance by endophytic bacteria isolated from guava (*Psidium guajava*) and mango (*Mangifera indica*) leaves. Int Ann Sci 11:16–23

Rosenblueth M, Martínez-Romero E (2006) Bacterial endophytes and their interactions with hosts. Mol Plant Microbe Interact 19:827–837. https://doi.org/10.1094/MPMI-19-0827

Santana SRA, Tadeu VV, Gabiane RA, Valterlina MS, Welson LS, Carolina V, Paulo FJ (2020) Inoculation of plant growth-promoting bacteria attenuates the negative effects of drought on sorghum. Arch Microbiol 5:1015–1024

Schlaeppi K, Bulgarelli D (2015) The plant microbiome at work. Mol Plant Microbe Interact 28:212–221. https://doi.org/10.1094/MPMI-10-14-0334-FI

Schulz B, Boyle C (2006) What are endophytes? In: Microbial root endophytes. Springer, Berlin, Heidelberg, pp 1–13

Shah D, Mohammad SK, Shahkaar A, Haidar A, Pecoraro L (2021) Molecular and biochemical characterization, antimicrobial activity, stress tolerance, and plant growth-promoting effect of endophytic bacteria isolated from wheat varieties. Microorganisms 1:21

Shakya M, Gottel N, Castro H, Yang ZK, Gunter L, Labbé J, Muchero W, Bonito G, Vilgalys R, Tuskan G, Podar M, Schadt CW (2019) A multifactor analysis of fungal and bacterial community structure in the root microbiome of mature *Populus deltoides* trees. PLoS One 14:e0215425

Sharma A, Kaushik N, Sharma A, Marzouk T, Djébali N (2022) Exploring the potential of endophytes and their metabolites for bio-control activity. 3 Biotech 12:277. https://doi.org/10.1007/s13205-022-03321-0

Sood M, Kapoor D, Kumar V, Sheteiwy MS, Ramakrishnan M, Landi M, Araniti F, Sharma A (2020) Trichoderma: the 'secrets' of a multitalented biocontrol agent. Plan Theory 9:762. https://doi.org/10.3390/plants9060762

Souza A, Cruz JC, Sousa NR, Procópio AR, Silva GF (2014) Endophytic bacteria from banana cultivars and their antifungal activity. Genet Mol Res 13:8661–8670

Spaepen S, Vanderleyden J, Remans R (2007) Indole-3-acetic acid in microbial and microorganism-plant signaling. FEMS Microbiol Rev 31:425–448. https://doi.org/10.1111/j.1574-6976.2007.00072.x

Stajkovic O, De Meyer S, Milicic B, Willems A, Delic D (2009) Isolation and characterization of endophytic nonrhizobial bacteria from root nodules of alfalfa (*Medicago sativa* L.). Bot Serbica 33:107–114

Sugawara M, Shah GR, Sadowsky MJ, Paliy O, Speck J, Vail AW, Gyaneshwar P (2011) Expression and functional roles of *Bradyrhizobium japonicum* genes involved in the utilization of inorganic and organic sulfur compounds in free-living and symbiotic conditions. Mol Plant Microbe Interact 24:451–457. https://doi.org/10.1094/MPMI-08-10-0184

Suman J, Rakshit A, Ogireddy SD, Singh S, Gupta C, Chandrakala J (2022) Microbiome as a key player in sustainable agriculture and human health. Front Soil Sci 2:821589. https://doi.org/10.3389/fsoil.2022.821589

Tiwari P, Bae H (2023) Trends in harnessing plant endophytic microbiome for heavy metal mitigation in plants: a perspective. Plan Theory 12:1515. https://doi.org/10.3390/plants12071515

Trivedi P, Leach JE, Tringe SG, Sa TM, Singh BK (2020) Plant–microbiome interactions: from community assembly to plant health. Nat Rev Microbiol 18:607–621

Tsalgatidou PC, Thomloudi EE, Nifakos K, Delis C, Venieraki A, Katinakis P (2023) *Calendula officinalis*—a great source of plant growth promoting endophytic bacteria (PGPEB) and biological control agents (BCA). Microorganisms 11:206

Tsipinana S, Husseiny S, Alayande KA, Raslan M, Amoo S, Adeleke R (2023) Contribution of endophytes towards improving plant bioactive metabolites: a rescue option against red-taping of medicinal plants. Front Plant Sci 14:1248319. https://doi.org/10.3389/fpls.2023.1248319

Wahab A, Muhammad M, Munir A, Abdi G, Zaman W, Ayaz A, Khizar C, Reddy SP (2023) Role of arbuscular mycorrhizal fungi in regulating growth, enhancing productivity, and potentially influencing ecosystems under abiotic and biotic stresses. Plan Theory 17:3102. https://doi.org/10.3390/plants12173102

Wang Y, Dai CC (2011) Endophytes: a potential resource for biosynthesis, biotransformation, and biodegradation. Ann Microbiol 61:207–215

Wang LW, Xu BG, Wang JY, Su ZZ, Lin FC, Zhang CL, Kubicek CP (2012) Bioactive metabolites from Phoma species, an endophytic fungus from the Chinese medicinal plant *Arisaema erubescens*. Appl Microbiol Biotechnol 93:1231–1239

Wang YJ, Deering AJ, Kim HJ (2021) Effects of plant age and root damage on internalization of Shiga toxin-producing Escherichia coli in leafy vegetables and herbs. Horticulturae 7:68. https://doi.org/10.3390/horticulturae7040068

Wen J, Okyere SK, Wang J, Huang R, Wang Y, Liu L, Nong X, Hu Y (2023) Endophytic fungi isolated from *Ageratina Adenophora* exhibits potential antimicrobial activity against multidrug-resistant *Staphylococcus aureus*. Plan Theory 12:650. https://doi.org/10.3390/plants12030650

Weyens N, van der Lelie D, Taghavi S, Vangronsveld J (2019) Phytoremediation: plant–endophyte partnerships take the challenge. Curr Opin Biotechnol 21:556–560

Wu Y, Xiao S, Qi J, Gong Y, Li K (2022) *Pseudomonas fluorescens* BsEB-1: an endophytic bacterium isolated from the root of *Bletilla striata* that can promote its growth. Plant Signal Behav 17:2100626. https://doi.org/10.1080/15592324.2022.2100626

Yan A, Wang Y, Tan SN, Mohd Yusof ML, Ghosh S, Chen Z (2020) Phytoremediation: a promising approach for revegetation of heavy metal-polluted land. Front Plant Sci 11:359. https://doi.org/10.3389/fpls.2020.00359

Yu Y, Ying G, Zijie L, Chunhao J, Jianhua G, Niu D (2022) Induced systemic resistance for improving plant immunity by beneficial microbes. Plan Theory 3:386

Zhang H, Sun H, Zhou S, Bai N, Zheng X, Li S, Zhang J, Lv W (2019) Effect of straw and straw biochar on the community structure and diversity of ammonia-oxidizing bacteria and archaea in rice-wheat rotation ecosystems. Sci Rep 9:9367. https://doi.org/10.1038/s41598-019-45877-7

Zhang Z, Tingting L, Xiao Z, Jing X, Ya W, Riming Y, Yumei J, Zhu D (2021) Cultivable endophytic bacteria in seeds of Dongxiang wild rice and their role in plant-growth promotion. Diversity 13:665

Zheng CJ, Li L, Han T, Qin LP (2012) Identification of a quinazoline alkaloid produced by *Penicillium vinaceum*, an endophytic fungus from *Crocus sativus*. Pharm Biol 50:129–133

Zouari I, Jlaiel L, Tounsi S, Trigui M (2016) Biocontrol activity of the endophytic *bacillus amyloliquefaciens* strains CEIZ-11 against *Pythium aphanidermatum* and purification of its bioactive compounds. Biol Control 100:54–62

Chapter 15
Biotic Stress to Plants: Fungal Pathogen as a Major Biotic Stress

Udit Yadav and Poonam C. Singh

Abstract Plants are subjected to a variety of stressors during their growth cycle. These stressors had an impact not just on general plant health but also on crop yield. Despite the extensive use of agrochemicals in agriculture, each year crop yield decreases due to losses in crops caused by a variety of stress factors like biotic and abiotic factors. Abiotic stress includes soil pollution, temperature, drought, humidity, and soil water content, among others. Parasites such as viruses, bacteria, insects, and fungi are examples of biotic stressors. Due to their rapid spreading rate and disruptive infection through seeds or by whole plant body, fungal diseases are regarded as one of the leading causes of crop losses among these biotic agents. Fungal diseases can potentially result in a crop yield loss of up to 100%. Several natural, chemical, and biological control techniques for fungal illnesses have been documented. Wilt, Blight, Rust, powdery mildew, and chocolate spot are the most frequent fungal diseases. The current article focuses on diverse biotic stress inducers of plants, with a particular emphasis on fungal infections.

Keywords Biotic stress · Crop plants · Fungal disease · Bacteria · Viruses

U. Yadav
Department of Botany, IFTM University, Moradabad, India

P. C. Singh (✉)
Microbial Technology Division, CSIR-National Botanical Research Institute, Rana Pratap Marg, Lucknow, India

Academy of Scientific and Innovative Research (AcSIR), Ghaziabad, India
e-mail: pc.singh@nbri.res.in

P. Singh Chauhan et al. (eds.), *Plant-Microbe Interaction and Stress Management*, Rhizosphere Biology,
https://doi.org/10.1007/978-981-97-4239-4_15

15.1 Introduction

Stress is defined as a change in physiological, morphological, and genetic conditions induced by a variety of causes that tend to disrupt the equilibrium state (Yadav et al. 2020). Stress in plants refers to external factors that have a negative impact on agricultural plant growth, development, and productivity (Chakraborti et al. 2022). Crop production is influenced by many abiotic factors such as drought, frost, freezing, heat, salt waterlogging, and mineral toxicities, as well as important biotic stresses such as fungal, bacteria, and viral diseases (Gupta et al. 2021). These stresses result in massive crop losses. However, fungal diseases cause a significant decline in crop productivity. Fungal infections have a qualitative and quantitative impact on crop yield, causing 30–100% yield losses in plants (Gaur et al. 2020). Plant-associated diseases are defined as the appearance of morphological and physiological problems caused by biotic or abiotic stresses. Fungi are the most important biotic stressors impacting crop yield. Parasites, viruses, bacteria, and weeds, on the other hand, can reduce crops output (Mahmoud 2021). Fungal spores can spread rapidly throughout broad areas of the world, resulting in major yield losses and a significant drop in the crop quality and quantity compared to other microbial diseases (Chakraborty et al. 2021). Therefore, fungal pathogens have been considered as the principal biotic factors affecting crop productivity. Biotrophic infections such as downy mildews, rusts, and powdery mildews are common foliar fungal diseases that harm crops (Kakoti et al. 2020). Fusarium wilt is one of the most prevalent and devastating fungal infections impacting a variety of crops (Kumari and Katoch 2020).

The majority of biotic stresses are managed by creating and generating resistant and tolerant crop types rather than the present crop variations (Pérez-Méndez et al. 2021). Disease management options include crop rotation, integrated pest management, tillage, and weed control (Stark and Thornton 2020). Intercropping with other crops has been demonstrated to lower microbial counts in the rhizosphere and phyllosphere, indicating a decrease in fungal disease occurrence (Mahmoud et al. 2018). Natural oils were also discovered like garlic oil to lower the number of the most plant pathogen (Mansour et al. 2021); essential oils of many medicinal plants like Ocimum, thyme as well as their extracts, have inhibitory actions against many pathogenic bacteria and fungi (El-Mohamedy et al. 2017). By inhibiting their fungal hydrolytic enzymes, new antifungal medicines based on metal have shown effectiveness against many fungal pathogenic species infections. The metal ion causes fungal morphological changes that can limit fungal growth and function (Ibrahim et al. 2020) or disrupt cell metabolism by altering the antioxidant system (Mahmoud et al. 2021). Biological control of fungus with the use of several new biological agents such as *Rhizobacteria*, *Bacillus*, *Pseudomonas*, *Aspergillus*, *Aspergillus niger*, and *Trichoderma harzianum* constitute the most environmentally friendly microorganisms of disease management (Suprapta 2012).

15.2 Biotic Stress

Biotic stress in plants are induced by the living organisms, such as viruses, bacteria, and fungi. These living organisms are called plant pathogen. These plant pathogen causes variety of illnesses, infections, and crop plant damage, reducing crop productivity in their hosts, which can finally lead to plant death (Ahmad et al. 2021). The following are examples of biotic stress in plants:

15.2.1 Nematodes

Nematodes are basically roundworms that looks like animal parasites seen in cattles and pets. Many nematodes can attack crop plants by root hair, causing planta stabilization or endoparasite migration into the stem and bulb. *Chenopodium album*, which is weeds, have also been observed to be infected by nematodes. Nematode-infected seeds have the ability to entirely destroy the crop. However, considerable nematode infection avoidance in resistant crop varieties has been documented (Abd-Elgawad 2022). *Meloidogyne* spp. is a lethal parasite that is expanding over the world. It is a huge species of nematode that can infect a variety of crops (Tapia-Vázquez et al. 2022). Van der Linde et al. (1956) discovered the first nematode parasitizing a local crop in the peanut root infected with *Meloidogyne arenaria*. Infected plants exhibit signs such as gall formation, which is distinguished by elongated or roundish, knot-like protuberances (Sikandar et al. 2020). The size of galls is determined by a variety of parameters, including species type and host state (Sikandar et al. 2020). These galls are like nodules formed by the *Rhizobium* found in legumes plants, but these galls can be distinguished because nodules easily rubbed off the plant roots, whereas galls establish very firm contact with integral parts of the root and are difficult to remove without damaging the roots (Fitoussi et al. 2021). To reduce output losses caused by nematode infection, genetically modified plant development is a cost-effective and environmentally benign management method (Sivasubramaniam et al. 2020). The use of synthetic nematicides to destroy pathogenic nematodes is a widespread and popular method (Sivasubramaniam et al. 2020).

15.2.2 Viruses

Viruses are the tiniest obligatory parasites of plants. The basic viral particle is made up of a protein coat that surrounds either deoxyribonucleic acid (DNA) or ribonucleic acid (RNA) genetic material (Blystad et al. 2020). Viruses can have a wide host range, including tomato and potato plants, as well as nonleguminous plants (Mkhize et al. 2023). Because viruses rely on live cells for life and growth, the symptoms vary significantly and are frequently very confusing due to resemblances with abiotic and biotic symptoms. Mosaic virus produces various kinds of symptoms like varying degrees of rolling or narrowing of leaf, yellowing, necrosis, and, eventually, death of the plant (Jones 2021). However, the symptoms of other viruses infecting many may also be identical (Kaur et al. 2020). These plant viruses have a broad host range and require a connection between crop seasons for their transmission (Singh et al. 2020). There is no well-known specialized strategy for healing the plants of viral infection in legumes. The current viral control method includes preventing or reducing infection rates by cultivating resistance varieties of plants and removing the sources of infection (Punja et al. 2021).

15.2.3 Bacteria

Bacteria are the most common and simplest prokaryotic bacteria, with a size of less than 10 m and no nucleus or complicated organelles (Gracelin 2021). These bacteria can be beneficial like Plant growth-promoting bacteria (PGPR) and pathogenic like *Xanthomonas* sp. Plant pathogenic bacteria present in the soil can cause significant damage to the crops, such as the gram-negative *Xanthomonas* sp., that causes bacterial blight disease which is a significant bean-borne disease. *Xanthomonas* genus is the part of the Xanthomonadacea family which has 27 species causing various disease with a host range of up to 400 distinct plants and showing very high plant host specificity (Goettelmann 2023). In case of legumes plants, the most common bacterial pathogens of legumes are *Xanthomonas axonopodis pv. Phaseoli* and *Xanthomonas fuscans subsp. fuscans* (Chen et al. 2021), while *X. phaseoli* has been observed to be particularly aggressive (Chen et al. 2021). These two diseases spread around the world anytime beans are grown (Goettelmann 2023). These bacteria also caused major loss in crop in Ethiopia's Rift Valley and Eastern Ethiopia (Belete and Bastas 2021). Warm temperatures combined with humid weather increase disease severity, resulting in yield and quality losses (Girma et al. 2022). Depending on the climatic circumstances and plant genotype, output losses ranging from 10% to 45% have been documented (Shahriar et al. 2020). To cause infection in plants, pathogenic bacteria enter inside the plant and create inoculums through stomata. These inoculums multiply and cause the formation of water-soaked patches on leaves, which enlarge these patches. Infected leaf becomes flaccid. Turns lemon-yellow, and undergoes necrosis, resulting to defoliation and stem girdling (Belete and Bastas 2021). Lesions form quickly in warm conditions; they enlarge and combine in varied in size that depends on disease and plant stage. Highly diseased leaves tatter in the wind and eventually fall off the plant (Manjunatha et al. 2022). Infected seeds had patches on the seed coat that look like butter yellow to brown or were only found in the hilum area of the seed. Infected seeds showed poor germination or vigor in severe infection (Jadon et al. 2020). Bacteria reproduce rapidly in the cell gaps once they enter the plant, taking 10 to 14 days to establish the initial infection (Luneau et al. 2022). Field sanitation, bactericide application, elimination of weed, crop rotation, healthy seed plantation, burning of crop residues, seed treatment, and resistant cultivars use are all methods for managing *Xanthomonas* disease (Sharma et al. 2022). The foliage infection was controlled by spraying it every 7–14 days during damp cold weather with copper hydroxide, copper sulphate, and potassium methyl dithiocarbamates, which reduced the bacterial population (Ramadass and Thiagarajan 2021). Biological control has been seen as an environmentally acceptable method of reducing plant disease through microbial activity. By applying it to the seeds, *Pseudomonas* sp. lowers *Xanthomonas* disease by up to 39% (Dimkić et al. 2022). *Rahnella aquatilis* has shown promising disease suppression by antagonistic activity (Abdelmeguid et al. 2023). Kashyap and Siddiqui (2021) exploited *Rhizobium leguminosarum* ability to suppress *Xanthomonas* disease as a seed treatment.

15.2.4 Fungus

Fungi are unicellular or filamentous organisms having a chitin cell wall and an ergosterol cell membrane. They are well-known organic matter decomposers and recyclers, and they interact with plants both positively and negatively (Abo Nouh et al. 2021). They can thrive as free saprophytic organisms or in parasitic or symbiotic interactions with other organisms. Fungi are well-known for being the primary cause of more than 70% of major plant diseases. Some very common fungal diseases and their causative agents are shown in Table 15.1. Rice blast, rice sheath blight, wheat rust, powdery mildew of pulses, and other prevalent fungal diseases in crops include the following. The disease-induced physiological abnormalities are caused by the interaction of plants and fungi. According to Kulik et al. (2020), the top 10 fungal pathogens are *Magnaporthe oryzae*, *Botrytis cincerea*, *Puccinia* spp., *Fusarium graminearum*, *Fusarium oxysporum*, *Blumeria graminis*, *Mycosphaerella graminicola*, *Colletotrichum* spp., *Ustilago maydis*, and *Melampsora lini*. Wilting, cell disintegration or collapse, abscission, increased respiration, aberrant cell division, cell enlargement, and chlorophyll degeneration are all indications of infected plants (Chelladurai et al. 2023). Fungal disease cycle have various steps like inoculation, pre-penetration, penetration, infection, colonization, growth, and pathogen reproduction that lead to infection symptoms (Sharma 2021). The interaction between the plant and fungi occurs at certain areas to commence pathogen infection. Primary inoculum generates original fungal infections, whereas secondary inoculum is created as a result of the primary infections. Prior to penetration, fungal adhesion to the plant's exterior surface occurs. In powdery mildew, for example, cutinase enzyme releases at adhesion sites, making fungi more hydrophilic and thereby cementing the spore to the plant surface (Jaswal et al. 2020). Spores begin to germinate by utilizing stored nutrients such as lipids and carbohydrates to build germ tubes, which then expand to numerous locations within the plant (Li et al. 2022). The plant cell wall is breached by appressoria, which obtain nutrition via lysosomal secretions (Zhu et al. 2021). Natural plant aperture on the plant surface are the entering point of fungal pathogens inside the plant to begin the infection

Table 15.1 Common fungal disease occurs in crop plants

Plant disease	Host plant	Causal fungal pathogen	Symptoms	References
Powdery mildew	Cucumber, strawberry, grapevine	*Podosphaera xanthii*, *Erysiphe*, *Leveillula*, *Microsphaera*, *Sphaerotheca*, and *Oidium*	White, powdery growth of the fungus on infected portions of the plant host	Kiss et al. (2004), Pertot et al. (2008) and El Sharkawy et al. (2018)
Downy mildew	Lettuces, onions, pansies	*Pseudoperonospora cubensis*, *Hyaloperonospora parasitic*, *Plasmopara viticola*, *Peronospora belbahrii*	Discolored blotches on the upper leaf surface, these may be pale green, yellow, purple, or brown, depending on the plant affected	Zhang et al. (2017), Shahin (2017) and Wright (2004)
Early Blight disease	Tomato, potato, eucalyptus	*Alternaria solani*, *Alternaria tomatophila*	Brownish-black spots, concentric rings in the darkened area	Da Silva et al. (2021), Imran et al. (2022) and Kishore et al. (2008)
Late blight of potato	Tomato, potato	*Phytophthora infestans*	Small, light to dark green, circular to irregular-shaped water-soaked spots	Yao et al. (2016), Shanthiyaa et al. (2013) and Júnior et al. (2006)
Leaf spot	Soyabean, capsicum, ground nut	*Cercospora sojina*, *Cercospora capsici*, *Colletotrichum truncatum*	Spots on the leaves, surrounding areas of the leaf start to turn yellow	Simonetti et al. (2012), Jagtap et al. (2012) and Hasan et al. (2014)
Rust	Wheat, cereals, sugarcane, corn	*Cronartium ribicola*, *Puccinia coronata*, *Puccinia graminis*, *Puccinia triticina*	Pale leaf spots develop into spore-producing structures called pustules	Anderson et al. (2016), Tanner et al. (2015) and Shiomi et al. (2006)
Damping off	Tomato, cucumber	*Alternaria*, *Botrytis cinerea*, *Phytophthora*, *Sclerotium rolfsii*	Rotting stems at or near the soil line, root decay	Eid (2014), Asaka and Shoda (1996) and Zagade et al. (2012)
Loose smut	Barley, wheat	*Ustilago nuda*, *Ustilago tritici*	Smutted grain heads, infected plants to grow slightly taller and mature slightly sooner	Vishunavat (2013) and Meena and Ramyabharathi (2012)

(continued)

Table 15.1 (continued)

Plant disease	Host plant	Causal fungal pathogen	Symptoms	References
Fusarium Wilt	Onion, potato, tomato, brassica	*Fusarium oxysporum f.* sp. *lycopersici, F. oxysporum f.*sp. *cepae, F. oxysporum f.* sp. *tuberosi*	Causes severe root and crown rots or wilt diseases by attacking roots and basal stems; cucurbit fruit and potato tubers can be affected in storage	Bubici et al. (2019) and Gang et al. (2013)
Black root rot	Salvia, petunia, Persian cyclamen, snapdragon, begonia	*Thielaviopsis basicola, Berkeleyomyces basicola, Berkeleyomyces rouxiae*	Stunted and slow-growth, dark brown to black lesions form on the middle portion of the root	Schoina et al. (2011) and del Pilar Martínez-Diz et al. (2021)

process. Leaf browning, necrosis, stem dwarfism, and other signs characterize successful infections till plant death (Koshariya et al. 2023). Plant invasion is caused by the growth of fungal mycelia between the plant's cuticle and epidermis. After invasion of the plant, most phytopathogens, such as fungi, penetrate the plant tissues, multiply, and spread inside the plant until plant cells die (Sharma et al. 2021).

15.3 Major Fungal Diseases of Plants

In general, diseases caused by fungi inflict greater harm than diseases caused by other pathogens. The following are some fungal infections that cause considerable crop loss:

15.3.1 Powdery Mildew

Powdery mildew is a fungal disease that affects a wide variety of plants and is caused by a variety of ascomycete fungi of the Erysiphales order. Powdery mildew is an obligatory biotrophic disease that requires nutrients from the plant host to develop and reproduce. To produce enough pressure to tear the plant cell wall, the powdery mildew spore (conidium) forms an infection structure (appressorium).

15.3.1.1 Disease Cycle and Symptoms

Conidia of these fungi germinate on the leaf surface and use appressoria to penetrate the epidermal cells (Chethana et al. 2021). Following the passage from the surface to the inside, these fungi cross the mesophyll (rusts) or epidermis (powdery mildew) cell wall and generate haustoria, which are distinct infection structures that act as nutrient acquisition mediators. The haustoria develop beneath the plant cell wall but do not damage the cell membrane. Instead, they push and invade the cell membrane, forming a long-lasting "cell within cell" complex. The plant cell cytoplasm and tonoplast, a polysaccharide-rich extra-haustorial matrix, and the haustorial wall and membrane comprise this fungus-plant hybrid structure (Chethana et al. 2021). Following establishment within the host cells, the haustoria use a variety of transporters to push nutrients into the haustorium from the infected host cells (Polonio et al. 2021). Simultaneously, they produce effectors that dampen host defense and maintain invaded cells alive (Polonio et al. 2021). Downy mildew disease symptoms include the formation of white patches of fungal growth on the lower surface of the leaf. Under good conditions, the patches grow huge and unite to cover the entire lower surface. Other symptoms include an upward curl of the leaf edge and purplish to red spots on the leaves. On the underside of the leaves, tiny, spherical, black fungal structures (cleistothecia) may be present. Pathogens can colonize the fruit and form white mycelial growth on the seeds, giving them a powdery appearance. Fruit surfaces harden and may crack.

15.3.2 Downy Mildew

Downy mildews are largely foliage blights caused by obligate biotrophic plant parasites that cause significant economic losses on a variety of crops. It refers to any of numerous oomycete microorganisms that are obligate plant parasites. Downy mildews are only found in the Peronosporaceae family. They are a particular challenge in commercial agriculture for growers of crucifers, grapes, and vine-grown vegetables.

15.3.2.1 Disease Cycle and Symptoms

Plasmopara viticola is exclusively biotrophic, which means it can only develop in connection with living cells of its host, rather than in vitro, making it difficult to research. Its oospores can land in the soil and remain dormant until suitable conditions are met. It will then spread to a host plant and begin to grow (Koledenkova et al. 2022). This pathogen's life cycle involves both primary and secondary infections (Vigneron et al. 2023). It overwinters in leaf litter, shoots, and soil as oospores (resting sexual spores) and, in rare cases, as mycelium in infected but not dead twigs (Vigneron et al. 2023). In the spring, oospores germinate to form macrosporangia, which, when wet, release zoospores (asexual spores) that encyst and infect when they encounter host tissue. The encysted zoospores form germ tubes that enter the host through stomata and colonize the leaf parenchyma intercellularly with a diploid, non-segmented tubular mycelium that uptakes nutrients from host cells via intracellular haustoria (Koledenkova et al. 2022). Large, angular, or blocky yellow spots visible on the upper surface are the first signs of disease (Salcedo et al. 2021).

They are also characterized by the emergence of occasional yellow patches. Lesions grow quickly and turn brown as they age. The underside of diseased leaves looks wet. A purple-brown mold (see arrow) becomes visible upon closer investigation. With a 10× hand lens, little spores shaped like footballs can be seen among the molds. If the illness spreads further, leaves may develop a blighted appearance because of the numerous infectious spots. Downy mildew spreads quickly in disease-friendly conditions (cold nights with long dew periods), killing leaf tissue but not stems or petioles.

15.3.3 Fusarium Wilt

Fusarium is one of the most complex and adaptable genera in the Eumycota (Gordon 2017). Fusarium species are common soil-borne pathogens that cause vascular wilts, rots, and damping-off illnesses in a variety of horticultural and food crops (Bodah 2014). Fusarium wilt is a widespread plant disease caused by many strains of the soil-dwelling fungus *F. oxysporum*. Tomato (*Solanum lycopersicum* L.), banana (*Musa* spp. L.), melon (*Cucumis melo* L.), cotton (*Gossypium* spp. L.), and legumes are among the many plant species affected by *F. oxysporum*. Grain and forage legumes are responsible for 27% of global primary crop production, trailing just cereals and oilseeds (Jha et al. 2020). Their cultivation covers between 12% and 15% of the world's arable land, having a significant impact on agronomy, the environment, and human and animal nutrition and health.

15.3.3.1 Fusarium Wilt of Tomato

Tomatoes (*Lycopersicon esculentum*) are one of the most widely produced vegetables in the world. The major pathogen of tomato vascular wilt is *F. oxysporum f.* sp. *lycopersici* (FOL).

Disease Cycle and Symptoms

Tomato wilt is caused by FOL (Borisade and Uwaidem 2018). The FOL enters the epidermis of the root, vascular tissue helps in its spreading, and it inhabits inside the plant xylem vessels, causing xylem vessel blockage and severe water stress, culminating in wilt-like symptoms (Singh et al. 2017). The progression of *F. oxysporum* infection of plant vascular tissues is a complex event, with the following sequential processes involved in the infection process: (1) root recognition via signaling mechanism, (2) attachment to pathogen by root hair surfaces and proliferation of the hyphae, (3) root cortex penetration, entering vascular tissue, and differentiation within xylem arteries, and (4) spewing of poisons and virulence factors. Disease development and the typical withering of the host plant result from vascular colonization (Di et al. 2017). The disease's earliest symptoms show gradually in the lower leaves, followed by plant withering. FOL penetrates the tomato plant, colonizing and discoloring the vascular tissue, which extends to the apex, causing the plants to wilt, collapse, and die. The disease is characterized by wilted plants with yellow-colored leaves and little or no crop production. In the absence of a host, the latent chlamydospore of FOL in contaminated soil can live indefinitely (Khan et al. 2017).

15.3.3.2 Fusarium Wilt of Legumes

Legumes are one of the world's most significant crops. However, volatility in yield due to sensitivity to biotic stress has a significant impact on its productivity. Global production of legumes limited by Fusarium wilt, which is caused by the soil-borne pathogen *F. oxysporum* (Fo). In heavily affected fields of legume species, this disease can result in total losses. For example, *Fo f.* sp. *pisi* is a damaging pathogen of field pea (*Pisum sativum* L.) that has been identified in every country where pea is grown (Bani et al. 2018). The production of the world's most consumed food grain legume, common bean (*Phaseolus vulgaris* L.), is affected by the *Fo f.* sp. *phaseoli* (Xue et al. 2015). Annual losses owing to *Fo f.* sp. *ciceris* in chickpea (*Cicer arietinum* L.), the second most important global grain legume (Muche and Yemata 2022), can exceed 100% under favorable disease circumstances (Muche and Yemata 2022). The two principal fungal infections responsible for yield losses in lentil (*Lens culinaris Medik*) are vascular wilt caused by *Fo f.* sp. *lentis* (Jadhav et al. 2023) and Ascochyta blight (Henares et al. 2023). Fusarium wilt, caused by *Fo f.* sp. *medicaginis*, affects alfalfa (*Medicago sativa* L.) and annual medics such as barrel medic *(Medicago truncatula L.)*.

Disease Cycle and Symptoms

The infection cycle begins in the presence of a host. Under favorable condition, fungal spores start to germinate and elongate towards the host plant roots (Bani et al. 2018). These germinated spores penetrate the root without the creation of specialized structures through the natural holes found at intercellular connections of the cortical cells or through wounds (Xue et al. 2015). When hyphae enter inside the root, it invade the root cortex, pierce the endodermis, and reach up to the xylem vessels (Xue et al. 2015). The fungus then proceeds upwards through the xylem tissue, where it moves and multiplies, invading the host until the plant wilts and completely die (Muche and Yemata 2022). Following plant death, the fungus begins sporulation on the surface of the plant, distributing their spores on the soil for the next infection cycle (Henares et al. 2023). Symptoms of the Fusarium wilt disease includes browning, leaf epinasty, stunting, wilting, defoliation, and death of the plant (Muche and Yemata 2022). When roots and stems are broken vertically or cross-sectioned, they produce a dark-brown staining of xylem tissues. Wilting is caused by a combination of pathogen activity, the formation of fungal mycelium, and host defense responses, such as the development of gels and gums, which block or clog the xylem vessels and cause water stress like symptoms (Muche and Yemata 2022).

15.3.4 Early Blight Disease

Early blight is the most frequent potato disease observed in most potato-growing locations (Brouwer et al. 2023). The fungus *Alternaria solani* causes this disease, which causes progressive deterioration of host tissue by reducing the photosynthetic process. However, when the disease's symptoms are severe, yield loss exceeds 50%.

15.3.4.1 Disease Cycle and Symptoms

The main infection of potato foliage is produced by inoculum from another diseased plant or by infection that has over-seasoned on the infected plant detritus. Following initial infection, lesions undergo sporulation, and spores are expelled under favorable environmental circumstances (Hou et al. 2021). Following sporulation, spores are disseminated by wind on the leaves of vulnerable plants, resulting in germination and direct or indirect penetration of host epidermal cells via stomata or wound.

Early blight foliar symptoms first develop as small, irregular to round dark brown spots on the bottom surface of older leaves. Later in the growing season, multiple lesions on the upper leaves may emerge, and leaves may drop as symptoms worsen (Meno et al. 2021). Early blight infection in tubers manifests as black and sunken lesions on the surface. Tuber lesions can be circular or irregular in shape, with a raised dark brown border.

15.3.5 Damping Off

Damping off is a seedling disease caused by fungal and fungi-like organisms. Damping off is an emerging disease, which affects all agricultural and forestry crops. It may harm 5% to 80% of seedlings, resulting in economic ramifications for farmers. Damping off is caused by *Rhizoctonia solani*, *Aphanomyces euteiches*, *Phytophthora* spp., and *Verticillium* spp., which are typically found in Northern European and North American locations (Arora et al. 2021; Bedine Boat et al. 2020; Zaki et al. 2021).

15.3.5.1 Disease Cycle and Symptoms

Pythium forms motile spores that can infect plant roots when environmental conditions are appropriate, and a vulnerable host is present. These spores germinate to form hyphae, which penetrate the roots and damage plant tissue (Khrieba 2020). Pythium also generates sexual spores in the root or soil that are resistant to unfavorable environmental circumstances and can remain dormant for months or years. In the presence of a suitable host and environmental circumstances, the hyphae penetrate the host and initiate the infection process, resulting in plant tissue death. The diseases cause seedling rot prior to emergence, a black water-soaked lesion on the seedling stem, wilting, collapse, and seedling mortality (Arora et al. 2021).

15.4 Conclusion

Agriculture is one of the most essential components of India's economic growth. Because our need for these agricultural products has grown throughout time, so have the producing areas to meet it. These increases in growing areas are still insufficient due to crop losses caused by fungal infections, which might result in total crop loss. Common fungal diseases that affect agricultural crops include wilt, powdery mildew, rust, root rot, and so forth. Environmental factors and plant defense system influence the severity of infection. Many natural disease control approaches, such as intercropping, crop rotation, draining wet soils, fertilizing, tillage, and weed management, were considered. Chemical control with traditional and novel antifungal agents, such as calcium and copper-based compounds, demonstrated efficacy against fungal infections. Natural oils have also recently been employed as a safe controlling agent. Biological controls utilizing *Bacillus pumilus*, *Pseudomonas putida*, *T. harzianum*, and other microorganisms effectively prevent Fusarium root rot and other legume diseases. Despite all of the regulating tactics listed, researchers must continue to look for additional sustainable controlling agents in order to protect the environment and combat disease resistance development.

References

Abd-Elgawad MM (2022) Understanding molecular plant–nematode interactions to develop alternative approaches for nematode control. Plan Theory 11(16):2141

Abdelmeguid KA, Soliman ER, Hamada MA, El-Hendawy HH (2023) Biological control of pepper soft rot disease caused by *Pectobacterium carotovorum* using *Rahnella aquatilis*. Egypt J Bot 64(1):451–466

Abo Nouh FA, Gezaf SA, Abo Nahas HH, Abo Nahas YH, Vargas-De-La-Cruz C, Solorzano Acosta RA, Landa-Acuña D, Luis-Alaya B, Abdel-Azeem AM (2021) Bioprospecting for biomolecules from different fungal communities: an introduction. In: Industrially important fungi for sustainable development. Bioprospecting for biomolecules, vol 2, pp 1–71

Ahmad M, Ali Q, Hafeez MM, Malik A (2021) Improvement for biotic and abiotic stress tolerance in crop plants. Biol Clin Sci Res J 2021(1):e004

Anderson FE, López SPS, Sánchez RM, Fuentealba CGR, Barton J (2016) *Puccinia araujiae*, a promising classical biocontrol agent for moth plant in New Zealand: biology, host range and hyperparasitism by *Cladosporium uredinicola*. Biol Control 95:23–30

Arora H, Sharma A, Sharma S, Haron FF, Gafur A, Sayyed RZ, Datta R (2021) Pythium damping-off and root rot of *Capsicum annuum* L.: impacts, diagnosis, and management. Microorganisms 9(4):823

Asaka O, Shoda M (1996) Biocontrol of *Rhizoctonia solani* damping-off of tomato with *Bacillus subtilis* RB14. Appl Environ Microbiol 62(11):4081–4085

Bani M, Pérez-De-Luque A, Rubiales D, Rispail N (2018) Physical and chemical barriers in root tissues contribute to quantitative resistance to *Fusarium oxysporum f. sp. pisi* in pea. Front Plant Sci 9:199

Bedine Boat MA, Sameza ML, Iacomi B, Tchameni SN, Boyom FF (2020) Screening, identification and evaluation of *Trichoderma* spp. for biocontrol potential of common bean damping-off pathogens. Biocontrol Sci Technol 30(3):228–242

Belete T, Bastas KK (2021) Anti-quorum sensing effects of some medicinal and aromatic plant extracts on *Xanthomonas axonopodis* pv. Phaseoli. Turk J Agric Food Sci Technol 9:2537–2546

Blystad DR, Kvarnheden A, Valkonen J (2020) Plant viruses. Plant pathology and plant diseases. CABI, Wallingford, pp 107–131

Bodah ET (2014) Genetic mechanisms of resistance to Fusarium root rot in pea (*Pisum sativum L.*). Doctoral dissertation, Washington State University

Borisade OA, Uwaidem YI (2018) Priming F1-resistant tomato hybrid (Lindo-F1) seedlings with copper-I-oxide Metalaxyl composite fungicide before infection enhanced resistance to Fusarium wilt. Int J Pathog Res 1(3):1–13

Brouwer SM, Wolters PJ, Andreasson E, Liljeroth E, Vleeshouwers VG, Grenville-Briggs LJ (2023) Double trouble: co-infection of potato with the causal agents of late and early blight. Plant Pathol 72(8):1365–1379

Bubici G, Kaushal M, Prigigallo MI, Gómez-Lama Cabanas C, Mercado-Blanco J (2019) Biological control agents against Fusarium wilt of banana. Front Microbiol 10:616

Chakraborti S, Bera K, Sadhukhan S, Dutta P (2022) Bio-priming of seeds: plant stress management and its underlying cellular, biochemical and molecular mechanisms. Plant Stress 3:100052

Chakraborty A, Singh J, Upadhyay H (2021) Fungal diversity and ecology: major challenges for crop production in present climate scenario. In: Recent trends in mycological research: agricultural and medical perspective, vol 1, pp 89–107

Chelladurai K, Sujatha N, Valli LN (2023) Fungal infection in plant leaves—a review. In: E3S web of conferences, vol 399. EDP Sciences, Les Ulis, p 02010

Chen NW, Ruh M, Darrasse A, Foucher J, Briand M, Costa J, Studholme DJ, Jacques MA (2021) Common bacterial blight of bean: a model of seed transmission and pathological convergence. Mol Plant Pathol 22(12):1464–1480

Chethana KT, Jayawardena RS, Chen Y, Konta S, Tibpromma S, Phukhamsakda C, Abeywickrama PD, Samarakoon MC, Senwanna C, Mapook A, Tang X (2021) Appressorial interactions with host and their evolution. Fungal Divers 110(1):75–107

Da Silva HAO, Teixeira WD, Borges ÁV, Silva Junior AL, Alves KS, Rodrigues Junior OM, de Abreu LM (2021) Biocontrol of potato early blight and suppression of *Alternaria grandis* sporulation by *Clonostachys* spp. Plant Pathol 70(7):1677–1685

del Pilar Martínez-Diz M, Díaz-Losada E, Andrés-Sodupe M, Bujanda R, Maldonado-González MM, Ojeda S, Yacoub A, Rey P, Gramaje D (2021) Field evaluation of biocontrol agents against black-foot and petri diseases of grapevine. Pest Manag Sci 77(2):697–708

Di X, Gomila JO, Takken FL (2017) Involvement of salicylic acid, ethylene and jasmonic acid signalling pathways in the susceptibility of tomato to *Fusarium oxysporum*. Mol Plant Pathol 18(7):1024–1035

Dimkić I, Janakiev T, Petrović M, Degrassi G, Fira D (2022) Plant-associated *Bacillus* and *Pseudomonas* antimicrobial activities in plant disease suppression via biological control mechanisms—a review. Physiol Mol Plant Pathol 117:101754

Eid K (2014) Biological control of bean damping-off caused by *Sclerotium rolfsii*. Egypt J Phytopathol 42(1):179–191

El-Mohamedy RSR (2017) Plant essential oils for controlling plant pathogenic fungi. Volatiles and Food Security: Role of Volatiles in Agro-Ecosystems 171–198

El Sharkawy HHA, Abo El Wafa TSA, Ibrahim SAA (2018) Biological control agents improve the productivity and induce the resistance against downy mildew of grapevine. J Plant Pathol 100:33–42

Fitoussi N, Borrego E, Kolomiets MV, Qing X, Bucki P, Sela N, Belausov E, Braun Miyara S (2021) Oxylipins are implicated as communication signals in tomato–root-knot nematode (*Meloidogyne javanica*) interaction. Sci Rep 11(1):326

Gang G, Bizun W, Weihong M, Xiaofen L, Xiaolin Y, Chaohua Z, Jianhong M, Huicai Z (2013) Biocontrol of Fusarium wilt of banana: key influence factors and strategies. Afr J Microbiol Res 7(41):4835–4843

Gaur A, Kumar A, Kiran R, Kumari P (2020) Importance of seed-borne diseases of agricultural crops: economic losses and impact on society. In: Seed-borne diseases of agricultural crops: detection, diagnosis & management, pp 3–23. https://doi.org/10.1007/978-981-32-9046-4_1

Girma F, Fininsa C, Terefe H, Amsalu B (2022) Distribution of common bacterial blight and anthracnose diseases and factors influencing epidemic development in major common bean growing areas in Ethiopia. Acta Agric Scand Sect B Soil Plant Sci 72(1):685–699

Goettelmann F (2023) Resistance gene identification through advanced host-pathogen genomics in the Xanthomonas-ryegrass pathosystem. Doctoral dissertation, ETH Zurich

Gordon TR (2017) *Fusarium oxysporum* and the Fusarium wilt syndrome. Annu Rev Phytopathol 55:23–39

Gracelin DHS (2021) Bacteria: an overview. Front Life Sci 1:101

Gupta A, Bano A, Pandey N, Rai S, Sharma S, Pathak N (2021) Recent advances in biotic and abiotic stresses of *Solanum melongena* L. In: *Solanum melongena*: production, cultivation and nutrition, pp 151–171

Hasan MM, Islam MR, Hossain I, Shirin K (2014) Biological control of leaf spot of groundnut. J Biosci Agric Res 1(2):66–78

Henares BM, Blake SN, Farfan-Caceres L, Tahghighi H, Debler JW, Russ MH, Farquharson EA, Rose JA, Khani M, Davidson JA, Kamphuis LG (2023) Virulence profiles and genome-wide association study for *Ascochyta lentis* isolates collected from Australian lentil-growing regions. Phytopathology 113(8):1515–1524

Hou C, Zhuang J, Tang Y, He Y, Miao A, Huang H, Luo S (2021) Recognition of early blight and late blight diseases on potato leaves based on graph cut segmentation. J Agric Food Res 5:100154

Ibrahim SS (2020) Essential oil nanoformulations as a novel method for insect pest control in horticulture. Horticultural crops 195–209

Imran M, Abo-Elyousr KA, Mousa MA, Saad MM (2022) Screening and biocontrol evaluation of indigenous native *Trichoderma* spp. against early blight disease and their field assessment to alleviate natural infection. Egypt J Biol Pest Control 32(1):40

Jadhav S, Kumar A, Lal SK, Akhtar J, Aski M, Mishra GP, Singh AK, Javeria S (2023) Effect of *Fusarium oxysporum f. sp. lentis* on seed quality parameters in lentil. Indian J Plant Genet Resour 36:70–76

Jadon KS, Thirumalaisamy PP, Kumar R (2020) Major seed-borne diseases in important pulses: symptomatology, aetiology and economic importance. In: Seed-borne diseases of agricultural crops: detection, diagnosis & management, pp 469–542

Jagtap GP, Gavate DS, Dey U (2012) Control of *Colletotrichum truncatum* causing anthracnose/pod blight of soybean by aqueous leaf extracts, biocontrol agents and fungicides. Sci J Agric 1(2):39–52

Jaswal R, Kiran K, Rajarammohan S, Dubey H, Singh PK, Sharma Y, Deshmukh R, Sonah H, Gupta N, Sharma TR (2020) Effector biology of biotrophic plant fungal pathogens: current advances and future prospects. Microbiol Res 241:126567

Jha UC, Bohra A, Pandey S, Parida SK (2020) Breeding, genetics, and genomics approaches for improving Fusarium wilt resistance in major grain legumes. Front Genet 11:1001

Jones RA (2021) Global plant virus disease pandemics and epidemics. Plan Theory 10(2):233

Júnior VL, Maffia LA, da Silva RR, Mizubuti ES (2006) Biocontrol of tomato late blight with the combination of epiphytic antagonists and rhizobacteria. Biol Control 38(3):331–340

Kakoti P, Gogoi P, Yadav A, Singh BP, Saikia R (2020) Foliar fungal diseases in pulses: review and management. In: Management of fungal pathogens in pulses: current status and future challenges, pp 131–142

Kashyap D, Siddiqui ZA (2021) Effect of silicon dioxide nanoparticles and *Rhizobium leguminosarum* alone and in combination on the growth and bacterial blight disease complex of pea caused by *Meloidogyne incognita* and *Pseudomonas syringae* pv. Pisi. Arch Phytopathol Pflanzenschutz 54:499–515

Kaur SI, Kashyap PL, Kang SS, Sharma A (2020) Detection and diagnosis of seed-borne viruses and virus-like pathogen. In: Seed-borne diseases of agricultural crops: detection, diagnosis & management, pp 169–199

Khan B, Akash Z, Asad S, Javed N, Rajput NA, Jabbar A, Din WU, Atif RM (2017) Antagonistic potential of *Trichoderma harzianum* against *Fusarium oxysporum f. sp. cubense* associated with Panama wilt of banana. Pak J Phytopathol 29:111–116

Khrieba MI (2020) Damping-off caused by Pythium species: disease profile and management. In: Pythium: diagnosis, diseases and management. CRC Press, pp 257–269

Kishore PV, Gandhi SK, Singh S (2008) Biological control of *Alternaria solani*, the causal agent of early blight of tomato. J Biol Control 22:67–72

Kiss L, Russell JC, Szentiványi O, Xu X, Jeffries P (2004) Biology and biocontrol potential of *Ampelomyces* mycoparasites, natural antagonists of powdery mildew fungi. Biocontrol Sci Tech 14(7):635–651. https://doi.org/10.1080/09583150410001683600

Koledenkova K, Esmaeel Q, Jacquard C, Nowak J, Clément C, Ait Barka E (2022) *Plasmopara viticola* the causal agent of downy mildew of grapevine: from its taxonomy to disease management. Front Microbiol 13:889472

Koshariya AK, Mahant MM, Afsana C, Reddypriya P (2023) Introduction to plants pathology. AG Publishing House (AGPH Books), Bhopal

Kulik T, Bilska K, Żelechowski M (2020) Promising perspectives for detection, identification, and quantification of plant pathogenic fungi and oomycetes through targeting mitochondrial DNA. Int J Mol Sci 21(7):2645

Kumari N, Katoch S (2020) Wilt and root rot complex of important pulse crops: their detection and integrated management. In: Management of fungal pathogens in pulses: current status and future challenges, pp 93–119

Li C, Jia S, Rajput SA, Qi D, Wang S (2022) Transcriptional stages of conidia germination and associated genes in *Aspergillus flavus*: an essential role for redox genes. Toxins 14(8):560

Luneau JS, Cerutti A, Roux B, Carrère S, Jardinaud MF, Gaillac A, Gris C, Lauber E, Berthomé R, Arlat M, Boulanger A (2022) Xanthomonas transcriptome inside cauliflower hydathodes reveals bacterial virulence strategies and physiological adaptations at early infection stages. Mol Plant Pathol 23(2):159–174

Mahmoud GA, Hamd-Alla WA, Hefzy M (2018) Intercropping faba bean with some crops and its impact on the production, phyllosphere and rhizosphere microorganisms under modern irrigation systems. Assiut Journal of Agricultural Sciences 49(4);44–63

Abd-Elmonsef Mahmoud G, Zidan AS, Aly AA, Mosbah HK, Ibrahim AB (2019) Calcium and strontium anthranilato complexes as effective Fusarium moniliforme controlling agents. Applied Organometallic Chemistry 33(2);e4740

Mahmoud A, AbdElgawad H, Hamed BA, Beemster GT, El- Shafey NM (2021) Differences in cadmium accumulation, detoxification and antioxidant defenses between contrasting maize cultivars implicate a role of superoxide dismutase in Cd tolerance. Antioxidants 10(11);1812

Mahmoud GAE (2021) Biotic stress to legumes: fungal diseases as major biotic stress factor. In: Sustainable agriculture reviews 51: legume agriculture and biotechnology, vol 2, pp 181–212

Manjunatha V, Bhattacharjee D, Flores C (2022) Disease management of Faba Beans. In: Faba bean: chemistry, properties and functionality. Springer International, Cham, pp 357–394

Mansour R, Biondi A (2021) Releasing natural enemies and applying microbial and botanical pesticides for managing Tuta absoluta in the MENA region. Phytoparasitica 49(2);179–194

Meena B, Ramyabharathi SA (2012) Effect of fungicides and biocontrol agents in the management of sugarcane smut disease. J Todays Biol Sci Res Rev 1(1):96–103

Meno L, Escuredo O, Rodríguez-Flores MS, Seijo MC (2021) Looking for a sustainable potato crop. Field assessment of early blight management. Agric For Meteorol 308:108617

Mkhize P, Malatji J, Mangena P (2023) Indexing for bacterial, fungal and viral pathogens in legume plants. In: Advances in legume research: physiological responses and genetic improvement for stress resistance, p 37

Muche M, Yemata G (2022) Epidemiology and pathogenicity of vascular wilt of chickpea (*Cicer arietinum* L.) caused by *Fusarium oxysporum f. sp. ciceris*, and the host defense responses. S Afr J Bot 151:339–348

Pérez-Méndez N, Miguel-Rojas C, Jimenez-Berni JA, Gomez-Candon D, Pérez-de-Luque A, Fereres E, Catala-Forner M, Villegas D, Sillero JC (2021) Plant breeding and management strategies to minimize the impact of water scarcity and biotic stress in cereal crops under Mediterranean conditions. Agronomy 12(1):75

Pertot I, Zasso R, Amsalem L, Baldessari M, Angeli G, Elad Y (2008) Integrating biocontrol agents in strawberry powdery mildew control strategies in high tunnel growing systems. Crop Prot 27:622–631

Polonio Á, Pérez-García A, Martínez-Cruz J, Fernández-Ortuño D, de Vicente A (2021) The haustorium of phytopathogenic fungi: a short overview of a specialized cell of obligate biotrophic plant parasites. Prog Bot 82:337–355

Punja ZK (2021) Emerging diseases of *Cannabis sativa* and sustainable management. Pest Manag Sci 77(9):3857–3870

Ramadass M, Thiagarajan P (2021) Current trends and emerging technologies for pest control management of rice (*Oryza sativa*) plants. Environ Biotechnol 4:125–179

Salcedo AF, Purayannur S, Standish JR, Miles T, Thiessen L, Quesada-Ocampo LM (2021) Fantastic downy mildew pathogens and how to find them: advances in detection and diagnostics. Plan Theory 10(3):435

Schoina C, Stringlis IA, Pantelides IS, Tjamos SE, Paplomatas EJ (2011) Evaluation of application methods and biocontrol efficacy of *Paenibacillus alvei* strain K-165, against the cotton black root rot pathogen *Thielaviopsis basicola*. Biol Control 58(1):68–73

Shahin SI (2017) Effect of some biocides on development of the onion downy mildew disease, caused by Peronospora destructor (Berk.). Egypt J Biol Pest Control 27(1):71–77

Shahriar SA, Imtiaz AA, Hossain MB, Husna A, Eaty M, Khatun N (2020) Rice blast disease. Annu Res Rev Biol 35(1):50–64

Shanthiyaa V, Saravanakumar D, Rajendran L, Karthikeyan G, Prabakar K, Raguchander T (2013) Use of *Chaetomium globosum* for biocontrol of potato late blight disease. Crop Prot 52:33–38

Sharma P, Roy M, Roy B (2021) Determining the Effect of Storage, Moisture and Temperature on Vigour and Viability of Rapeseed and Mustard. Current Topics in Agricultural Sciences 5;39–51

Sharma I (2021) Phytopathogenic fungi and their biocontrol applications. In: Fungi bio-prospects in sustainable agriculture, environment and nano-technology. Academic Press, San Diego, CA, pp 155–188

Sharma A, Abrahamian P, Carvalho R, Choudhary M, Paret ML, Vallad GE, Jones JB (2022) Future of bacterial disease management in crop production. Annu Rev Phytopathol 60:259–282

Shiomi HF, Silva HSA, Melo ISD, Nunes FV, Bettiol W (2006) Bioprospecting endophytic bacteria for biological control of coffee leaf rust. Sci Agric 63:32–39

Sikandar A, Zhang MY, Wang YY, Zhu XF, Liu XY, Fan HY, Xuan YH, Chen LJ, Duan YX (2020) *Meloidogyne incognita* (root-knot nematode) a risk to agriculture. Appl Ecol Environ Res 18(1):1679

Simonetti E, Carmona MA, Scandiani MM, Garcia AF, Luque AG, Correa OS, Balestrasse KB (2012) Evaluation of indigenous bacterial strains for biocontrol of the frogeye leaf spot of soya bean caused by *Cercospora sojina*. Lett Appl Microbiol 55(2):170–173

Singh VK, Singh HB, Upadhyay RS (2017) Role of fusaric acid in the development of Fusarium wilt symptoms in tomato: physiological, biochemical and proteomic perspectives. Plant Physiol Biochem 118:320–332

Singh S, Awasthi LP, Jangre A (2020) Transmission of plant viruses in fields through various vectors. In: Applied plant virology. Academic Press, San Diego, CA, pp 313–334

Sivasubramaniam N, Hariharan G, Zakeel MCM (2020) Sustainable management of plant-parasitic nematodes: an overview from conventional practices to modern techniques. In: Management of phytonematodes: recent advances and future challenges, pp 353–399

Stark JC, Thornton M (2020) Field selection, crop rotations, and soil management. In: Potato production systems, pp 87–100

Suprapta DN (2012) Potential of microbial antagonists as biocontrol agents against plant fungal pathogens

Tanner RA, Pollard KM, Varia S, Evans HC, Ellison CA (2015) First release of a fungal classical biocontrol agent against an invasive alien weed in Europe: biology of the rust, *Puccinia komarovii* var glanduliferae. Plant Pathol 64(5):1130–1139

Tapia-Vázquez I, Montoya-Martínez AC, De los Santos-Villalobos S, Ek-Ramos MJ, Montesinos-Matías R, Martínez-Anaya C (2022) Root-knot nematodes (*Meloidogyne* spp.) a threat to agriculture in Mexico: biology, current control strategies, and perspectives. World J Microbiol Biotechnol 38(2):26

Van der Linde WJ (1956) The Meloidogyne problem in South Africa. Nematologica 1;177–183

Vigneron N, Grimplet J, Remolif E, Rienth M (2023) Unravelling molecular mechanisms involved in resistance priming against downy mildew (*Plasmopara viticola*) in grapevine (*Vitis vinifera* L.). Sci Rep 13(1):14664

Vishunavat K (2013) Smut fungi: potential pathogens and biocontrol agents. Centre of Advanced Faculty Training in Plant Pathology, New Delhi, pp 131–135

Wright PJ (2004) Resistance of Peronospora destructor (onion downy mildew) to metalaxyl fungicides. N Z Plant Prot 57:347–347

Xue R, Wu J, Zhu Z, Wang L, Wang X, Wang S, Blair MW (2015) Differentially expressed genes in resistant and susceptible common bean (*Phaseolus vulgaris* L.) genotypes in response to *Fusarium oxysporum f. sp. phaseoli*. PLoS One 10(6):e0127698

Yadav S, Modi P, Dave A, Vijapura A, Patel D, Patel M (2020) Effect of abiotic stress on crops. In: Sustainable crop production, vol 3, p 228

Yao Y, Li Y, Chen Z, Zheng B, Zhang L, Niu B, Meng J, Li A, Zhang J, Wang Q (2016) Biological control of potato late blight using isolates of Trichoderma. Am J Potato Res 93:33–42

Zagade SN, Deshpande GD, Gawade DB, Atnoorkar AA, Pawar SV (2012) Biocontrol agents and fungicides for management of damping off in chilli. World J Agric Sci 8(6):590–597

Zaki SA, Ouf SA, Aly AA, Abd-Elsalam KA (2021) Fungi involved in damping-off of cotton seedlings and their differential pathogenicity on two cotton cultivars. Egypt J Bot 61(3):911–921

Zhang X, Zhou Y, Li Y, Fu X, Wang Q (2017) Screening and characterization of endophytic bacillus for biocontrol of grapevine downy mildew. Crop Prot 96:173–179

Zhu X, Li L, Wang J, Zhao L, Shi H, Bao J, Su Z, Liu X, Lin F (2021) Vacuolar protein-sorting receptor MoVps13 regulates conidiation and pathogenicity in rice blast fungus *Magnaporthe oryzae*. J Fungi 7(12):1084

The manufacturer's authorised representative in the EU is Springer Nature Customer Service Centre GmbH, Europaplatz 3, 69115 Heidelberg, Germany. If you have any concerns regarding our products, please contact ProductSafety@springernature.com

Printed and bound by CPI Group (UK) Ltd, Croydon, CR0 4YY
30/09/2026
02225300-0014